IET POWER AND ENERGY SERIES 79

Vehicle-to-Grid: Linking Electric Vehicles to the Smart Grid

Other volumes in this series:

Volume 1	**Power circuit breaker theory and design** C.H. Flurscheim (Editor)
Volume 4	**Industrial microwave heating** A.C. Metaxas and R.J. Meredith
Volume 7	**Insulators for high voltages** J.S.T. Looms
Volume 8	**Variable frequency ac motor drive systems** D. Finney
Volume 10	**SF_6 switchgear** H.M. Ryan and G.R. Jones
Volume 11	**Conduction and induction heating** E.J. Davies
Volume 13	**Statistical techniques for high voltage engineering** W. Hauschild and W. Mosch
Volume 14	**Uninterruptible power supplies** J. Platts and J.D. St Aubyn (Editors)
Volume 15	**Digital protection for power systems** A.T. Johns and S.K. Salman
Volume 16	**Electricity economics and planning** T.W. Berrie
Volume 18	**Vacuum switchgear** A. Greenwood
Volume 19	**Electrical safety: A guide to causes and prevention of hazards** J. Maxwell Adams
Volume 21	**Electricity distribution network design, 2nd Edition** E. Lakervi and E.J. Holmes
Volume 22	**Artificial intelligence techniques in power systems** K. Warwick, A.O. Ekwue and R. Aggarwal (Editors)
Volume 24	**Power system commissioning and maintenance practice** K. Harker
Volume 25	**Engineers' handbook of industrial microwave heating** R.J. Meredith
Volume 26	**Small electric motors** H. Moczala *et al.*
Volume 27	**AC-DC power system analysis** J. Arrillaga and B.C. Smith
Volume 29	**High voltage direct current transmission, 2nd Edition** J. Arrillaga
Volume 30	**Flexible ac transmission systems (FACTS)** Y-H. Song (Editor)
Volume 31	**Embedded generation** N. Jenkins *et al.*
Volume 32	**High voltage engineering and testing, 2nd Edition** H.M. Ryan (Editor)
Volume 33	**Overvoltage protection of low-voltage systems, Revised Edition** P. Hasse
Volume 36	**Voltage quality in electrical power systems** J. Schlabbach *et al.*
Volume 37	**Electrical steels for rotating machines** P. Beckley
Volume 38	**The electric car: Development and future of battery, hybrid and fuel-cell cars** M. Westbrook
Volume 39	**Power systems electromagnetic transients simulation** J. Arrillaga and N. Watson
Volume 40	**Advances in high voltage engineering** M. Haddad and D. Warne
Volume 41	**Electrical operation of electrostatic precipitators** K. Parker
Volume 43	**Thermal power plant simulation and control** D. Flynn
Volume 44	**Economic evaluation of projects in the electricity supply industry** H. Khatib
Volume 45	**Propulsion systems for hybrid vehicles** J. Miller
Volume 46	**Distribution switchgear** S. Stewart
Volume 47	**Protection of electricity distribution networks, 2nd Edition** J. Gers and E. Holmes
Volume 48	**Wood pole overhead lines** B. Wareing
Volume 49	**Electric fuses, 3rd Edition** A. Wright and G. Newbery
Volume 50	**Wind power integration: Connection and system operational aspects** B. Fox *et al.*
Volume 51	**Short circuit currents** J. Schlabbach
Volume 52	**Nuclear power** J. Wood
Volume 53	**Condition assessment of high voltage insulation in power system equipment** R.E. James and Q. Su
Volume 55	**Local energy: Distributed generation of heat and power** J. Wood
Volume 56	**Condition monitoring of rotating electrical machines** P. Tavner, L. Ran, J. Penman and H. Sedding
Volume 57	**The control techniques drives and controls handbook, 2nd Edition** B. Drury
Volume 58	**Lightning protection** V. Cooray (Editor)
Volume 59	**Ultracapacitor applications** J.M. Miller
Volume 62	**Lightning electromagnetics** V. Cooray
Volume 63	**Energy storage for power systems, 2nd Edition** A. Ter-Gazarian
Volume 65	**Protection of electricity distribution networks, 3rd Edition** J. Gers
Volume 66	**High voltage engineering testing, 3rd Edition** H. Ryan (Editor)
Volume 67	**Multicore simulation of power system transients** F.M. Uriate
Volume 68	**Distribution system analysis and automation** J. Gers
Volume 69	**The lightening flash, 2nd Edition** V. Cooray (Editor)
Volume 70	**Economic evaluation of projects in the electricity supply industry, 3rd Edition** H. Khatib
Volume 78	**Numerical analysis of power system transients and dynamics** A. Ametani (Editor)
Volume 79	**Vehicle-to-grid: Linking electric vehicles to the smart grid** J. Lu and J. Hossain (Editors)
Volume 905	**Power system protection, 4 volumes**

Vehicle-to-Grid: Linking Electric Vehicles to the Smart Grid

Edited by Junwei Lu
and Jahangir Hossain

The Institution of Engineering and Technology

Published by The Institution of Engineering and Technology, London, United Kingdom

The Institution of Engineering and Technology is registered as a Charity in England & Wales (no. 211014) and Scotland (no. SC038698).

First published 2015

The Institution of Engineering and Technology
Michael Faraday House
Six Hills Way, Stevenage
Herts, SG1 2AY, United Kingdom

www.theiet.org

British Library Cataloguing in Publication Data
A catalogue record for this product is available from the British Library

ISBN 978-1-84919-855-4 (hardback)
ISBN 978-1-84919-856-1 (PDF)

Typeset in India by MPS Limited

Contents

Preface ix

1 Introduction to the Smart Grid using PEVs **1**
Junwei Lu and Jahangir Hossain

Abstract 1
1.1 Introduction 1
1.2 The Smart Grid and Microgrid 2
1.3 Impact of PEVs on Distributed Energy Resources in the Smart Grid 4
1.4 V2G Technology and PEVs Charging Infrastructures 6
References 9

2 Impact of EV and V2G on the Smart Grid and Renewable Energy Systems **11**
Taha Selim Ustun

Abstract 11
2.1 Introduction 11
2.2 Types of Electric Vehicles 12
2.3 Motor Vehicle Ownership and EV Migration 17
2.4 Impact of Estimated EVs on Electrical Network 25
2.5 Impact on Drivers and the Smart Grid 30
2.6 Standardization and Plug-and-Play 32
2.6.1 IEC 61850 Communication Standard and IEC 61850-7-420 Extension 33
2.6.2 Extending IEC 61850 for Electric Vehicle Modeling 34
2.7 Conclusion 38
References 39

3 Distributed Energy Resource with PEV Battery Energy Storage in the Smart Grid **43**
Jianguo Zhu, Md Rabiul Islam, Jiefeng Hu, Sonki Prasetya, Yuedong Zhan, Mohammad Jafari, Hung Dung Pham and Youguang Guo

Abstract 43
3.1 Introduction 43
3.2 Distributed Energy Sources 44
3.2.1 Solar 45
3.2.2 Wind 48

3.2.3 Fuel Cells 50
3.2.4 Electric Vehicles 52
3.2.5 Backup Power Supplies 54
3.2.6 MPPT Strategies for PV Based Microgrids 56
3.3 The Smart Grid and Microgrids 58
3.3.1 Microgrid Topologies 60
3.3.2 Microgrid Control Strategies 63
3.3.3 Simulation Results and Discussion 67
3.3.4 Experimental Results and Discussion 69
3.4 Conclusions 74
References 74

4 Power Conversion Technology in the Smart Grid and EV 81
M.A. Mahmud

Abstract 81
4.1 Introduction 81
4.2 Dynamical Modelling of EVs 84
4.2.1 Dynamical Modelling of EV Connected to Single-Phase Smart Grid Node 85
4.2.2 Dynamical Modelling of EV Connected to Three-Phase Smart Grid Node 87
4.3 Power Conversion Problem Formulation in Smart Grids with EVs 88
4.4 Feedback Linearisation and Feedback Linearisability of Smart Grids with EVs 89
4.4.1 Overview of Feedback Linearisation 90
4.4.2 Feedback Linearisability of Smart Grids with EVs 91
4.4.3 Partial Feedback Linearisation of Smart Grids with EVs 91
4.5 Distributed Controller Design for Smart Grids with EVs 92
4.6 Performance Evaluation 94
4.6.1 Controller Performance Evaluation During the Charging Mode of EVs 97
4.6.2 Controller Performance Evaluation During the Charging Mode of EVs 97
4.7 Conclusions 100
Appendix I: Definition of Lie Derivative and Relative Degree 101
Appendix II: Stability of Internal Dynamics of Smart Grids with EVs 101
Appendix III: System Parameters 102
References 103

5 Power Control and Monitoring of the Smart Grid with EVs 107
M.S. Rahman, F.H.M. Rafi, M.J. Hossain and J. Lu

Abstract 107
5.1 Introduction 107
5.2 Impacts of EV Penetration on Grid Power Profile and Requirements of Its Control and Monitoring 108

5.2.1 Voltage and Frequency Regulation 109
5.2.2 Supporting and Balancing of Intermittent RES Powered Smart Grid 110
5.2.3 Controlling and Monitoring via '*Smart Charging/Discharging*' 110
5.3 Hybrid EV Powertrain Architectures 111
5.4 Industrially Used Control, Monitoring and Management Strategies 117
5.4.1 EV Induction Motor (IM) Control 117
5.4.2 Requirement of EV Battery Management and Monitoring 119
5.4.3 Techniques for EV Battery Management 120
5.4.4 Techniques for EV Battery Capacity Monitoring 124
5.5 V2G Communication System 129
5.6 System Model 132
5.6.1 Wind Turbine Modelling 132
5.6.2 PV System Modelling 134
5.6.3 EV Energy Storage System (EV-ESS) Modelling 137
5.7 Problem Formulation and Control Strategy 143
5.8 Simulation Results 145
5.8.1 Severe Three-Phase Fault 146
5.8.2 Intentional Islanding 147
5.9 Conclusions 150
References 151

6 PEV Charging Technologies and V2G on Distributed Systems and Utility Interfaces 157
Domagoj Leskarac, Chirag Panchal, Sascha Stegen and Junwei Lu

Abstract 157
6.1 Introduction 157
6.1.1 Overview 157
6.1.2 Vehicle-to-Grid Concept and PEV Communication Requirements 163
6.1.3 Distributed Generation and the Smart Grid 165
6.1.4 Charging Diversity and Utility Interfaces 167
6.1.5 Local, Central and Distributed Generation 168
6.2 Current PEV Charging Standards 170
6.2.1 Socket Types 171
6.3 Contact-Based PEV Charging 172
6.3.1 Rectifier Topologies for G2V 173
6.3.2 Inverter Topologies for V2G 173
6.3.3 DC/DC Converters 176
6.3.4 Full-Bridge Converters with STATCOM Capabilities 176
6.3.5 Internal STATCOM 178
6.3.6 PEV On-board Charging Designs 178
6.4 Smart Transformer for Charging Station in a Microgrid 180

6.5 Contact Charging Safety Considerations 181
6.5.1 AC Charging 181
6.5.2 DC Charging 182
6.6 Wireless Charging Systems for PEV 183
6.6.1 Wireless Power Characteristics 183
6.6.2 Methods of Wireless Power Transfer 185
6.6.3 Standards of Wireless Charging Systems for PEVs 197
6.6.4 Application of Wireless Charging Systems for PEVs 201
6.6.5 Wireless Safety Considerations 208
6.7 Conclusion 209
References 209

7 Economic, Social and Environmental Dimensions of PHEV in the Smart Grid 223
Christopher J. Bennett, Markos Katsanevakis and Rodney A. Stewart

Abstract 223
7.1 Introduction 224
7.2 Economic Dimension 225
7.2.1 Smart Electric Vehicle Charging 226
7.2.2 PHEVs as Distributed Energy Resources 229
7.3 Social Dimension 233
7.3.1 Convenience 233
7.3.2 Uncertainty, Trust and Equity 234
7.3.3 Perception of Poor Safety 235
7.3.4 Public Perceptions and Overcoming Resistance to Change 235
7.4 Environmental Dimension 236
7.4.1 LCA for PHEVs and Consideration of Further Automotive Technologies 236
7.4.2 Environmental Impact Comparative Assessment 237
7.4.3 Global Metal Reserves, Production and Pricing Data Related to Vehicle Batteries and Electric Motors 239
7.5 General Method for Developing PHEV Scheduling Systems 243
7.5.1 Introduction 243
7.5.2 Developing Load Forecast Models 245
7.5.3 Initial Scheduling 248
7.5.4 Online Control System 249
7.6 Conclusion 249
References 250

Index 253

Preface

The electric power system is undergoing a profound change, driven by the increasing penetration of renewable energy resources (RES) and integration of electrification for modern transportation. The need for environmental compliance and energy conservation becomes a very important issue for the future. We need better grid reliability while dealing with an aging infrastructure, and we need improved operational efficiencies and customer service. The changes that are happening are particularly significant for the electricity distribution grid, where 'blind' and manual operations, along with the electromechanical components, will need to be transformed into a 'smart grid'. The smart grid, regarded as the next generation power grid, uses two-way flows of electricity and information to create a widely distributed automated energy delivery network. The smart grid will meet environmental targets to accommodate a greater emphasis on demand response. It will also support the plug-in electric vehicles (PEVs), including all-electric vehicles and plug-in hybrid electric vehicles (PHEVs), as well as distributed generation and storage capabilities. It is obvious that these needs and changes present the electric power industry with the biggest challenge it has ever faced. On the one hand, the PEVs will play a significant role in the electric power system. The connection of PEVs can have a serious impact on the future smart grid. Although PEVs can provide a new opportunity to reduce oil consumption by drawing on power from the electric power grid, PEVs can also be used as energy storage to provide energy into the power grid. The power flow of the grid connection of PEVs can be bi-directional if PEVs have the function of vehicle-to-grid (V2G), which can be either flexible loads (charging mode) or storage sources (discharging mode). To maximize the benefits of V2G, the emerging PEV infrastructure must provide access to electricity from the smart grid, satisfy driver expectations and ensure safety.

The aim of this book is to cover a broad area of the V2G and the smart grid, and provide an in-depth coverage of specific topics related to PEVs and battery charging technologies and infrastructures. It will also cover the impact of the V2G on the smart grid and renewable energy systems. Chapter 1 presents a short introduction to the smart grid using PEVs. Chapter 2 discusses the impact of PEVs and the V2G on the smart grid and renewable energy systems. Chapter 3 is devoted to distributed energy resources with PEV battery energy storage in the smart grid. Chapter 4 investigates power conversion technology in the smart grid and PEVs. Chapter 5 covers power control and monitoring of the smart grid with PEVs. Chapter 6 introduces PEV charging technologies and V2G on distributed energy resources and utility interfaces. Chapter 7 investigates the economic, social and environmental impacts on the smart grid using PEVs.

Chapter 1

Introduction to the Smart Grid using PEVs

Junwei Lu and Jahangir Hossain**

Abstract

The smart grid, regarded as the next generation of power grid, uses two-way flows of electricity and information to create a widely distributed automated resilient energy delivery network. The smart grid will meet environmental targets to accommodate a greater emphasis on demand response, and to support plug-in electric vehicles (PEVs) as well as distributed generation and storage capabilities. The PEVs including all-electric vehicles and plug-in hybrid electric vehicles (PHEVs) will play a significant role in the electric power system. The connection of PEVs can have serious impact on the future smart grid. Although PEVs can provide a new opportunity to reduce oil consumption by drawing on power from the electric power grid, they can be used as energy storage to provide the energy into power grid. The power flow between PEVs and the grid can be bidirectional if PEVs have the function of vehicle-to-grid (V2G), which can be either as flexible loads (charging mode) or storage sources (discharging mode). To maximize the benefits of V2G, the emerging PEV infrastructure must provide access to electricity from the smart grid, satisfy driver expectations, and ensure safety. This chapter introduces a basic concept of the smart grid and its "building blocks," microgrids, and impact of PEVs on distributed energy resources in the smart grid and V2G technology and PEVs charging infrastructures.

1.1 Introduction

Plug-in electric vehicles (PEVs), including all-electric vehicles and plug-in hybrid electric vehicles (PHEVs), will provide a new opportunity to reduce oil consumption and play a central role in decarbonising road transport. This could have a significant effect on future grid performances and load demands. Electricity suppliers and business investors will need to anticipate the long-term investments that will be needed to respond to this emerging trend, both in the electricity network and associated PEV charging infrastructure. Since the use of PEVs will increase,

*School of Engineering, Griffith University, QLD, 4222, Australia

PEVs could have a significant impact on the future smart grid and load demand, as this new type of electricity load will need careful management in order to minimize the impact on peak electricity demand. The integration of the smart grid and microgrid technology with vehicle-to-grid (V2G) can enable PEVs to be used as distributed energy storage devices, feeding electricity stored in their battery tanks and sending it back into the grid when electric power is needed. Microgrids and smart grids using PEVs can help to reduce electricity system costs by offering a cost-effective means of providing regulation services, spinning reserves, and peak-shaving capacity. However, there are several technical, environmental, and economic barriers to such a development, including PEV charging infrastructure and battery technology. Overcoming these limitations will be critical to the future of V2G supplies and could have a significant impact on future grid and load demanding. This chapter presents a brief introduction to the smart grid using PEVs, and provides some general information related to this book. More detailed technical information on each topic will be discussed in each chapter.

1.2 The Smart Grid and Microgrid

The smart grid uses advanced information and communications technologies to improve the reliability, security, and efficiency of the electric system from large generation of power, through to the delivery systems to electricity consumers and a growing number of distributed-generation and storage resources. The smart grid vision is to make lower voltage networks more visible and to enable the participation of customers in the operation of the power system, particularly through smart meters and smart homes. The definition and operation of smart grids have been described in the book "Smart Grid – Technology and Application" [1–2]. There are a number of different types of the smart grid technology, all of which make use of information and communications technologies including both hardware and software. The smart grid will play an important role in encouraging the integration of PEVs, and PEVs could reinforce the benefits of smart grids.

A microgrid is defined by the U.S. Department of Energy (DOE) as "… a group of interconnected loads and DERs within clearly defined electrical boundaries that act as a single controllable entity with respect to the grid. A microgrid can connect and disconnect from the grid to enable it to operate in both grid-connected and island-mode" [3]. Microgrids are "building blocks of smart grids," which consist of several basic technologies for operation. These include: distributed generation, distributed storage, interconnection switches, and control systems, as illustrated in Figure 1.1. Distributed generation (DG) units are small sources of energy located at or near the point of use. DG technologies typically include photovoltaic (PV), wind, fuel cells, microturbines, and reciprocating internal combustion engines with generators. Distributed storage (DS) technologies are used in microgrid applications where the generation and loads of the microgrid cannot be exactly matched. The interconnection switch is the point of connection between the microgrid and the rest of the distribution system or the smart grid. The control

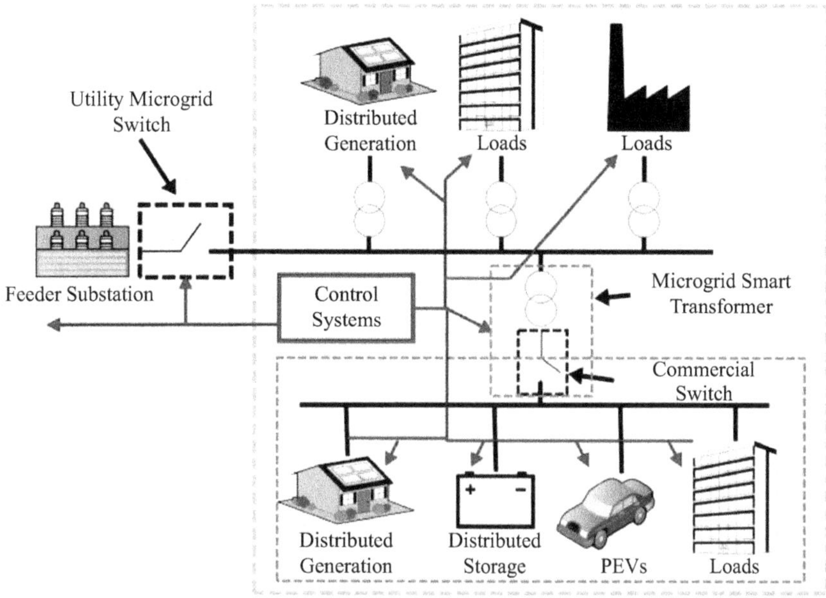

Figure 1.1 Basic concept of the microgrid introduced in IEEE 1547.4 Standard [4]

system of a microgrid is designed to safely operate the system in grid-parallel and stand-alone modes.

Microgrids are electric power systems (EPSs) that have distributed resources (DRs) and load. They also have the ability to disconnect from and parallel with the EPS, and they include the local EPS. Microgrids may also include portions of the area EPS, and are intentional and planned. The modes of operation of the microgrids include normal parallel mode, transition-to-island mode, island mode, and reconnection mode. The planned microgrids include the local EPS island (facility island), secondary island, lateral island, circuit island, substation bus island, substation island, and an adjacent circuit island. Microgrid standards are intentional islands formed at a facility or in an electrical distribution system that contains at least one distributed energy resource and associated loads. Microgrid standards will become more realistic and easier to meet the power-quality requirements of the IEEE 1547 series of standards [4]. IEEE 1547.4 covers key considerations for planning and operating microgrids. This includes: impacts of voltage, frequency, power quality, inclusion of single point of common coupling (PCC) and multiple PCCs, protection schemes and modifications, monitoring, information exchange and control, understanding load requirements of the customer, knowing the characteristics of the DER, identifying steady state and transient conditions, understanding interactions between machines, reserve margins, load shedding, demand response, cold load pickup, additional equipment requirements, and additional functionality associated with inverters.

1.3 Impact of PEVs on Distributed Energy Resources in the Smart Grid

Recent technological advances in electricity distribution and load demand management, which make use of information and communications technologies and distributed control systems, promise to facilitate the integration of PEVs into the electricity load and to lower costs. The impact of PEVs on the smart grid demand response has been moving from a load-following (LF) to a load-shaping (LS) strategy [2]. Therefore, PEVs can be used as distributed storage devices, feeding electricity stored in their batteries back into the system when needed (vehicle-to-grid, or V2G, supply). V2G technology can help to reduce electricity costs by providing a cost-effective means of providing regulation services, spinning reserves, and peak-shaving capacity. The smart grid and microgrid technologies can enable the PEVs' charging load to be shifted to off-peak periods, thereby flattening the daily load curve and significantly reducing both generation and network investment needs. In this way, PEVs could both benefit from and promote investment in smart grids. However, there are a number of technical, practical, and economic barriers to this development, including low battery discharge rates and limited storage capacity. Both have the potential to significantly impact the future of V2G supply [5].

The integration of PEVs and clean electricity generated from renewable sources, such as solar PV arrays and wind turbines, offers the greatest near-term potential for net-zero energy transportation. Today, approximately 12 percent of electricity comes from renewable sources, but the variable nature of the wind and sun may pose operational challenges [6]. V2G technology makes it possible to store energy in and source energy from vehicle batteries to balance intermittent renewable resources, enhance grid stability, and/or reduce demand for costly grid-fed electricity at peak hours. With V2G technology, the grid and other infrastructure will allow the transportation system of the future to use dramatically less fossil fuel and significantly cut greenhouse gas emissions while also addressing grid demand and electric vehicle consumer adoption issues. There are also concerns about the impact of a nationwide migration to a fleet of electric-drive vehicles on electricity demand. The utility infrastructure has a finite capacity to meet ever-growing electricity needs, and the majority of this system is powered by coal and natural gas generators. The rapid growth in the number of PEVs in use would have a significant impact on the need for investment in electricity network capacity and the smart grid technologies. Depending on the rate of penetration of the light-duty vehicle fleet, PEVs could account for a substantial share of total electricity consumption and, more importantly, peak load. The greater the increase in consumption, the larger the potential benefits from the smart grid technologies that improve the ability of the electricity utility to manage load in order to schedule charging as much as possible outside of peak hours [7–10].

The connection of batteries of PEVs can have significant impacts on microgrids and the future smart grid. The power flow of the PEV connection can be bidirectional, as vehicles can function either as flexible loads (charging) or storage sources (discharging). This dual procedure must be controllable in order to achieve a more

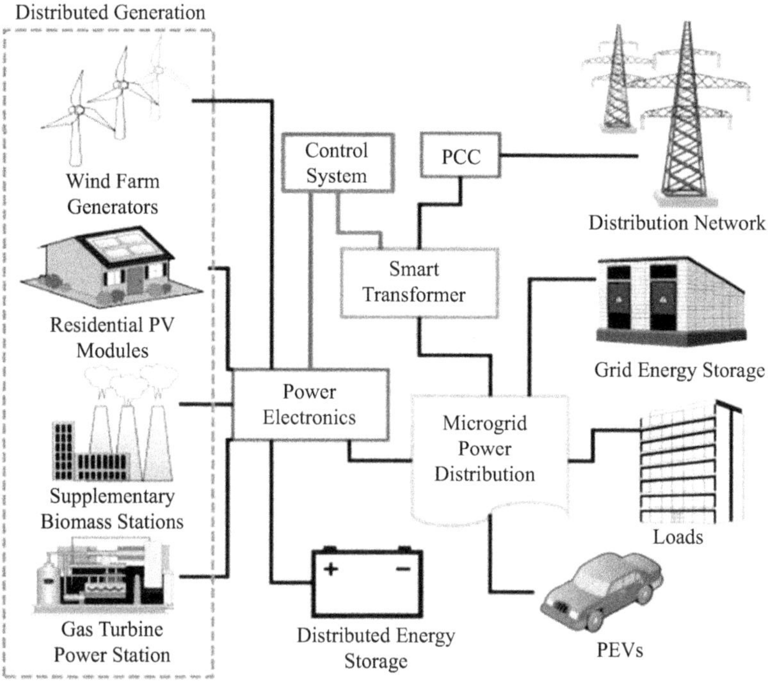

Figure 1.2 Impacts of PEV penetration in a microgrid

economic and optimal grid operation. As shown in Figure 1.2, within microgrids, the presence of EVs has more advantages. They contribute to the optimization procedure allowing a higher and more efficient renewable energy system penetration, while smoothing the load curve. Microgrids comprise low voltage (LV) distribution systems with distributed energy sources, storage devices and controllable loads, operated through a connection to the main power network or island, in a controlled, coordinated way. The uncontrolled connection of PEVs in the grid increases the peak demand, producing technical and economical issues. Considering the PEV batteries as the microgrid's controllable components, charging and discharging (V2G procedure) can be executed in an optimal way. This is because PEV battery management can be used for balancing load-generation, islanding, black start and regulation in a microgrid. PEVs used as energy storage devices and flexible loads in a microgrid or a commercial building with distributed generators (DGs) can help to take advantage of the surplus energy in periods of high renewable energy production and low load demand. Besides the technical and economic benefits, this approach is environmentally beneficial, as the charging procedure is executed by using energy sources of low CO_2 emissions, reducing the CO_2 footprint of PEVs. Several PEVs plugged into a low voltage network in the residential area, or a larger number of PEVs in a parking garage connected to a lateral feeder, could cause a localized overload on the distribution network and transformers. Many distribution networks

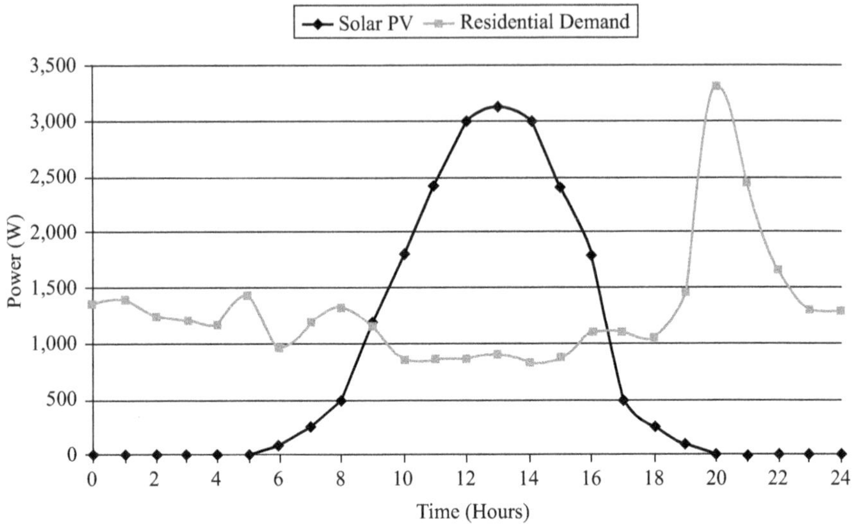

Figure 1.3 Solar energy sources can increase the availability of peak-period midday charging and powering PEVs can be made more affordable

have been operating close to their operating limits, and the additional load may push them above their emergency operating limits. Furthermore, the potential unbalanced conditions created by such loads could cause problems on main feeders and other laterals. Therefore, the load forecasting and scheduling will play an important role in the microgrid and future smart grid with V2G. V2G technology makes it possible for vehicles to become valuable grid resources while conserving energy and supporting CO_2 emissions reduction. As illustrated in Figure 1.3, if solar energy sources can increase the availability of peak-period midday charging, powering PEVs can be made more affordable. At the same time, PEVs can store PV energy during periods of low demand. A more detailed discussion on peak demand management can be found in Chapters 2, 3, 4, 5, and 7.

1.4 V2G Technology and PEVs Charging Infrastructures

The use of PEVs has increased significantly in recent years due to concerns regarding the environment and increasing fossil fuel prices. These vehicles possess dual dynamic characteristics as a load while in G2V (grid to vehicle) mode and as a generator while in V2G mode. If it is designed carefully, the novel concept of V2G with four-quadrant topologies can improve a utility's grid performance in the areas of efficiency, stability, and reliability by offering reactive power support, active power regulation and the tracking of various renewable energy resources, load balancing by valley filling, peak load shaving, and current harmonic filtering [11–12]. Power-factor corrected (PFC) unidirectional chargers only transmit power from a utility to a vehicle battery and operate with an input power factor (PF) of almost unity. In other words, they are not designed to exchange reactive power with the grid. Today, most

PEVs on the market use these types of chargers but, considering their limitations, they are not promising in comparison with other type of topologies for reactive power operation. As four-quadrant topologies are suitable for reactive power compensation, the second innovation of this project is the design and commercialization of a four-quadrant energy-efficient charging system for a V2G system, which will contribute to stabilizing grid voltage.

PEVs provide a new opportunity to reduce oil consumption by drawing on power from the electric grid. To maximize the benefits of PEVs, the emerging PEV infrastructure must provide access to clean electricity generated from renewable sources, satisfy driver expectations, and ensure safety. PEVs have shown great promise in that they have the potential to curb emissions and reduce the cost of transportation. Although wide-scale adoption of PEVs is still a few years away, politicians, electric utilities, and auto companies are eagerly awaiting the opportunities that may arise from reduced emissions and gasoline consumption, new services and increased revenues, and new markets that would create new jobs. This is particularly true for electric utility companies, which could see substantial revenue growth through the electrification of the transportation market segment. For consumers, PEVs will significantly lower operational costs when compared with traditional gasoline cars. While PEVs will provide economic and environmental benefits, they can also offer a potential source of energy storage which has a significant impact on microgrids or future smart grids. A number of papers have discussed several topologies and control methods including reactive power compensation using bidirectional PEV chargers that can perform bidirectional power transfer using a PEV as a distributed energy resource [13–16].

The PEV-charging infrastructure could form an important part of the microgrid or future smart grid. This includes physical charging facilities (connectors and meters), as well as billing, scheduling, and other intelligent features for smart charging during off-peak periods or storing PV energy during periods of low demand. As the share of PEV charging in the overall electricity load increases, the grid would need to incorporate other assets in order to enhance the capacity to provide power-system ancillary services (reserve generating capacity and peak-shaving facilities), and, potentially, power discharging hardware and software to enable PEV batteries to be used as storage devices. Three other features of this technology are also promising: smart charging, V2G communications, and fast charging. Automated smart charging systems use information about the grid frequency, voltage, availability of renewably generated power, and pricing to determine when and how much power to deliver to PEVs [5]. V2G communications pair smart charging capabilities with features allowing PEVs to feed energy back to the grid. Bidirectional on-board charging systems give PEVs the potential to balance variable production from intermittent renewable sources, supply emergency backup power, and act as grid-tied energy storage devices. PEV charging needs to become more convenient and cost effective for a larger number of people to buy PEVs. Wireless power transfer (WPT) takes this V2G communication one step further. While PEV owners might have the means to plug in vehicles at home or at work, charging stations can be far apart once drivers get on the road. Table 1.1 presents

Table 1.1 Comparison of various charging systems for PEVs

Charging infrastructures	**Charge level**		**DC charge**	**V2G function**	**Four quadrant topologies**
	Single phase	**Three phase**			
PEV on board charging system	Level 1 Level 2 Domestic charge	Level 2 Domestic/Industrial/Public charge	N/A	V2G using bi-directional DC/DC converter with transformer isolation	Yes
Conventional charging station (DC only)	N/A	Level 2 Domestic/Industrial/Public charge	Level 3 Public Fast charge	N/A	N/A
Smart transformer based charging station (DC and AC)	N/A	Level 2/Level 3 Industrial/Public charge	Level 3 Public Fast charge	V2G using built-in bi-directional DC/DC converter with transformer isolation	Yes
Three-phase charge using internal motor drive system (DC/AC inverter)	N/A	Level 2/Level 3 Industrial/Public charge On board fast charge	N/A	V2G using PEV internal bi-directional DC/AV motor drive inverter	Yes
Wireless charging station and PEV on board wireless charging system	Level 1 Level 2 Domestic charge	Level 2 Domestic/Industrial/Public charge	N/A	V2G using bi-directional DC/DC converters	Yes

various charging systems where bidirectional on-board charging systems with four quadrant topologies can be used. A more detailed discussion on PEVs charging infrastructures can be found in Chapter 6.

References

[1] J. Ekanayake, K. Liyanage, J. Wu, A. Yokoyama, and N. Jenkins, *Smart Grid – Technology and Application*, Singapore: John Wiley & Sons, 2012.

[2] Trevor MORGAN, "Smart grids and electric vehicles: Made for each other?" Discussion Paper No. 2012-02, Menecon Consulting, United Kingdom (Further information about the International Transport Forum is available at www.internationaltransportforum.org).

[3] Peter Asmus, "Why microgrids are moving into the mainstream," *IEEE Electrification Magazine*, March, 2014.

[4] IEEE Standards Coordinating Committee 21, "IEEE guide for design, operation, and integration of distributed resource island systems with electric power systems," IEEE Std 1547.4™-2011, July 20, 2011.

[5] Qiuwei Wu (Editor), *Grid Integration of Electric Vehicles in Open Electricity Market*, Wiley, 2013.

[6] C. Gearhart, J. Gonder, and T. Markel, "Connectivity and convergence," *IEEE Electrification Magazine*, June, 2014.

[7] F. Ahourai and M. A. Al Faruque, "Grid impact analysis of a residential microgrid under various EV penetration rates," Technical Report 13-08, July 30, 2013.

[8] Wencong Su, Habiballah Rahimi-Eichi, Wente Zeng, and Mo-Yuen Chow, "A survey on the electrification of transportation in a smart grid environment," *IEEE Transaction on Industrial Informatics*, Vol. 8, No. 1, February 2012.

[9] C. Pang, P. Dutta, and M. Kezunovic, "BEVs/PHEVs as dispersed energy storage for V2B uses in the smart grid," *IEEE Transaction on Smart Grid*, Vol. 3, No. 1, March 2012.

[10] M. C. Kisacikoglu, B. Ozpineci, and L. M. Tolbert, "EV/PHEV bidirectional charger assessment for V2G reactive power operation," *IEEE Transaction on Power Electronics*, Vol. 28, No. 12, December 2013.

[11] M. C. Kisacikoglu, B. Ozpineci, and L. M. Tolbert, "Examination of a PHEV bidirectional charger system for V2G reactive power compensation," Twenty-Fifth Annual IEEE Applied Power Electronics Conference and Exposition (APEC), February 2010.

[12] S. Schey, D. Scoffield, and J. Smart, "A first look at the impact of electric vehicle charging on the electric grid in the EV project," EVS26 International Battery, Hybrid and Fuel Cell Electric Vehicle Symposium, May 2012.

[13] M. Meiqin, S. Shujuan, and L. Chang, "Economic analysis of the microgrid with multi-energy and electric vehicles," 8th International Conference on Power Electronics – ECCE Asia May 30–June 3, 2011.

[14] Y. Ota, H. Taniguchi, T. Nakajima, K. M. Liyanage, J. Baba, and A. Yokoyama, "Member autonomous distributed V2G (vehicle-to-grid) satisfying scheduled charging," *IEEE Transaction on Smart Grid*, Vol. 3, No. 1, March 2012.

[15] N. C. Kar, K.L.V. Lyer and *et al.*, "Courting and Sparking", *IEEE Electrification Magazine*, September 2013.

[16] T. S. Ustun, C. Ozansoy, and A. Zayegh, "Electric vehicle potential in Australia, its impact on smart grids," *IEEE Industrial Electronics Magazine*, December 2013.

Chapter 2

Impact of EV and V2G on the Smart Grid and Renewable Energy Systems

*Taha Selim Ustun**

Abstract

This chapter is aimed at examining different electric vehicle (EV) technologies along with different electric vehicles such as cars, motorbikes, service mobiles, etc. Having covered these EV types, the chapter will further focus on the potential impact in the existing electric system. This will be done based on estimated migration to EV, the vehicle ownership in different countries and at generation, transmission and distribution levels. The introduction of EV and vehicle-to-grid (V2G) has impacts on the smart grid/microgrid management, and these will be discussed. Finally, standardization and plug-and-play requirements for wide-spread EV acceptance shall be covered.

2.1 Introduction

The idea of powering cars with electricity has been around for quite some time. When Nikola Tesla removed the gasoline engine of his Pierce Arrow, replacing it with an electric motor, and drove for hours, some called him possessed while others called him crazy. Thankfully, after a century, we have less mystic and more scientific explanations about vehicles that can run without the use of fossil fuels. The scary experience for the old is becoming the daily routine for some environmentalists who opt to go for green cars.

Of course, it did not happen all of a sudden. The recent improvements in electric motor technology as well as batteries made it commercially possible to produce electric cars. Considering the concerns on global warming and rising oil prices, governments are offering incentives for electric vehicle market. All of these, when backed by more efficient engines and saving on fuel costs, increased the interest of drivers in electric vehicles.

In parallel with the increasing number of electric vehicles, the interaction with the electric grid has also become a popular subject. There are different aspects of

*School of Electrical and Computer Engineering, Carnegie Mellon University, Pennsylvania, USA

this field such as aggregated load of electric vehicles, the changing load profile, proposed load management schemes as well as V2G opportunities. These factors are equally interesting and under the radar of power engineers for research and development.

In addition to technical research that focuses on the impact on electrical grid, there are socio-economic researches which focus on the impact of electric vehicles on drivers, and their driving patterns and life styles. Also, similar approaches are taken to estimate the migration to electric vehicles and the socio-economic groups which might be more welcoming them.

The cumulative impact of electric vehicles on electrical networks as well as drivers is investigated. First, types of different electric vehicles are examined due to their differing impacts on the grid. Second, an estimation of migration to electric vehicles and the potential of the vehicle fleet are worked out from statistical information of various countries. Then, the impacts of the estimated potential on the grid as well as the driving patterns of the owners and operation of smart grids are studied from various aspects. Finally, standardization considerations are laid out to facilitate mass migration to EVs.

This chapter serves as a bridge between various research fields. It contributes to knowledge by linking data from a distinct field with its consequences on another field. The technical features of different EVs are coupled with socio-economic and spatial data of various countries to extract the expected migration and the resultant electrical impact. This result is interpreted with load profiles of different users and the feasibility of V2G implementation is established on real-life information. Furthermore, the unprecedented requirements of EV usage, such as constant need for charge, and the smart technologies utilized to tackle them are investigated. This is particularly important considering that these smart technologies have impacts on users, electrical networks, and the network operators. All of these considerations show that there is large data to be assessed in EV technology. Considering the number of vehicles and manufacturers present in the market, it is easily appreciated how hard it is to realize communication between them. Standardization efforts and a model developed by the author are examined in detail for automated deployment of EVs in the electrical network.

In summary, a wide scope of information pertaining to EVs and their impact on electrical networks is provided to equip the readers with necessary knowledge to analyze EV migration and their impacts at large.

2.2 Types of Electric Vehicles

Different types of electric vehicles have different characteristics and, thus, different impacts on the electric grid. Therefore, it is beneficial to have a look at types of electric vehicles before estimating their potential effects. There are several ways of classifying electric vehicles based on their sizes and the employed technology.

As far as the size of vehicles that run on electrical energy is concerned, it is possible to talk about ships, which can be considered as moving electrical grid, as well as electric scooters with very small battery sizes. For the scope of this chapter, we shall focus on the electric cars (i.e., automobiles, service cars, etc.) and electric motors since they are connected to the electrical grid and have a continuous interaction with it.

The electrification of vehicles can be discussed as a gradual change from vehicles using conventional internal combustion engines. The first step is to equip these vehicles with electrical auxiliaries such as electric motors, power electronics interfaces, and batteries. This is called hybridization of cars and is preferred due to its increased efficiency which is only 30 percent in internal combustion engines [1]. This hybridization can be realized at different factor such as 5–10 percent for micro-hybrids and 10–25 percent for mild hybrids. Regardless of the factor, hybridization has two major benefits [2]:

- Higher operation efficiency due to higher efficiency and speed of electric propulsion systems.
- The ability to charge batteries during braking through regenerative braking.

Figures 2.1 and 2.2 show Hybrid electric vehicles (HEVs) with a series power train and a parallel power train, respectively. As the name implies, a series power train necessitates a cascade operation. The fuel tank feeds the internal combustion engine which, in turn, charges the batteries with a generator over power electronics interface. The electric motor that gives traction to the wheels is only fed by the batteries. Since the engine is decoupled from the wheels mechanically, there is freedom to relocate it. Series power trains are popular for large vehicles, easy to develop and control, yet have large manufacturing costs [2].

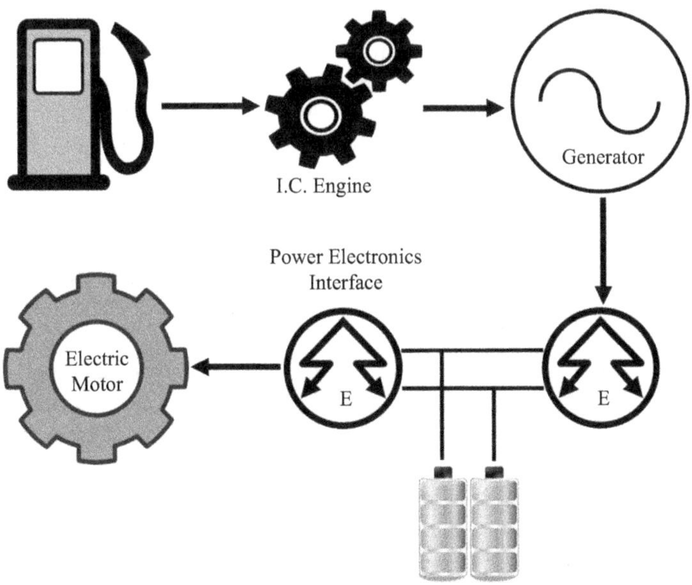

Figure 2.1 Schematic of an HEV with a series power train

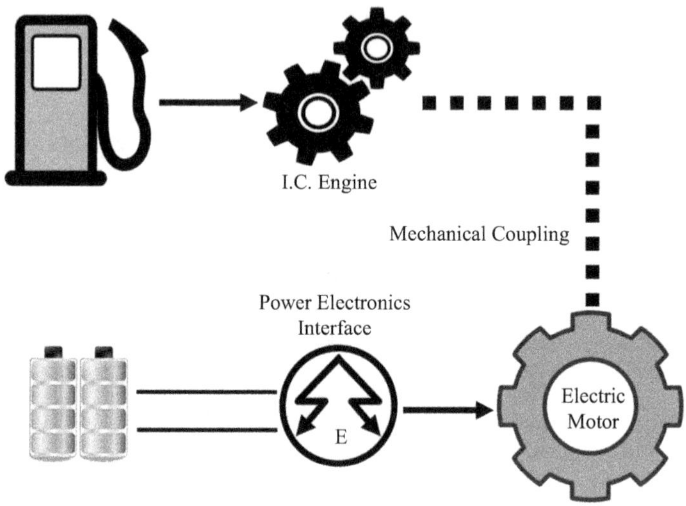

Figure 2.2 Schematic of an HEV with a parallel power train

In contrast to above, with the parallel power train, it is possible to provide traction to the wheels through fuel tank–internal combustion engine couple or through battery–electric motor. While it is true that the electric motor increases the efficiency and decreases carbon emissions, parallel power train does not operate in all-electric mode at high speeds. This significantly limits the implementation opportunities.

Figure 2.3 shows a plug-in HEV (PHEV) with both of these technologies combined, i.e., series-parallel power train. This topology possesses the advantages of both of the previous ones, includes more components (mechanical coupling and a generator at the same time), and is more expensive. But there is a more fundamental difference between HEVs and PHEVs which is the external connection for the battery system that allows charging independently from the engine. The interaction between PHEVs and the electrical grid is provided through this connection. In short, HEVs do not have any interaction with, hence impact on, electrical grid. Only PHEVs and pure EVs (obtained by completely eliminating the fuel tank and the internal combustion engine) interact with the grid, enable V2G technology and are taken into account accordingly.

Table 2.1 summarizes different types of electric vehicles based on hybridization technology and their impacts on the vehicle and the electric grid. Both HEV and PHEV provide higher efficiency and lower carbon emissions in the vehicle while pure EV provides maximum efficiency and zero emission. HEVs do not have any interaction with the grid. On the other hand PHEVs and EVs charge from the grid and in case of necessity may provide power to the grid through V2G technology. Pure EVs can only provide the energy stored in the battery (i.e., act as a storage) while PHEVs can also generate electricity with internal combustion engine – generator topology (i.e., act as both storage and generator).

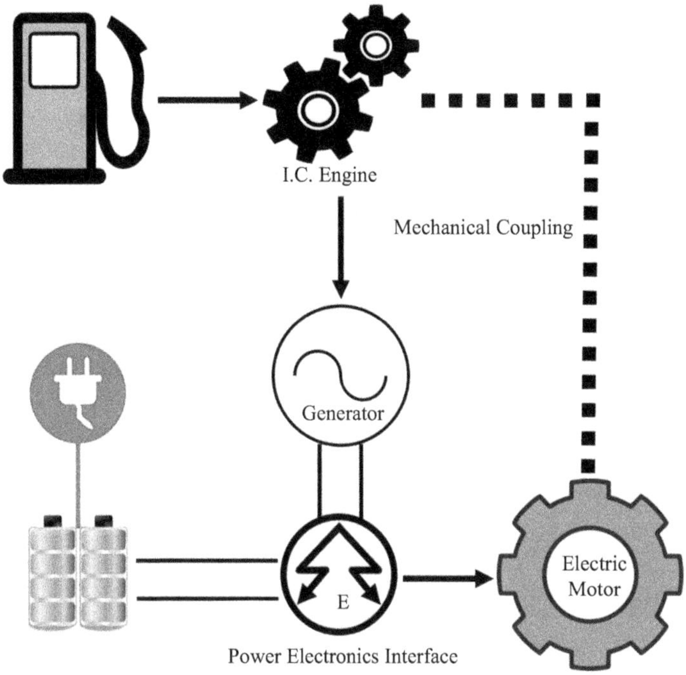

Figure 2.3 Schematic of PHEV with a series-parallel power train

Table 2.1 Types of electric vehicles based on technology

Technology	Overall efficiency	Carbon emissions	Impact on the grid
Hybrid electric vehicle	Higher	Lower	None
Plug-in hybrid electric vehicle	Higher	Lower	Charging, V2G, Emergency case generation
Pure electric vehicle	Much higher	Eliminated	Charging, V2G

Another way of discussing the types of electric vehicles is based on their battery sizes. This is especially the case for plug-in vehicles (PEVs), including both PHEVs and pure EVs, due to their interaction with the grid. As expected, a higher hybridization will imply larger battery size and, consequently, a larger impact on the grid operation. In addition, there are all-electric motorbikes that are developed as the fastest sports bikes [3], high comfort riders [4, 5] and even electric version of the classic Harley Davidson [6]. The electric motorbikes have a more dynamic market with a myriad of technologies used. There are models with gasoline tanks as well as fuel cells [7], while Zero Motorcycles has a model which is designed in a way that permits battery swapping. As it will be discussed in the next section, there are cases, such as Macau, where use of electric motorbikes is more preferable.

Table 2.2 shows different models of electric cars and motorbikes along with their electrical specifications [3–13].

The electric vehicles in the market and/or under development have electrical characteristics which can be represented as a significant residential load. The average battery size for electric cars is around 20 kWh while motorbikes, not considering high performance bikes such as Lightning, have batteries around 10 kWh. As it will be discussed in the next sections, these capacities constitute serious increase in residential loads and in distribution networks. The electric range is a serious parameter for estimating mass migration to electrical vehicles and its impact on drivers' behaviors which will be discussed below.

The final point to consider in this section is the types of chargers. Charging is a multifaceted issue as it involves the manufacturers, customers and other parties such as electric service providers. It involves special stations, devices, and connectors. The most important concern is the time required to fully charge a battery.

Table 2.2 Electrical specs of some electric vehicles

Vehicle type	Manufacturer	Model	EV type	Electric range (km)	Battery size (kWh)
Motorbike	Zero	DS ZF8.5	All electric	City: 152 Highway: 92	8.5
	Harley Davidson	Lightwire		53	~10
	Brammo	Empulse		City: 206 Highway: 93	10.2
	Lito	Sora		City: 200 Highway: 100+	12
	Mission	Mission R		City: 370 Highway: 225	17
	Lightning	LS-218		275	20
Automobile	Toyota	Prius	PHEV	8	4
	Chevrolet	Volt	EREV	64	16
	BMW	i3	EV	160	18.8
	Nissan	LEAF	EV	160	24
	Toyota	RAV4 EV	EV	190	27
	Cooper	Mini E	EV	251	28
	Tesla	Roadster	EV	354	53

Table 2.3 Charging options for EVs

Option	Utility	Field of use	Charging power (kW)
Level 1	110 V, 15 A	Opportunity	1.4
Level 2(a)	220 V, 15 A	Residential	3.3
Level 2(b)	220 V, 30 A	Home/public	6.6
Level 3	480 V, 167 A	Public/private	50–70

This may range between 30 minutes and 18 hours, depending on the electrical grid parameters and availability of special charging equipment [2].

Currently, there are three charging levels defined: Level 1, Level 2 (a and b), and Level 3. Table 2.3 summarizes the characteristics of these levels. All of these levels require an upgrade of the existing electrical network except for Level 1. Level 1, also called opportunity, takes very long time to charge but costs less and uses off-peak times. Level 2 can be used in houses for private use (Level 2a) or in public charging stations located at schools, train stations, or shopping centers (Level 2b with a larger charging ability). Level 3, which is fast charging, requires completely separate charging stations and special charging equipment [14]. EV users can leave their vehicles in these charging stations, similar to conventional petrol stations, and get recharged in a quick manner. This facility is especially required for vehicles with large battery packs such as Tesla Roadster, Cooper Mini, Lightning, and Mission motorbikes.

It is obvious that the electrical burden posed by EVs on the electrical grid depends on the charging option used. Level 1 can be easily deployed in residential distribution networks and thanks to its small charging power it will not have a large impact on the operation of the grid. It may sound appealing to use Level 1 for all charging applications at the first sight. However, as shown in Figure 2.4, faster charging options are required for vehicles with a thirst for power. For instance, BMW Active E needs almost a complete day to fully charge. Although preferred for their mobility and availability, even electric motorbikes had to be tied up to the grid for more than half of the day.

In relation with the above, it is expected that there will be public fast charging stations for several vehicles. Aggregating many vehicles over a fast charging option means that these stations will be a focal load point and, as discussed below, have a significant impact on the grid.

2.3 Motor Vehicle Ownership and EV Migration

As much as knowing different technologies and their interaction with the electrical grid, estimating the number of electric vehicles is also important. Electric vehicles are distributed and small-scale prosumers, i.e., an entity capable of both generation and consumption. The significant level of impact on the electrical grid is due to

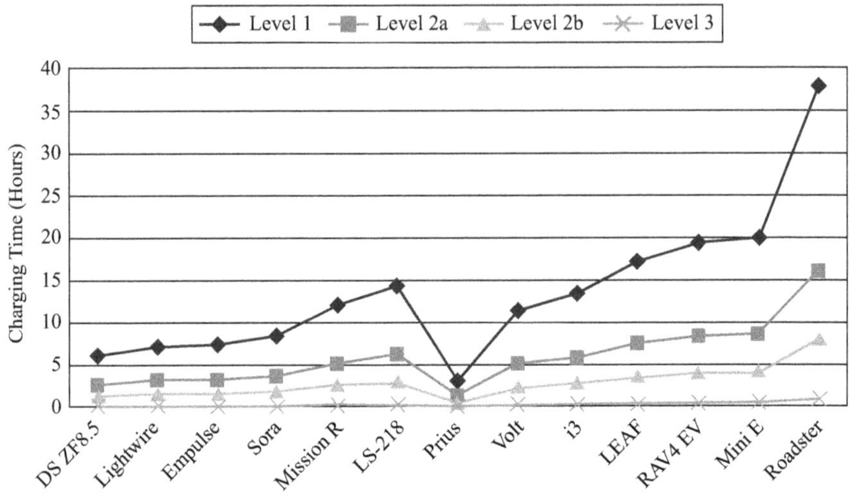

Figure 2.4 Charging times of various EVs with different charging options

massive number of vehicles present in a neighborhood. Although an individual electric vehicle does not constitute a very large component for the electric network, the cumulative effect can reach levels such that network planners and operators need to consider.

For this reason, researches have been undertaken to document the motor vehicle ownership and possible migration to electric vehicles in different countries. There are three main groups that are taken into account. The first group is comprised of countries with large surface area such as Australia [15] and the United States [16]. The common feature of these countries is the low density of population due to sparse settlement. The vastness of the residential areas, large distances that should be covered and the absence of reliable public transportation are the key reasons for the common desire to own a personal vehicle.

The second group includes countries from Western Europe. The residents of these countries have sufficient financial means to own several vehicles. However, in contrast with the first group, these countries have smaller cities with clustered population. Therefore, public transport is more reliable and preferable.

The last group focuses on some Asian cities with high incomes. These cities possess high financial means, large clusters of population and cultural codes that are different than the first two groups. In a general sense, these countries are more inclined toward public transport and the service quality is much higher. An international comparison of public and private transport usage levels is shown in Figure 2.5 (Data Source [17]).

As mentioned, the first group countries, i.e., the United States, Australia/New Zealand, and Canada, have very high portion of private motor vehicle use. Western European Union (WEU) countries, the second group, have significantly

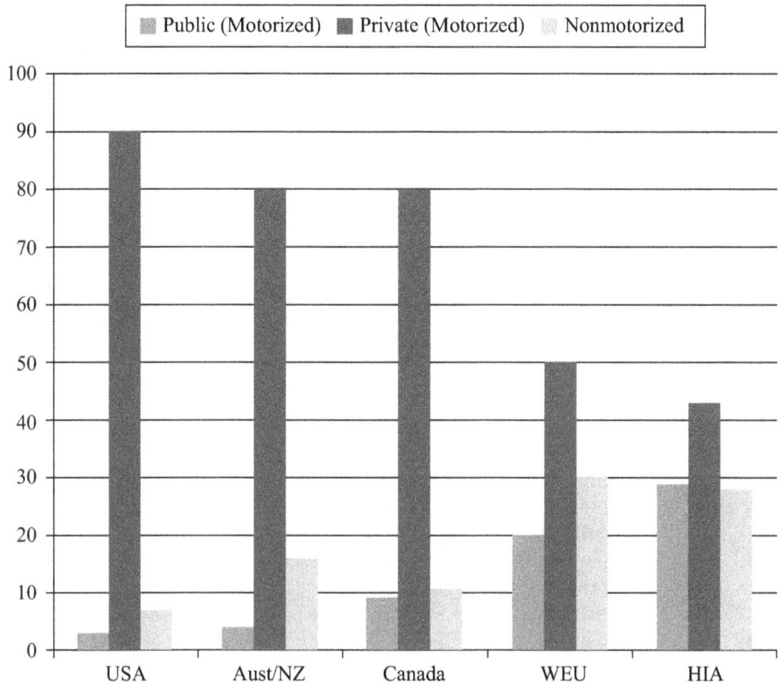

Figure 2.5 Public and private transport use in different countries [17]

higher motorized public mode and nonmotorized mode such as bicycle use. Representing the third group, High Income Asian (HIA) countries, such as Singapore, have the lowest motorize public transport use.

There are different reasons for this large contrast between these groups. First and foremost, the structure of the cities plays a major role. People living in the first group countries have a taste for stand-alone buildings and even the growing population does not change this view. In contrast, flats and apartments are more common in WEU and HIA. The natural result of this sparse and clustered living is the difference in population densities. Table 2.4 shows surface areas and populations of some major world cities. The list is arranged in descending order based on the land area. However, when special attention is paid on the population density an interesting fact can be observed. The cities in first group countries, regardless of their population, have lower population density than cities located in other groups. For instance, Washington, Paris, and Osaka have very similar land area. The population density of Washington is less than half of population density of Paris and is equal to one fifth of population density of Osaka. Alternatively, we can have a look at Chicago, London, and Beijing which had populations around 8 million in 2007. When we compare the land areas we observe that Chicago spreads over an area that is four times that of London and, surprisingly, seven times that of Beijing. A final comparison can be done with the population density, where the city with the highest

Table 2.4 Surface area and population densities of major cities [18]

City/Urban area	Country	Population	Land area (km^2)	Density (per km^2)
Tokyo/Yokohama	Japan	33,200,000	6,993	4,750
Chicago	USA	8,308,000	5,498	1,500
Atlanta	USA	3,500,000	5,083	700
Washington	USA	3,934,000	2,996	1,300
Nagoya	Japan	9,000,000	2,875	3,150
Paris	France	9,645,000	2,723	3,550
Essen/Düsseldorf	Germany	7,350,000	2,642	2,800
Osaka/Kobe/Kyoto	Japan	16,425,000	2,564	6,400
Melbourne	Australia	3,162,000	2,080	1,500
Montreal	Canada	3,216,000	1,740	1,850
Sydney	Australia	3,502,000	1,687	2,100
London	UK	8,278,000	1,623	5,100
Kuala Lumpur	Malaysia	4,400,000	1,606	2,750
Milan	Italy	4,250,000	1,554	2,750
Seoul/Incheon	South Korea	17,500,000	1,049	16,700
Berlin	Germany	3,675,000	984	3,750
Madrid	Spain	4,900,000	945	5,200
Lisbon	Portugal	2,250,000	881	2,550
Cologne/Bonn	Germany	1,960,000	816	2,400
Barcelona	Spain	3,900,000	803	4,850
Beijing	China	8,614,000	748	11,500
Shanghai	China	10,000,000	746	13,400
Manchester	UK	2,245,000	558	4,000
Rotterdam	Netherlands	1,325,000	531	2,500
Stockholm	Sweden	1,400,000	518	2,700
Singapore	Singapore	4,000,000	479	8,350
Taipei	Taiwan	5,700,000	376	15,200

value is Seoul, a HIA city and with a relatively small land area, while the city that has the lowest density is Atlanta which ranks third in terms of its land mass.

The daily commute is a significant part of life in developed countries. For instance, in the United States the impact on the population change is very high where some counties experience 94.7 and 111.4 percent population change increase, such as New York County, NY and Redmond City, Washington, while some others lose 41.4 percent of its population [19].

In spite of this daily movement, the share of public transport is very limited. The U.S. statistics indicate that Americans prefer private vehicles the most [20]. The daily work commute means distribution given in Figure 2.6 shows that 86.2 percent of the U.S. population uses private automobile for this purpose. It is also noteworthy to mention that the share of public transportation and other means (including walking) does not show real change. It is observed that the mobility need of the increasing population is met by the private automobiles. (Data Source [21]).

The share of public transport in 50 largest metropolitan statistical areas is also examined in the research. New York, New Jersey, and Long Island region lead the pack with a little more than 30 percent public transportation use. However, the

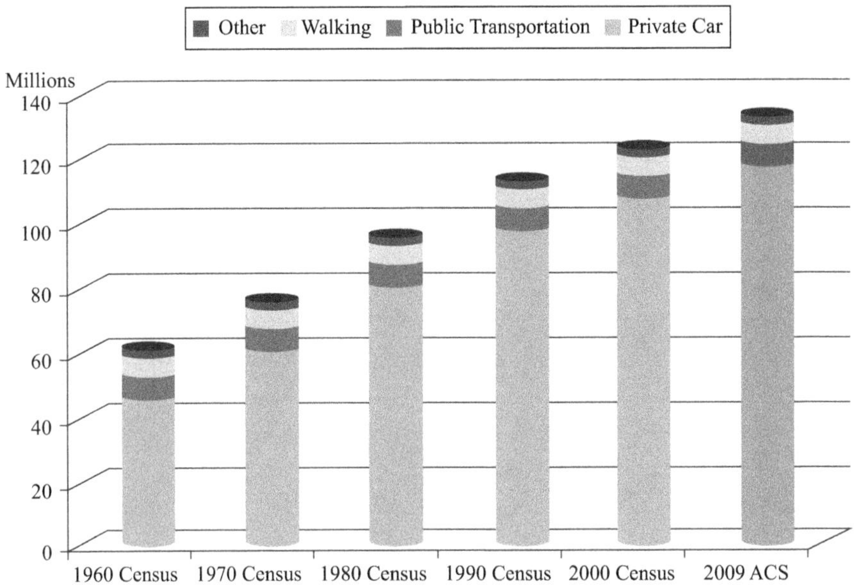

Figure 2.6 Transportation modes for daily commute in the United States

figures drop dramatically after this. The next two regions only have around 15 percent and while eighth region only has 6 percent public transport usage [22]. This low use of public transport affects the quality and frequency of public transport vehicles such as buses or trains. With less frequency, people tend to use their own vehicle more and this loop results in poor service and low preference of public transportation.

Similar trends are observed in Australia as well. The public transportation network is very weak in Australia and it is not likely to change according to Australian Bureau of Statistics. The research states that the wide-spread nature of cities makes it very difficult to extend the public transportation services to every corner of the city [23]. The ratio of public transport to other means in Australia is very lop-sided. It is worthy to note that nine out of every ten passenger-kilometers are covered by cars while bus or rail services only represent 10 percent of the total travel. The historical growth trends as well as the future projections indicate that rapidly increasing population does not change the passenger load met by the bus and rail services. The real portion still relies on privately owned passenger cars for all Australian cities.

All of the above discussion becomes relevant when it is reflected on the number of motor vehicles owned per capita. The above factors make driving a necessity rather than a luxury for the countries in the first group. U.S.-wide figures indicate the highest motor-vehicle ownership per capita in the world; 828 motor vehicles per 1,000 people [25]. As shown in Figure 2.7, in 2011, only 9 percent of American households did not have access to a vehicle while 34 percent had one vehicle, 57 percent had two or more vehicles. It is not a

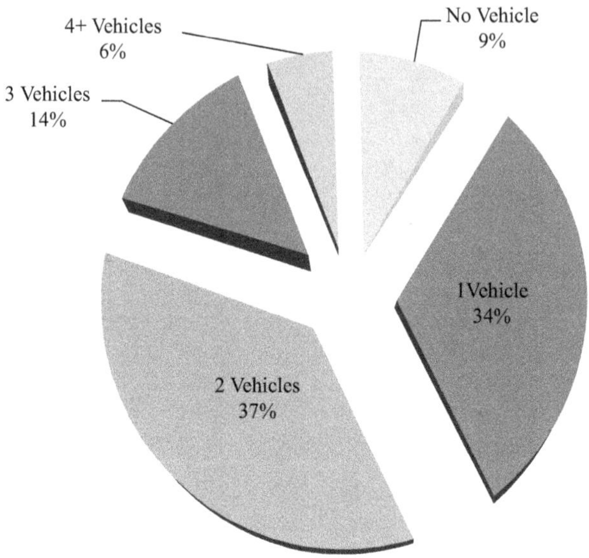

Figure 2.7 Motor Vehicle Ownership per household in the United States [25]

Table 2.5 State motor vehicle registration 1990–2009 in the United States

Type	1990	2009
All motor vehicles	188,798,000	246,283,000
Private and commercial	185,541,000	242,058,000
Public	3,257,000	4,225,000
Automobiles	133,700,000	134,880,000
Private and commercial	132,164,000	133,438,000
Public	1,536,000	1,442,000
Trucks	54,470,000	110,561,000
Private and commercial	53,101,000	108,269,000
Public	1,369,000	2,292,000

surprise that 70.2 percent of total petroleum consumption in the United States is due to transportation and this corresponds to more than 1.7 billion annual metric tons of CO_2 [25].

According to census data, the number of motor vehicles is rising in the United States. State motor vehicle registration data pertaining to 1990–2009 given in Table 2.5 indicate two key information:

(i) The overwhelming majority of the vehicles are privately owned.
(ii) About 50 percent of all vehicles registered are privately owned automobiles [26].

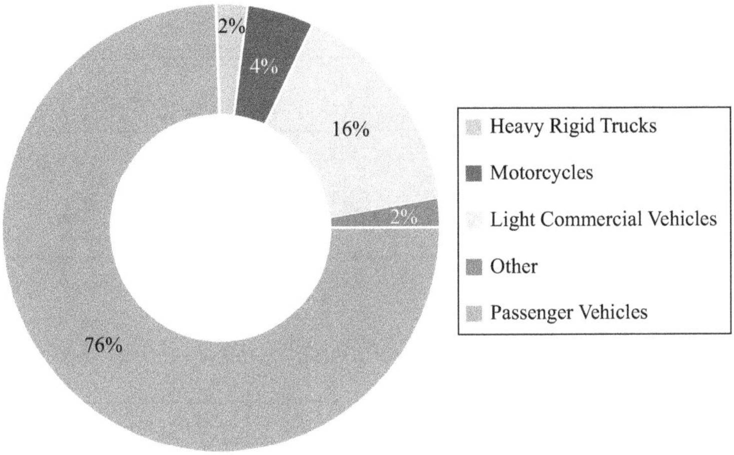

Figure 2.8 Motor vehicle types in Australia [29]

These figures do not include motorcycle registrations that reached 7,883,000 in 2009.

Passenger vehicle ownership in Australia is not different; where 85 percent of the population at least owns one car [27] and 60 percent of the households have two or more cars [28]. Figure 2.8 shows that in Australia, majority of the registered vehicles are passenger vehicles. In 2011, this number reached 16,368,383 with a 2 percent increase from 2010. Furthermore, in 2010, 72 percent of the total kilometers traveled, i.e., 2,666 billion kilometers, was covered by the same vehicle group [29].

The above data clearly show that countries in the first group have an enormous passenger vehicle fleet. Due to unreliable nature and insufficiency of the public transport, low quality of the service, safety concerns and high ticket cost, future estimations show that passenger vehicle fleet will grow in parallel with the population. This is an opportunity for electric vehicle introduction into the market. The size of the market is large and the demand is increasing with different customers from a wide range of backgrounds. As long as electric vehicles can provide the same driving range with a comparable price, there will be drivers interested in them. However, this might be the Achilles' heel for the electric vehicles some of which suffer from short ranges. In order to estimate possible migration, it is crucial to investigate daily vehicle use patterns of drivers and link this information to current electric vehicle characteristics discussed in the previous section.

Annual and daily average kilometers traveled by passenger vehicles in Australia are given in Table 2.6 [30]. The figures corresponding to capital cities and urban areas are welcoming electric vehicle technology even with the current battery technology. The average travel time to work in Australia is 39 minutes while average travel time to home is 40 minutes. On average, a passenger vehicle is used on for 69 minutes in a day [27].

On a slightly different angle, Federal Highway Administration's (FHWA) National Household Travel Survey in 2009 shows that in the United States daily

Table 2.6 Passenger vehicle use in Australia (2010)

	Capital city	Other urban areas	Other areas	Total interstate	Interstate	Australia
Total (million km)	95,619	30,787	31,400	157,806	5,555	163,360
Daily use (million km)	29.6	21.1	24.7	37.0	16.7	38.1

Table 2.7 Urban and rural travel in the United States

Travel type	Trip length (miles)		Trip time (minutes)	
	Urban	Rural	Urban	Rural
Work	11.11	15.36	22.36	24.34
Work-related	15.53	20.99	26.66	29.41
Shopping	5.49	9.50	13.47	17.14
Family/Personal	6.07	9.17	14.27	16.33
Other	10.98	16.06	20.16	25.13
Social	10.72	12.85	19.69	20.90
All	8.79	12.59	17.96	20.67

travel per person is 3.8 trips and 36.12 miles [31]. On average, daily commutes make up 25 percent of the total daily travel. Table 2.7 shows urban and rural vehicle travel based on the travel purpose, time, and the distance. On average, the urban trip of a vehicle is around 10 miles and a rural trip does not exceed 20 miles. In terms of travel time, urban and rural trips have an approximate average travel time of 20 minutes. Although rural areas have longer trips, especially for work-related trips, ranges of current electric vehicle are sufficiently long. This is crucial in capitalizing on vehicle usage and using it for electric vehicle migration.

In the light of the above discussion based on real motor vehicle data, travel time and usage pattern of drivers, it is safe to say that countries of the first group have large number of vehicles which is bound to increase in the future. This opportunity can be used to introduce electric vehicles into the market and realize vehicle migration. The daily use ranges and times lend themselves perfectly to current electric vehicle technology. With the foreseen developments in electric vehicle technology, it can be expected that even rural and inter-state travels can be covered in the near future. In this fashion, drivers can cut down on fossil fuel consumption, decrease carbon emissions, and enjoy their driving freedom all the same.

Knowing the number of the vehicle fleet, battery characteristics of electric vehicles that are likely to appear (or have already appeared) in the market is very crucial from power engineering perspective. This electrical entity is reflected on the grid as a load or a storage/generator while charging and discharging/V2G, respectively. The average driving time shows that electric vehicles will be idle approximately 95 percent of the time. This allows for different management patterns by the grid which is discussed in the following sections.

2.4 Impact of Estimated EVs on Electrical Network

Considering the potential of electric vehicles and the rate of migration to them, there is bound to be a strong impact on different levels. It is true that the electric vehicle uptake is estimated to be slow and the impact would not be experienced abruptly. However, given the current trend toward electric vehicles and the willingness to cut down dependence on cars that run on petrol, it is safe to say that numbers and impacts of electric vehicles will surge in the near future.

Even if is assumed that the share of electric vehicles is 5 percent in the entire vehicle fleet, the resulting electrical load (and generation/storage capacity) is at considerable levels. Figure 2.9 shows the estimated aggregated capacity of electric vehicles based on Lito, Leaf, and Roadster models. Lito and Leaf are chosen as representative cases for electric motorbikes and automobiles while Roadster with a large battery is chosen as a representative case for future when average vehicles have larger capacities.

The most trivial impact of electric vehicles on the grids is the increase in the load. The good news is that the impact is not as much as feared. Due to small number of electric vehicles and the expected low migration rate, the impacts of electric vehicles on the grid is manageable and sometimes automatically solved with regular upgrade and maintenance works. Southern California Edison stated in its recent report that of almost 400 upgrades only 1 percent was specifically

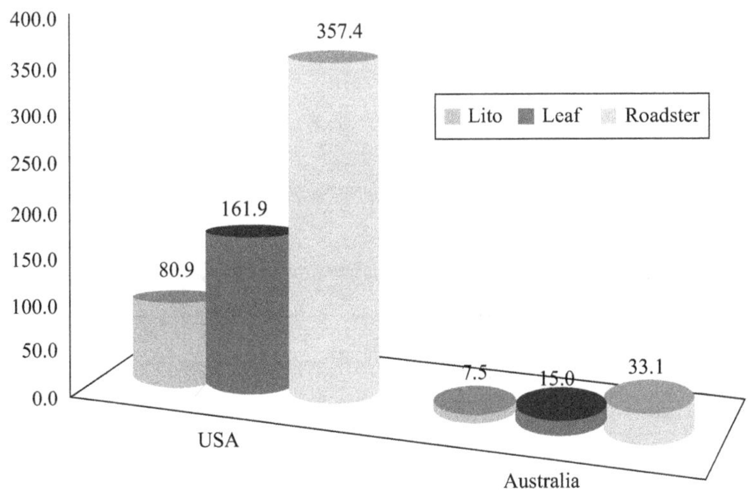

Figure 2.9 Electrical capacity of EVs (in GWh)

required for electric vehicle power demand [32]. These upgrades are only required at distribution level, since the impact of electric vehicles does not reach transmission and generation level as yet [33].

Figure 2.10 shows the national energy demand of Australia for the period 1999–2010 (Data Source [34]). Contrasting with the load constituted by the electrical vehicles given in Figure 2.9, it is observed that the impact on the national grid is negligible. Even if it is assumed that the entire vehicle fleet in Australia is replaced with electric vehicles, such as Nissan Leaf, the resulting load would be 300 GWh. This is around 0.15 percent of the national energy in 2009–2010 in Australia.

The national grid in the United States has a similar situation as well. U.S. Energy Information Administration publishes generation, transmission, market bidding, and other regulatory information about the national energy market including the electricity sector. Table 2.8 shows the generation in total electric power industry by electric utilities as well as independent power producers (IPPs) for the year 2012 [35].

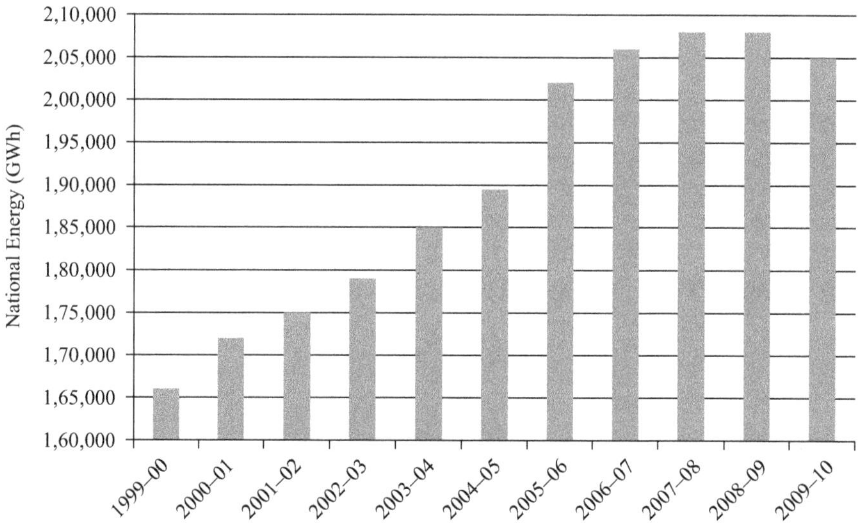

Figure 2.10 National energy demand for Australia

Table 2.8 National electric power industry in the United States

Net generation	Total (GWh)	Utilities	IPPs
Coal	1,514,043	1,146,480	354,076
Natural gas	1,225,894	504,958	627,833
Nuclear	769,331	394,823	374,509
Hydro	276,240	252,936	20,923
Renewable energy	218,333	28,017	160,064
Total	4,047,765	2,339,172	1,551,186

The population is much higher in the United States than Australia, but so is the electricity market level. The estimated electric vehicle load in the United States, which is 161.9 GWh, is infinitesimally small when compared with the national grid generation levels. When a full electric vehicle migration is assumed in the United States based on Nissan Leaf, the resulting electric load would be only 0.8 percent of the national grid.

Therefore, power engineers are aware of the fact that the immediate impact of electric vehicles would be at the distribution level. The increasing load profile in residential neighborhoods as well as aggregated load caused by electric vehicle charging stations has to be taken into account by distribution network operators. Several simulation works have been undertaken to verify these results as well. Figure 2.11 shows a sample neighborhood developed to investigate the impact of the electric vehicle charging and power injection to the grid. Simulations have been performed for stand-alone operation and operation when connected to IEEE T14 and 34 Bus systems. The results listed in [15] show that the impact of electric vehicles in the grid is only at the distribution level and no change is required at transmission and generation level [35–37]. Also, the upgrades that are required can be made gradually as the electric vehicle migration progresses.

The impact of electric vehicles as a load is one aspect and as a potential distributed storage is another. The estimated storage capacity and the idle time during which storage can be realized have been discussed above. It is also vital to investigate and relate daily load profiles for distribution systems and daily vehicle usage patterns. Figure 2.12 shows distribution of vehicle trips over the 24-hour period (dark grey-commute, dashed line-family/personal, light grey-school, dotted line-social/recreational, black-total, Data Source [31]).

Regardless of the motivation for the trip, almost all trips tend to rise around 5 a.m. until 8 a.m., and 5 p.m. until 7 p.m. This pattern is especially true for daily commute to work and school. The reverse analysis of the situation indicates that the vehicles are

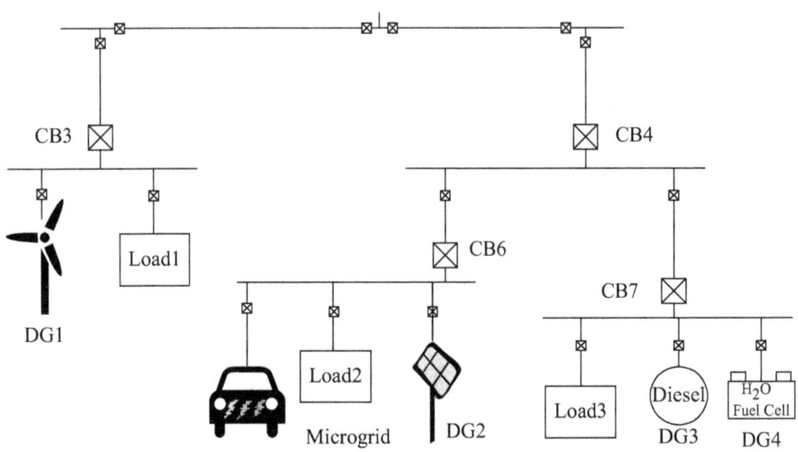

Figure 2.11 Sample microgrid with EV deployments

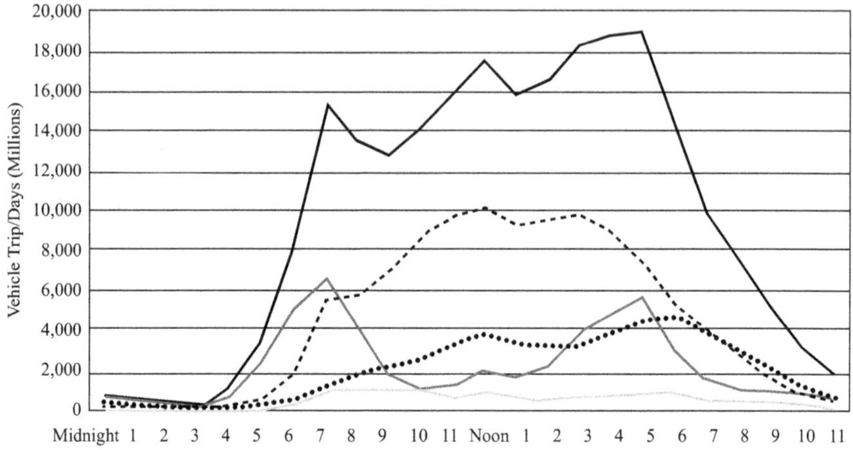

Figure 2.12 Daily driving pattern in the United States

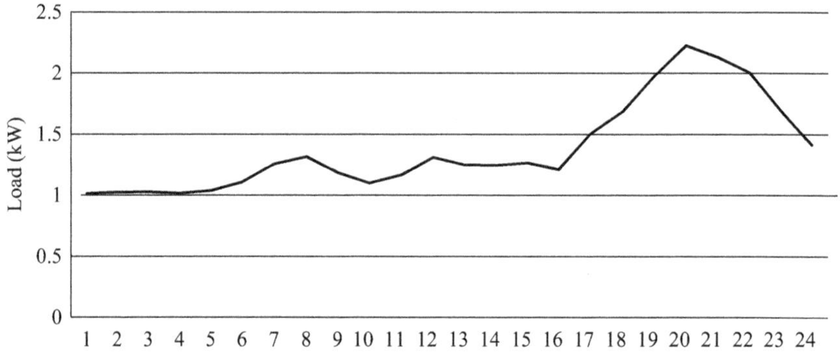

Figure 2.13 Load profile of a house in the United States

idle (i.e., parked) during other times and available for V2G purposes. There is ample time between two peak times both during the day time and during the night. Therefore, power operators will have the opportunity to use the charge in the batteries to support the grid and recharge the batteries before the next peak hour. In order to extract the possible effects on the grid, consider Figures 2.13 and 2.14 which depict daily load profiles for a household and a school as provided by Xcel Energy [38].

Contrasting Figures 2.12, 2.13, and 2.14 reveal the opportunity posed by electric vehicles in terms of grid support. Drivers leave their home and arrive at the school (for other cases, work place) before the peak hour starts. As a matter of fact, school load increases as people arrive. Therefore, the amount of vehicles arriving at the premises is directly proportional to the load increase. During day time, the energy is required at the school and the vehicles parked outside can energize the microgrid as distributed storage systems. The load is decreasing at

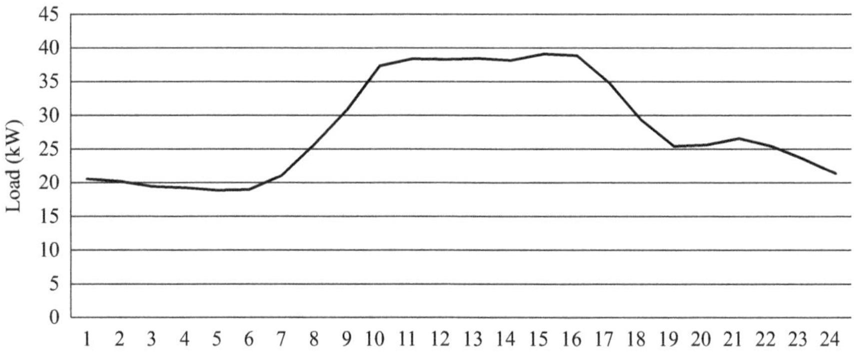

Figure 2.14 Load profile of a school in the United States

school when people start leaving for home. The peak load at home increases in line with the number of people, hence vehicles, reaching home. The vehicles will be parked at the houses during peak hours. Generally speaking, majority of the drivers would only need their vehicles in the morning to commute to work. Therefore, the charging time can be postponed to off-peak hours. The owners would not be affected as long as the car is fully charged in the morning. This can also help them save money in two ways:

1. By selling power back to grid during peak hours when the prices are high.
2. By delaying charging until off-peak hour when the prices are significantly lower.

This pattern is also true for Australia, the daily vehicle regime of which is shown in Figure 2.15 in terms of combined vehicle-kilometers traveled (VKT) (Data Source [24]). Considering isolated communities which rely on stand-alone microgrid [37, 39], this additional storage provided by V2G technology contributes much needed support for reliability and control.

Although grid operators are looking into ways of using batteries of electric vehicles, V2G technology is not implemented commercially yet. There have been some pilot projects in by PG&E and Xcel Energy, both in the United States. These are projects of very small scales and their impact on the larger grid has not been investigated. The main idea was to verify the applicability of V2G technology. These pilot systems were successfully implemented and the applicability of the technology has been verified.

As far as the power quality is concerned, electric vehicles can be classified as very "well-behaving" loads thanks to their power electronics interface and sophisticated control. Moreover, should the necessary hardware infrastructure be provided, it is possible to use electric vehicles to increase the power quality at the point of common coupling. According to a recent publication, reactive power compensation can be performed both during charging and discharging (V2G operation). It is also documented that reactive power compensation can be realized without affecting the charging time and/or state of the charge in the battery [40].

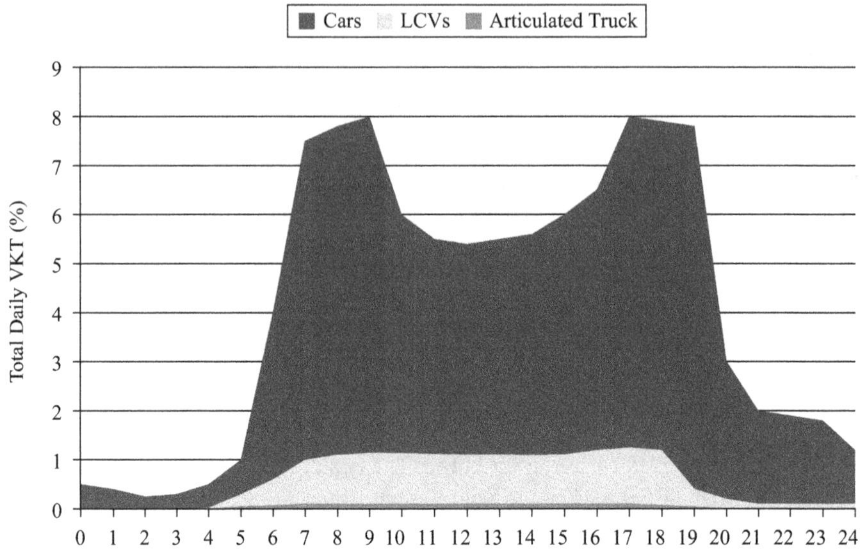

Figure 2.15 Daily commuting pattern in Australia for vehicle types

Despite being at pilot implementation stage, these features will ensure that electric vehicle deployments enhance the power quality and reliability in local networks and microgrids.

2.5 Impact on Drivers and the Smart Grid

Depending on the range and the capacity of the electric vehicle that is used, the drivers might need to plan their routes and car usage more wisely. This is not a very big issue of PHEVs that can run on petrol and electricity, but, in the first years of fleet migration, EVs are likely to force users to be more aware of their usage. As discussed above, statistical data show that majority of daily car use is less than the range of electric vehicles and there is not a change in a driver's daily routine. Long road trips require planning ahead. The number of available charging stations and available spots may constitute a problem in the beginning. As the EV market grows and auxiliary services improve, it is natural that there will be ample charging stations and drivers can charge their vehicles at will. Furthermore, fast charging infrastructures are expected to be deployed. During a short break, the charging can be completed even for a battery with large capacity. Nevertheless, as the market picks up, there will be certain growth and feasibility problems with these technologies.

Another effect on the drivers is the changing charging times with demand side management by the network operators. Vehicle owners may opt to postpone their charging into off-peak hours and save on their utility bills. The owners have the freedom to choose whether or not to participate in this scheme and normally it will not have an effect. However, should a need for emergency drive arise in the middle of the night (e.g., hospital) the battery may not have sufficient charge. This can be solved with the improvements in charging infrastructure. Although it is not realistic to expect this in the near future, houses may have super-fast chargers that minimize the charging time. Therefore, even with demand side management participation, the charging will take very short period of time.

The interface between the smart grid applications and electric vehicles offers exciting opportunities. One example of this is "smart charge" technology. This technology is developed to meet the differing, and sometimes conflicting, needs and preferences of vehicle owners and the grid operator. It goes without saying, the realization of such a scheme relies on extensive communication and coordination. For successful implementation, a myriad of devices such as smart meter; the vehicle, battery and the charger; vehicle owner and the grid operator should be synchronized. Therefore, smart charging is not only a power transfer operation but also a power management and optimization problem with many factors included.

This is where the smartness of the grid comes into play. The electric vehicle charging systems are equipped with IT-based communication and control technologies. The related parties, i.e., owner, the grid operator, have access to these systems through same IT technologies. The information is exchanged over communication lines (wired or wireless) in line with implemented communication standards. The next section discusses the standardization of electric vehicle communication.

As an example of the smart applications it is possible to mention ChargeIQ. Developed by an Australian Company, DiUS, the smart charging solution comprises a charger, smartphone app, and ZigBee-based communication [41]. The operation details of ChargeIQ are as follows. The network operator communicates with the smart meter via wide area network (WAN). This is required for power dispatch, charging management and V2G applications, if applicable. The smart meter is coupled with the smart meter over house area network (HAN). The parameter or preference changes from both grid operator and the vehicle owner are reported through this coupling. The smart charger is directly accessible by the user over the internet via a smartphone app. The user can control and input his preferences through this internet based connection.

ChargeIQ smartphone application provides necessary information to the user. The user has the ability to know the connection status, the time required to fully charge his vehicle, the time when the vehicle will be fully charged as well as the current state of the charge (SOC). These features are designed to minimize the impact of electric vehicles on the driver which are discussed above. Furthermore, the user can choose between "ChargeIQ" and "Charge Now," grid-friendly and owner-friendly charging options, respectively.

As it is used by DiUS, grid friendly charging corresponds to the concept of postponing vehicle charging to off-peak hours. This is a permission given by the user for the network operator to manage the charging time. In response to this assistance, the network operators offer incentives such as upfront payments or discounts in the bills. To implement this concept in ChargeIQ, a query screen is shown to the owner, asking whether he is willing to participate in grid-friendly charging. When agreed, ChargeIQ is managing the charging process and the power transfer is stopped during peak hours, as shown below. This information is reported to Network Operator for dispatch planning via smartphone – smart charger – smart meter link. This process increases the charge time and the user saves an extra $5.00 payment. The difference between on-peak and off-peak charging costs is also another gain for the user.

In addition to this individual coordination of chargers in houses, a tool to locate available charging stations and gather charging cost information is also required. Chargepoint [42] is an electric vehicle charging solution which lists near-by charging spots, their availabilities and costs. This solution is also internet-based and has a smartphone application.

When a specific location is selected, say Honolulu, the application can be used to compare prices and get directions to a selected charging station. Furthermore, Chargepoint allows users to create their own account, save their preferences, and keep record of their recent charging stations.

Chargepoint only focuses on the United States and Canada (with only two other charging stations in China). On the other hand, EV-charging, a more powerful tool, lists all charging stations in Europe, including Istanbul, Turkey [43]. EV-Charging solution has more embedded information such as plug type (AC, DC), charging Volt (400, 230, 36 V), operation times and cost as well as the type of accepted vehicles such as e-cars, e-bikes, and e-boats.

The smartphone application of EV-charging helps locate charging stations that are available for charging and suitable for the vehicle in terms of charging voltage, charger socket type, etc. The most important contribution of these applications is fast and easy detection of available charging stations even in places that are completely unknown to the driver and make the use of EVs more manageable by decreasing their impacts on the driver.

2.6 Standardization and Plug-and-Play

Previous sections discussed the communication required for the interaction between electric vehicles and the smart grids. The smart grid technologies that monitor the state of charge of the battery and devise charging plans need to communicate with the vehicles. Similar to all components of smart grids, electric vehicles need to have the ability to interface and operate with other components.

In order to achieve the above, only using standard chargers is not sufficient. Electrical vehicles should be modeled in communication systems. This is important to receive information such as the battery capacity of the vehicle, the state of the charge in the battery, the owner's preference to participate in V2G or not, and the time when the vehicle needs to be fully charged. These data are inevitable for proper planning in electrical networks.

Considering the number of different manufacturers in the car industry (also motorbikes), a nonstandard approach to communication modeling will result in different models that cannot talk to each other. Although companies usually prefer to keep their information under propriety rights, electrical vehicle models need to be transparent and standard.

In a properly functioning electrical grid, electrical vehicles from different manufacturers are likely to be connected to the same charging station and interact with the network operation components. The same charging station or the network planning device needs to be able to successfully detect and communicate with these different models. This can only be achieved by a universal standard modeling approach. There are several researches conducted in this field with specialized research centers such as Electric Vehicle-Smart Grid Interoperability Center [44, 45]. This chapter focuses on IEC 61850 communication standard which is very popular in power engineering circles.

2.6.1 IEC 61850 Communication Standard and IEC 61850-7-420 Extension

The need for a standard communication in power systems has been felt long before the introduction of electric vehicles. The communication required for supervisory control and data acquisition (SCADA) systems consist of several instruments from different vendors. The ability to operate these instruments seamlessly relies on a standard communication where, regardless of their manufacturer or model, all equipment speaks the same language.

IEC 61850 Substation Communication Standard was developed to address this need. It supports all parameters and functions in a substation by proposing data models [46]. Shortly after its publication, IEC 61850 created much interest among power engineers and is in use ever since. In parallel with the latest developments in microgrid and renewable energy areas, International Electrotechnical Commission (IEC) formed Workgroup 17 to prepare an extension to IEC 61850-7-4 which defines compatible logical nodes (LN) and data classes [47, 48]. This extension is aimed at defining new generation components:

1. solar panels
2. diesel generators
3. fuel cells
4. combined heat and power

In addition to function of switching DERs on and off, DER systems also involve:

- Managing of the interconnection between the DER units and the power systems they connect to, including local power systems, switches and circuit breakers, and protection.
- Monitoring and controlling the DER units as producers of electrical energy.
- Monitoring and controlling the individual generators, excitation systems, and inverters/converters.

- Monitoring and controlling the energy conversion systems, such as reciprocating engines (e.g., diesel engines), fuel cells, photovoltaic systems, and combined heat and power systems.
- Monitoring and controlling the auxiliary systems, such as interval meters, fuel systems, and batteries.
- Monitoring the physical characteristics of equipment, such as temperature, pressure, heat, vibration, flow, emissions, and meteorological information.

The system assumes a holistic sense in which the DER systems are modeled starting from their internal parameters (e.g., fuel type for diesel generators, battery test results for solar panels or hydrogen levels for fuel cells) to their grid connection types and parameters and even microgrid operator commands and control units. Having detailed characteristic variables and measurement values entrenched inside; this modeling system serves for a rigorous communication system. The extended discussion about this standard and its modeling can be found in [49]. For the sake of simplicity, this section would focus on standard approach taken for modeling the electric vehicles.

2.6.2 Extending IEC 61850 for Electric Vehicle Modeling

IEC 61850-7-420 extension is a very suitable and timely move to include renewable energy based generators. However, electric vehicles and functions pertaining to them are still not included. As discussed, electric vehicles are not mere loads that sink power from the grid. They can be used as distributed storage systems and can participate in management efforts through communication and control. Therefore, it is logical to introduce a new model in IEC 61850-7-420 for electric vehicles. This will ensure that messages and parameters are sent in a universal way so that electric vehicles, regardless of their models and manufacturers, can smoothly communicate and cooperate with other grid elements.

A logical node (LN) has been developed earlier for this purpose [50]. In IEC communication standard terminology an LN is a sub-function located in a physical node and exchanges data with other distinct logical entities. In simple terms, LNs are representations of real devices in the virtual world. To put it into perspective, they are real devices such as battery systems, energy converter, circuit breakers, and measurement units. On the other hand, 4-letter acronyms located above these rectangles are the names for their representations in the real world, i.e., names of the LNs. Several LNs might be used to model a single device. For instance, ZBAT is used to model battery and its parameters while ZBCT is used to model the charging interface and its features.

The controllers can also be modeled as a separate entity such as "DRCT" which is used as a standard controller model for all distributed energy resource (DER) controllers. This LN defines the control characteristics and capabilities of one DER unit or aggregations of one type of DER device with a single controller. If EVs are to be included as a separate DER, this class has to be updated. In the original document, IEC 61850-7-420 includes the four different types mentioned above. It is possible add another entry to DRCT Class, "DERtyp = 6" as shown in Table 2.9.

Table 2.9 DRCT class with electric vehicle addition

<table>
<tr><th colspan="5">DRCT Class</th></tr>
<tr><th>Data name</th><th>CDC</th><th>Explanation</th><th>T</th><th>M/O/C</th></tr>
<tr><td>LNName</td><td></td><td>Shall be inherited from logical-node class (see IEC 61850-7-2)</td><td></td><td></td></tr>
<tr><td colspan="5">Data</td></tr>
<tr><td colspan="5">System LN data</td></tr>
<tr><td></td><td></td><td>LN shall inherit all mandatory data from common LN class</td><td></td><td>M</td></tr>
<tr><td></td><td></td><td>Data from LLN0 may optionally be used</td><td></td><td>O</td></tr>
<tr><td colspan="5">Settings</td></tr>
<tr><td>DERNum</td><td>ING</td><td>Number of DER units connected to controller</td><td></td><td>M</td></tr>
<tr><td>DERtyp</td><td>ING</td><td>Type of DER unit:
<table>
<tr><td>Value</td><td>Explanation</td></tr>
<tr><td>0</td><td>Not applicable/unknown</td></tr>
<tr><td>1</td><td>Virtual or mixed DER</td></tr>
<tr><td>2</td><td>Reciprocating engine</td></tr>
<tr><td>3</td><td>Fuel cell</td></tr>
<tr><td>4</td><td>Photovoltaic system</td></tr>
<tr><td>5</td><td>Combined heat and power</td></tr>
<tr><td>6</td><td>Electric vehicle (with V2G technology)</td></tr>
<tr><td>99</td><td>Other</td></tr>
</table></td><td></td><td>M</td></tr>
<tr><td>MaxWLim</td><td>ASG</td><td>Nominal max output power</td><td></td><td>M</td></tr>
<tr><td>MaxVarLim</td><td>ASG</td><td>Nominal max output reactive power</td><td></td><td>M</td></tr>
<tr><td>StrDITms</td><td>ING</td><td>Nominal time delay before starting/restarting</td><td></td><td>M</td></tr>
<tr><td>StopDITms</td><td>ING</td><td>Nominal time delay before stopping</td><td></td><td>M</td></tr>
<tr><td>LoadRampRte</td><td>ING</td><td>Nominal ramp load or unload rate, power vs. time</td><td></td><td>M</td></tr>
</table>

This addition ensures that electric vehicles can be defined in an electrical grid, in the same way a diesel generator or fuel cell is. Being an identity card, DRCT will serve as an interface with network operator (such as SCADA system) and identify the electric vehicle. The other components of electric vehicles should be modeled with the components explained above. For instance battery and the charger can be modeled with ZBAT and ZBCT that are already present in the standard. The grid connection can be modeled through an inverter/converter and a switch. This topology is already modeled with previously developed LNs with ZRCT, ZINV, CSWI, and XCBR.

This holistic modeling includes all components of an electric vehicle and its connection to the grid. The only missing link is a control entity that holds key information about charging/discharging status about the vehicle and preferences inputted by the owner. Hence, a new LN class called electric vehicle control (EVCT) has been developed in [50] as shown in Table 2.3 to reflect the sub-function required for monitoring the critical functions and states of the V2G process. The EVCT LN node will hold the answers to the following questions:

- When to start the vehicle-to-grid (discharge) process?
- At what time during the day or night should the battery be fully charged?
- When to charge the car?
- Is demand side response in operation? Economy charging out of peak-hours or immediate charging at the time of connection?
- How much power has been supplied to the grid? How much imported?
- What are the settings for audible and visual signals?

The EVCT class defines the data about the charging and discharging (V2G) operations as discussed above in relation to ChargeIQ. The preferences of the owner, the incoming offers/commands from the network operator and the internal parameters of the vehicle are defined and held by EVCT in a standard manner. Table 2.10 shows the developed EVCT class and its contents [50]. It is split into four categories, i.e., "Settings," "Status information," "Controls," and "Measured values." The settings section holds the information that is required for successful interaction with the grid. Some of these are preferences of the owner such as ChrgReady, some other might be a parameter agreed on by the owner and the grid operator such as V2GStart while other might be built-in parameters set by the manufacturer such as Alim and Vlim. These current and voltage values shall be in compliance with the standard IEC 61851-22: ac electric vehicle charging station [51]. This standard stipulates upper and lower limits for voltage, current, and frequency for different countries. This way it is ensured that the proposed EVCT class can be used worldwide.

Another parameter that is tailored toward worldwide use is ChrgMode that holds the value for the selected charging mode among the modes defined by IEC 61851-1: Electric Vehicle Conductive Charging System, General Requirements [52]. This parameter is vital for providing a standard LN for EVs since different countries have different grid parameters and grid codes. For example, Mode 1 charging which is prohibited in the United States by the national regulations can be used in Japan and Sweden for residential charging. This way the same modeling can be used for different countries with necessary tweaks, such as disabling Mode 1 for vehicles in the United States. It is a solid advantage that both of the standards utilized in EVCT LN, i.e., IEC 61851 and IEC 61850, are prepared by the IEC. This means they are compatible and complementary with each other.

The mandatory parameters under "Status information" in Table 2.10 are required to transmit owner preferences to the grid operator. When the network operator connects to electric vehicle over EVCT and reads V2Gstatus and EconStatus, the preferences of the owner about whether or not to participate in V2G and economic charging (grid-friendly charging as used by ChargeIQ) are acquired. The last two optional variables show the status of visible and audible signals showing on-going and completed charging. Depending on the policies and regulations, two control inputs, V2GEnable and EconCharge, can be toggled either locally by the owner for demand response or by the central network control for demand side management.

Table 2.10 EVCT class for EV communication and control

EVCT Class				
Data Name	**CDC**	**Explanation**	**T**	**M/O/C**
LNName		Shall be inherited from logical-node class (see IEC 61850-7-2)		
Data				
System LN data				
		LN shall inherit all mandatory data from common LN class		M
		Data from LLN0 may optionally be used		O
Settings				
V2Gstart	ASG	V2G-Allowed period start time		O
V2Gend	ASG	V2G-Allowed period end time		O
ChrgReady	ASG	Time when EV should be fully charged		O
Alim	ASG	Input current limit		O
Vlim	ASG	Input voltage limit		O
ChrgMode	ING	Charging Mode (see IEC 61851 – 1) Value | Explanation 1 | Charging Mode 1 2 | Charging Mode 2 3 | Charging Mode 3 4 | Charging Mode 4		M
Status information				
ConnCount	INS	Count of grid connection		M
V2Gstatus	SPS	True: V2G participation is On False: V2G participation is Off		M
EconStatus	SPS	True: Economic charging is selected False: Immediate charging is selected		M
Charging signal	SPS	True: Charging indicator is On False: Charging indicator is Off		O
BattFullAudibleSignal	SPS	True: Battery full audible signal is On False: Battery full audible signal is Off		O
Controls				
V2Genable	DPC	Switch on/off V2G participation, On = True, Off = False		M
EconCharge	DPC	Toggle between economic and immediate charging, Economy = True, Immediate = False		M
Measured values				
Supplied power	MV	The amount of power supplied to the grid through V2G scheme		O
Received power	MV	Power received for charging the batteries		O

The optional measurements are aimed at keeping record of energy transfer between the grid and the EV. However, measurement can also be performed through smart meters and in that case separate modeling would be required.

Figure 2.16 shows how electric vehicles can be integrated into IEC 61850-7-420 blueprint given in IEC 61850-7-420 [50]. It is possible to utilize these models not only for an individual EV but also for aggregated V2G pools which consist of

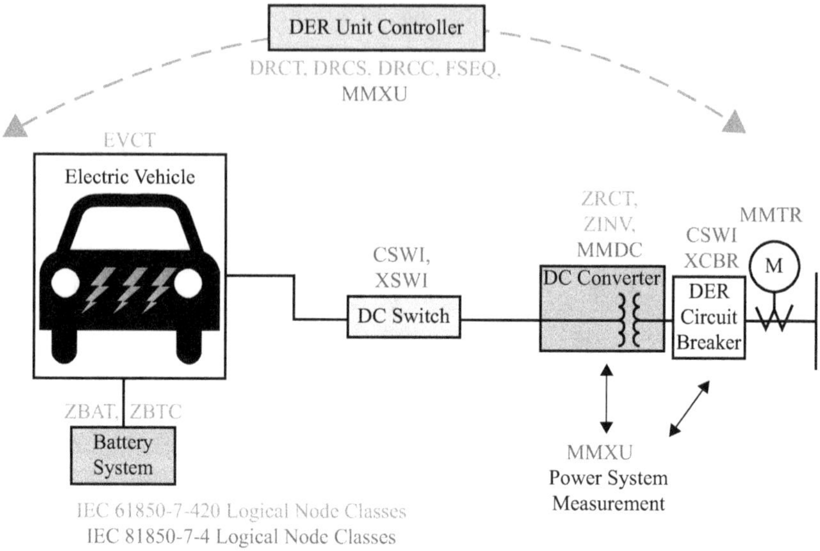

Figure 2.16 Universal electric vehicle modeling IEC 61850-7-420 [50]

several EVs. The generic nature of EVCT class is very versatile and accommodates different models. Thanks to universal approach assumed, various EVs produced by different manufacturers can be conveniently modeled and used separately or together in a pool. The seamless communication between different vehicles and the network operators is achieved with this communication model that provides EV parameters in a standard fashion.

2.7 Conclusion

There are several factors that make electric vehicles popular such as the will to reduce carbon emissions and dependency on imported oil due to security concerns. The availability of the technologies for wide spread electric vehicle implementation increased the interest in this market. The incentives given by the governments are pushing this sector forward and several manufacturers have already assembled their own electric vehicles in the form of an automobile or even motorbike. The expectation into the future is the steady growth of electric vehicles' share in vehicle fleets.

Due to the technology and infrastructure required, only developed countries are candidates for large electric vehicle roll-out. Among them, large countries with sparse population and low use of public transport pose a special opportunity for quicker migration to electric vehicles. These countries have households with high vehicle ownership. Therefore, the impacts of electric vehicles have to be taken into account by power engineers and network operators.

The impact studies undertake focus on both power sinking and sourcing features of electric vehicles. The extra load introduced with electric vehicles causes no issues at generation or transmission levels. The distribution networks may require some upgrades and these can be solved by regular upgrade works and in a gradual manner as the electric vehicle market matures. On the other hand, the unique opportunities posed by V2G technology require some serious effort both on technological and financial fronts before it becomes commercially viable. Research and development efforts show that V2G technology can be used to support grids in terms of power supply and power compensation. The fact that usual driving pattern lends itself to extensive use by the grid operators makes this technology promising for the future.

The interaction of electric vehicles with the smart grid can take quite innovative forms including smartphone apps, communication for demand side management and dynamic control as well as searching charging stations over the Internet. This communication interface requires a standard approach for different electric vehicle models for plug-and-play convenience.

In short, electric vehicles are becoming popular for technological, political, and financial reasons with multifaceted impacts on the grid operators, customers, and the infrastructure. In addition, this poses exciting opportunities for unprecedented applications both at macro and micro scales. For a smooth migration, this multidisciplinary area should be thoroughly investigated by researchers from different fields of expertise.

References

[1] C. C. Chan, "The state of the art of electric and hybrid vehicles," Proc. IEEE, vol. 90, no. 2, pp. 247–275, 2002.

[2] A. Emadi, "Transportation 2.0," *IEEE Power Energy Mag.*, vol. 9, pp. 19–29, July/August 2011.

[3] LS-218 [Online]. Available from http://lightningmotorcycle.com/product/specifications/.

[4] Empulse, Empulse R [Online]. Available from www.brammo.com.

[5] Zero DS models, Zero Motorcycles [Online]. Available from www.zeromotorcycles.com.

[6] Project Livewire [Online]. Available from http://www.gizmag.com/harley-davidson-project-livewire-electric-motorbike/32613/.

[7] Yamaha FC-AQEL [Online]. Available from http://global.yamaha-motor.com/news/2006/1019/fc-aqel.html.

[8] Zero Motorcycles Specs [Online]. Available from http://www.zeromotorcycles.com/zero-ds/specs.php.

[9] BMW i3, *The BMW i3 Brochure*, BMW AG, Munich, Germany, 2014.

[10] Tesla Motors [Online]. Available from http://www.teslamotors.com/.

[11] Holden–Volt [Online]. Available from http://www.holden.com.au/pages/volt-coming-soon.

[12] Toyota–Prius [Online]. Available from http://www.toyota.com/prius-plug-in/.

[13] Mini Cooper–Mini E [Online]. Available from http://www.miniusa.com/minie-usa/.
[14] A. Emadi, *Handbook of Automotive Power Electronics and Motor Drives*. Boca Raton, FL: CRC Press, p. 736, 2005.
[15] T. S. Ustun, A. Zayegh, and C. Ozansoy, "Electric vehicle potential in Australia: Its impact on Smartgrids," *IEEE Industrial Electronics Mag.*, vol. 7, no. 4, pp. 15, 25, December 2013.
[16] S. W. Hadley and A. A. Tsvetkova, "Potential impacts of plug-in hybrid electric vehicles on regional power generation," *Electricity J.*, vol. 22, no. 10, pp. 56–58, December 2009.
[17] J. Kenworthy, A study of 84 global cities: Transport Energy Use and Greenhouse Gases in Urban Passenger Transport Systems [Online]. 2003. Available from http://goo.gl/aFzm8d.
[18] Largest cities in the World, Ranked by Surface Area, Population and Density, 2006, City Mayr Statistics [Online]. Available from http://goo.gl/sKhV.
[19] B. McKenzie, W. Koerber, A. Fields, M. Benetsky, M. Rapino, "Commuter-Adjusted Population Estimates: ACS 2006-10," Journey to Work and Migration Statistics Branch, U.S. Census Bureau.
[20] U.S. Census Bureau, "American Community Survey, 2008–2012," Version 2, published on February 12, 2014.
[21] B. McKenzie and M. Rapino, "Commuting in the United States: 2009," American Community Survey Reports, September 2011.
[22] U.S. Census Bureau, American Community Survey, 2009.
[23] National Transport Commission (NTC), Smarttransport for a growing nation project, 2011.
[24] Australian Government and Department of Transport and Regional Services, Estimating urban traffic and congestion cost trends for Australian cities, Working Paper 71, 2007.
[25] U.S. Department of Transportation, Research and Innovative Technology Administration Bureau of Transportation Statistics, Transportation Statistics Annual Report, 2012.
[26] U.S. Census Bureau, Statistical Abstract of the United States: 2012.
[27] J. Hoye, I. Andreadakis, and S. Vercoe, National transport commission smart transport for a growing nation: Public attitudes to mobility and access social research report, GA Research, 2011.
[28] "Australian social trends," Australian Bureau of Statistics, Canberra, Australia, 4102.0-2008.
[29] "Motor vehicle census, Australia," Australian Bureau of Statistics, Canberra, Australia, 9309.0-2011.
[30] "Survey of motor vehicle use," Australian Bureau of Statistics, Canberra, 9208.0-2010.
[31] Federal Highway Administration, National Household Travel survey (NHTS), 2009.
[32] Available from http://cleantechnica.com/2013/08/13/southern-california-edison-offers-insight-into-impact-of-electric-vehicles-on-grid/.

[33] S. Rahman, "How can we minimize electric vehicle's impact on the electric power distribution network (Keynote speech)," in ISGT ASIA, Perth Australia, 15, November 2011.

[34] R. Garnaut, *The Garnaut Climate Change Review*. Cambridge University Press: Port Melbourne, Australia, 2008.

[35] U.S. Energy Information Administration, Total Electric Power Industry, 2011 and 2012 [Online]. Available from http://www.eia.gov/electricity/annual/html/epa_01_01.html.

[36] L. Dow, M. Marshall, X. Le, J. R. Aguero, and H. L. Willis, "A novel approach for evaluating the impact of electric vehicles on the power distribution system," in 2010 IEEE Power and Energy Society General Meeting, pp. 1–6, 2010.

[37] Horizon Power, Kalumburu power project (ARCPSP) [Online]. 2011. Available from http://www.horizonpower.com.au/384.html.

[38] Xcel Energy, Integrated Resource Plan, Appendix D, Hourly Load Profiles, 2012.

[39] Hydro Tasmania, "King Island renewable energy integration project," in CEIC, Storage Tech Clinic [Online]. 2010. Available from www.kingislandrenewableenergy.com.au/.

[40] M. Kesler, M. C. Kisacikoglu, and L. M. Tolbert, "Vehicle-to-grid reactive power operation using plug-in electric vehicle bidirectional offboard charger," Industrial Electronics, IEEE Transactions on, vol. 61, no. 12, pp. 6778, 6784, Dec. 2014.

[41] ChargeIQ, Smart Charging Solution, by DiUS [Online]. Available from http://percepscion.com/chargeiq/.

[42] ChargePoint, Smart Charging Station Application [Online]. Available from www.chargepoint.com.

[43] EV-Charging [Online]. Available from www.ev-charging.com.

[44] T. S. Ustun, C. R. Ozansoy, A. Zayegh, "Implementing vehicle-to-grid (V2G) technology with IEC 61850-7-420," IEEE Transactions on Smart Grid, vol. 4, no. 2, pp. 1180, 1187, June 2013.

[45] Keith, Hardy Electric Vehicle-Smart Grid Interoperability Center, Argonne National Laboratory. (March, 2012). Available from http://goo.gl/w9l6VY.

[46] IEC/TR 61850-1, "Communication networks and systems in substations," First Edition: International Electrotechnical Commission, 2003.

[47] IEC TC-57, "Communication networks and systems in substations – Part 7-420: Basic communication structure – Distributed energy resources logical nodes," IEC Standard IEC/TR 61850-7-420, Edition 1.0, 2009.

[48] I. T. WG17, "Introduction to IEC 61850-7-420: Distributed energy resources (DER) object modeling," White Paper, July 31, 2009.

[49] T. S. Ustun, C. Ozansoy, A. Zayegh, "Extending IEC 61850-7-420 for distributed generators with fault current limiters," Innovative Smart Grid Technologies Asia (ISGT), 2011 IEEE PES, pp. 1, 8, 13–16 November 2011.

[50] T. S. Ustun, C. R. Ozansoy, and Zayegh, A, "Implementing vehicle-to-grid (V2G) technology with IEC 61850-7-420," IEEE Transactions on Smart Grid, vol. 4, no. 2, pp. 1180, 1187, June 2013.

[51] IEC 61851-22: Electric Vehicle Conductive Charging System—Part 22: AC Electric Vehicle Charging Station. International Electrotechnical Commission, Geneva, Switzerland, IEC 2001, IEC TC-57.

[52] IEC 61851-1: Electric Vehicle Conductive Charging System—Part 1: General Requirements. International Electrotechnical Commission, Geneva, Switzerland, IEC 2010, IEC TC-57.

Chapter 3

Distributed Energy Resource with PEV Battery Energy Storage in the Smart Grid

Jianguo Zhu[1], Md Rabiul Islam[1], Jiefeng Hu[2], Sonki Prasetya[1], Yuedong Zhan[3], Mohammad Jafari[1], Hung Dung Pham[1] and Youguang Guo[1]

Abstract

This chapter presents various aspects of grid-connected distributed generation systems with electric vehicle, including the historical growth of two dominating renewable energy sources, i.e., wind and solar, the existing technologies for large-scale generation, transmission and distribution, power quality, reliability, technical challenges, and possible solutions. Recent development of grid integration technologies, converter topologies, and control techniques are the foremost intention of this chapter.

3.1 Introduction

The power from big power plants travels through the power transmission lines before distributed to the consumers. Most of conventional power stations, depending either on fossil fuels or hydro turbines, require high infrastructure supports for construction, operation as well as maintenance. Furthermore, a problem in a transmission subsystem will result in a power failure to the consumers.

Today, the world challenges in energy area are how to accommodate the needs of energy without endangering the environment by decreasing the dependency of polluting fossil fuels. Moreover, it has to be economically feasible and reliable. On the other hand, distributed generations (DGs) particularly generated from renewable energy (RE) sources are emerging in power systems. Sources like wind, solar, micro-hydro, and fuel cells (FCs) are highly implemented.

Normally DGs are connected in the distribution network voltage level. DGs should be placed near to the consumers to compensate the lower power ratings and

[1]Faculty of Engineering and Information Technology, University of Technology, Sydney, Australia
[2]School of Automation Science and Electrical Engineering, Beijing University of Aeronautics and Astronautics, Beijing, China
[3]Department of Automation, Kunming University of Science and Technology, Kunming, China

the high transmission losses. However, the RE based DGs suffer from the power fluctuation due to the nature's intermittency. Thus, the RE sources with proper control technologies are the solution for the grid problems.

Conventional electrical power system consisting of large power stations recently evolves to the so-called smart power grid system concept. It incorporates sensors and information exchange through communication among elements in the power system network [1]. Moreover, it also provides alternatives for customers to actively participate in the network by means of home based DGs penetration. Therefore, the smart power grid reduces the dependency as for big centralized power plants by introducing the aggregation of DGs. A clustered system using advanced technologies called microgrid can merge all the components above in an effective way. Thus, it improves the reliability and the quality of the grid power by controlling the grid regionally.

Connecting DGs as well as other elements such as the energy storage systems (ESS) or other smart power appliances in a proper way enhances the network instead of adding more electrical transmission infrastructures (power plants, transmission lines, substations, etc.) [2]. From the technical aspect, employing DGs have several positive benefits namely reducing line losses, upgrading the voltage profile, lowering emissions, improving overall energy efficiency, increasing the reliability and security, and also enhancing the power quality [3].

The concept of zero emission vehicles is emerged recently. These future vehicles employ electric motors to diminish the dependency of fossil fuels in order to preserve the environment as well as to increase the fuel flexibility. There are various types of electric vehicles (EVs); however, in principle an EV comprises of electric machine/motor, energy storage, and controller [4]. This chapter focuses on the energy storage in EVs.

The battery type used as the energy storage in EVs is commonly called as the secondary or rechargeable battery. This type of battery has flexibility in applications for instance standby power system, starting, lighting, and ignition (SLI) in automotive area and for power sources in EV applications [5]. Furthermore, a large capacity of the battery can be used to support the operation of RE generation systems. Batteries can be used to optimize the generated RE power by charging or discharging them during power excess or shortage condition respectively due to the intermittency of energy sources. Therefore, batteries in EVs are a potential element to be employed as power sources for EVs as well as energy storages for DGs.

3.2 Distributed Energy Sources

Distributed energy sources or DGs can be distinguished by the capacity of the power sources and the voltage level when paralleled to the grid. Moreover, high power demand requires higher power production. As mentioned earlier regarding the negative environmental impact of conventional energy sources, DGs support the smart power grid system by supplying clean and efficient energies for the consumers.

There are two types of DGs regarding the technology used, namely conventional and nonconventional DGs. The first type normally uses combustion or steam systems to generate electricity such as gas turbine and diesel generators. The second one may use newer energy source technologies and nonfossil fuels to produce electricity, for instance, wind turbines, solar photovoltaic (PV) systems, fuel cells (FCs), etc. Furthermore, the EV battery energy storage integration supports the penetration of RE based DGs since the batteries can compensate the fluctuation drawback of RE sources. Therefore, it is considered as the part of DGs. Some of the distributed energy sources are discussed in this chapter.

3.2.1 Solar

Solar energy is one of the main RE sources for future electricity supply. Solar PV generates electricity in well over 100 countries and continues to be the fastest growing renewable resource in the world. By the end of 2011, a total of 69.68 GW PV power capacity had been installed, sufficient to generate around 80 billion kWh/year and enough to cover the annual power supply needs for more than 20 million households as reported by European Photovoltaic Industry Association (EPIA). Solar PV installations have substantially increased over the last five years. The annual installation of new PV power capacity rose from 7.43 GW in 2009 to 16.80 GW in 2010 and to 29.66 GW in 2011. The globally installed PV power capacity is summarized in Table 3.1. According to statistical data the cumulative installed capacity of solar PV generation in 2007, 2009, and 2011 was 9.44, 23.21, and 69.68 GW, respectively. This means that the capacity of solar PV generation effectively tripled every two years.

Since 2007, medium and large-scale PV power plants have attracted significant interests and PV power plants of more than 10 MW in capacity have now become a reality. More than 200 PV power plants have already been installed in the world; each of them generating an output of more than 10 MW. Of these plants, 34 are located in Spain and 26 in Germany. The number of PV power plants will continue to rise. More than 250 PV power plants will be installed within the next few years [6]. Future PV power plants will have higher power capacity. Indeed, some are to have a capacity in excess of 250 MW. These multimegawatt PV power plants require large areas of land. Owing to this, they are usually installed in remote areas, far from cities.

When PV arrays are used to harvest solar energy, two important factors could limit the implementation of PV systems, i.e., high cost and low efficiency in energy conversion. The conversion efficiency of the current solar PV modules is typically only about 10–17 percent. In PV systems, the PV array represents about 57 percent

Table 3.1 Global solar PV installations

Year	Annual addition (MW)	Cumulative (MW)
2001	328	1,753
2003	578	2,798
2005	1,429	5,340
2007	2,528	9,443
2009	7,438	23,210
2011	29,665	69,684
2012	32,472	102,156

of the total cost of the system, and the battery storage system corresponds to 30 percent of the cost. Other system components such as inverters and maximum power point tracker (MPPT) contribute to only 7 percent of the total cost [7]. Due to the low conversion efficiency and high cost of solar array, it is very desirable to operate the PV panel at the maximum power point (MPP). Solar energy sources have variable daily and seasonal patterns. Daily bright sunshine hours have also variable daily and seasonal patterns. The output characteristics of the PV array are nonlinear and critically affected by the solar radiation, temperature, and load conditions [8, 9]; however, consumer power demand requirements have very different characteristics. If energy storage is not wanted, e.g., in large-scale systems then a grid-connected PV power plant is the only practical solution.

For power transmission, a medium voltage grid is usually used for medium and large-scale PV power plants. Although different power electronic converters have been developed in the last few decades using conventional topologies for solar PV systems, it is hard to connect the traditional converters to the grids directly as the distortion in generated output voltages is high and a single switch cannot stand at the grid voltage level. With the rapid growth of grid connected PV generations, the total harmonic distortions (THDs) generated from PV inverters is becoming a major concern. In this regard, conventional systems utilizing the power-frequency step-up transformer, filter and booster not only increase the size, weight, and loss but also increase the cost and complexity of the system installation and operation.

These heavy and large size power frequency step-up transformers significantly increase the weight and volume of the renewable power generation systems. Because of the heavy weight and large size of the power frequency transformer, the wind turbine generator and PV inverter system can be expensive and complex in terms of installation and maintenance especially in offshore and remote area applications. For example, the weight and volume of a 0.69/33 kV, 2.6 MVA transformer is typically in the range of 6–8 tons and 5–9 m^3 respectively, and the weight and volume of a 0.4/33 kV, 1 MVA power frequency transformer is typically in the range of 3–5 tons and 4–5 m^3 respectively [10]. A liquid-filled 2 MVA step-up transformer uses about 900 kg of liquid as the coolant and insulator, which requires regular monitoring and replacement. Table 3.2 summarizes some transformer data. These levels are critical in remote area applications where the costs of installation and regular maintenance are extremely high.

Today, to ensure that the goals of smart grids are met, the industrial trend is to move away from these heavy and large size passive components to compact and lightweight systems that use more and more semiconductor devices controlled

Table 3.2 Transformer data

Transformer type	Rating (MVA)	Size (m^3)	Weight (kg)	Liquid (kg)
Cast coil	2.50	8.50	6,200	–
Liquid filled	2.00	5.70	4,530	870
SLIM	2.30	4.00	5,040	900

by advanced digital controls. In comparison with conventional two-level converters, multilevel converters present lower switching losses, lower voltage stress on switching devices, and better harmonic performance. They also enable the development of medium voltage converters using mature and cheap power semiconductor devices. The development of multilevel medium voltage converters enables the connection of RE systems directly to the grid without using large, heavy, and costly power transformers. In 2011, different medium voltage multilevel inverter topologies were compared for the possible medium voltage grid connection of wind turbine and PV systems [11]. The modular multilevel cascaded (MMC) inverter topology was considered as a possible candidate for medium voltage inverter systems. The MMC inverter requires multiple isolated and balanced DC sources. In 2013, a high-frequency magnetic-link was proposed to generate multiple isolated and balanced DC sources for the MMC inverter from a single source [12]. Compared with the conventional transformers operated at the power frequency (50 or 60 Hz), the medium-frequency transformer-link has much smaller and lighter magnetic cores and windings, thus much lower costs. Fe based amorphous alloys have attracted a high degree of attention because of their significantly low magnetic core losses and very high saturation magnetic flux density [13]. However, the electromagnetic design of high-frequency common magnetic-links is a multi-physics problem and thereby affects the system efficiency and cost. An optimization technique was proposed and verified by prototype magnetic-links in [14].

In 2012, by combination of a quasi-Z source inverter into an MMC inverter a medium voltage PV inverter was proposed in [15]. The proposed inverter does not have isolation between PV array and medium voltage grid. Multiphase isolated DC/DC converter based MMC inverter topology was proposed in [16]. In the proposed configuration, the voltage balancing is the challenging issue, since each H-bridge cell is connected to a PV array through a DC/DC converter. A common DC-link may be one of the possible solutions to minimize the voltage imbalance problem, and a single DC-link based inverter was presented in [17]. Although this design may reduce the voltage balancing problem in the grid side, the generation of common DC-link voltage from different PV arrays makes the inverter operation complex and limits the range of MPPT operation.

As an alternative approach to minimize the voltage imbalance problem with a wide range of MPPT operation, a common magnetic-link, as shown in Figure 3.1, was proposed [6]. The boost converter was considered for the MPPT operation. The array DC power is converted to a high-frequency AC through a high-frequency inverter. The inverter also ensures constant output voltage. The inverter is connected to a primary winding of a multi-winding high-frequency magnetic-link. Each secondary winding works as an isolated source and is connected to an H-bridge cell through a bridge rectifier. The number of primary windings depends on the number of PV arrays and the number of secondary windings depends on number of levels of the inverter. In medium or large-scale PV power plants, several PV arrays are operated in parallel. The magnetic-link also provides electrical isolation between the PV array and the grid, which can thus inherently overcome the common mode and voltage imbalance problems and ensure a wide range of MPPT operation and safety of operating personnel.

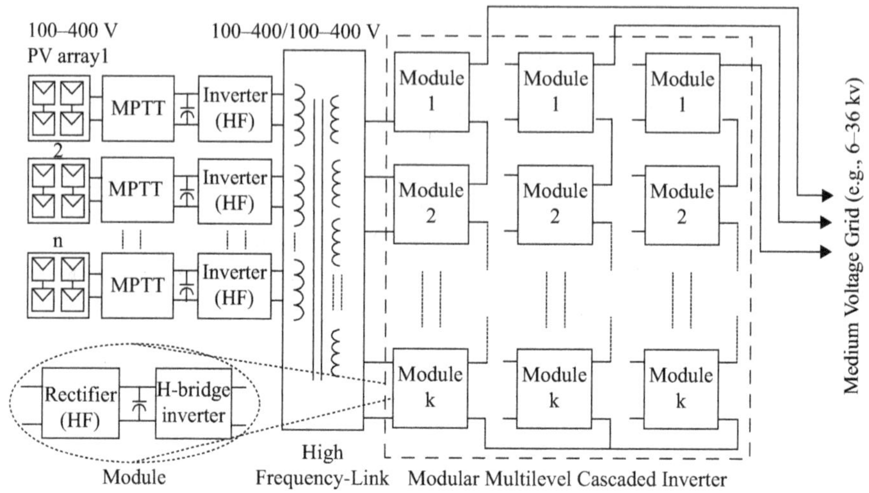

Figure 3.1 Common high-frequency magnetic-link based medium-voltage inverter [6]

Table 3.3 Global wind farms installed capacity

Year	Annual addition (MW)	Cumulative (MW)
2005	11,471	59,091
2007	19,951	93,889
2009	38,351	158,975
2011	40,125	238,126
2013	34,923	318,117

3.2.2 Wind

Over 318 GW of wind power generation has been installed by 2013. The cumulative installed capacity of wind power generation in 2008, 2010, and 2012 was at 120.26, 197.68 and 282.43 GW, respectively. The global annually installed capacity of wind farms is summarized in Table 3.3.

Wind speed varies continuously with time and height because of the changes in the thermal conditions of air masses. The motion of air masses is not only a global phenomenon but also a regional and local phenomenon. A wind turbine generator converts wind energy to electricity energy. Usually offshore winds tend to flow at higher speeds than onshore winds. This allows the turbine to produce more electricity as the possible energy produced from the wind is proportional to the cube of the wind speed. Also unlike onshore wind, offshore breezes can be strong in the afternoon, matching the time when load demands are at peak level. Moreover, wind farms cover large areas of land. The land area covered by a 3.6 MW turbine can be almost 0.37 km^2, such that 54 turbines would cover about a land area of 20 km^2 [18].

Since offshore wind farms can save land rental expense which is equivalent to 10–18 percent of the total operating and maintenance costs of a wind farm, offshore based wind farms have attracted great attention in the last few years. Considering the wind speed and capacitance effect of the transmission-line cables, offshore wind turbines are usually installed about 7–27 km from the shore. It is expected that the global offshore installed capacity will increase to approximately 20 GW by 2015 and rise sharply to 104 GW by 2025.

To integrate scattered wind turbine generators into a medium-voltage grid (e.g., 11–33 kV), a power frequency (50 or 60 Hz) transformer is commonly used to step-up the voltage for long distance transmission. In an offshore wind turbine power generation system, this transformer is usually installed at a height of about 80 meters inside the nacelle together with other equipments, such as the generator and power converter, which increases the weight of the nacelle and mechanical stress of the tower. Therefore, a reduction in mechanical loading represents an enormous saving of tower construction and turbine installation costs.

In 2008, an 11 kV, 25-level converter with 2.5 kW modular permanent magnet generator (72 isolated stator coils) was designed [19]. This multi-winding generator requires a special winding arrangement (which increases the weight and volume) and complicated control strategies [20]. The rotor diameter of a low-speed, direct-drive synchronous generator is several times larger than that of a conventional gearbox based high speed generator. The most critical drawback of the multi-coil generator and MMC converter based wind power generation system is the lack of electrical isolation between the grid and turbine generator. A transformer-less wind power generation structure with several parallel six-phase permanent magnet synchronous generators (PMSGs) has been proposed in [21], where a few six-phase generators are placed in the turbine nacelle. All the generators are driven by the same wind turbine and each stator winding generates an isolated source for an H-bridge inverter cell of the MMC converter. The MMC converter generates medium-voltage AC output, which can be connected to the medium-voltage network directly. The proposed structure requires multiple generators, depending on the number of levels of MMC converter. The multiple generators increase the system volume and weight, and make the system impractical especially in medium-voltage applications, where a high number of levels are required. Moreover, the grid isolation is the critical issue in the proposed converter system.

As an alternative approach, a high-frequency magnetic-link MMC medium-voltage converter was validated for step-up-transformer-less direct grid integration of wind turbine generators [18]. The common magnetic-link generates multiple isolated and balanced DC supplies for all the H-bridge inverter cells of the MMC converter from a single or multiple renewable sources. The grid electrical isolation and voltage imbalance problems are solved through the common high-frequency magnetic-link. Figure 3.2 shows the basic block diagram of the proposed high-frequency magnetic-link MMC converter for wind turbine generator systems [18]. Although the common magnetic-link may ensure equal voltages at the secondary terminals, the topology diminishes modularity of the power conversion system, which is very important for medium or high-voltage high-power system.

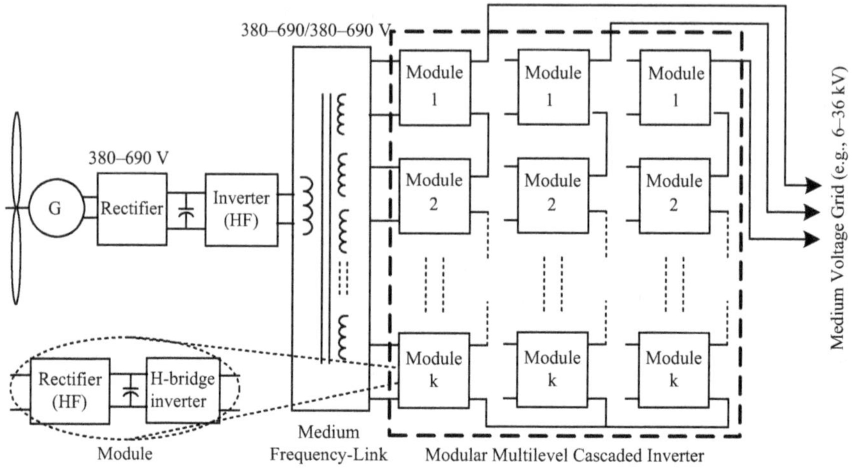

Figure 3.2 Wind turbine generator system for direct medium-voltage grid integration [18]

The implementation of high-power high-frequency inverter is the main challenging issue. Moreover, the leakage inductance may also limit the power capacity of the high-frequency magnetic-link. Recently, a new medium-voltage converter has been proposed in [22] with multiple-cascaded modules for wind farms, which has the following new features compared with the existing similar systems:

1. no requirement for special or multiple generators
2. an inherent voltage balance
3. completely modular in construction
4. possibility to use commercially available low-cost mature devices
5. no power limitation
6. an overall compact and lightweight system
7. an inherent minimization of the grid isolation problem

Figure 3.3 shows the circuit diagram of the medium-voltage power conversion system, where the available rectified wind power (from generator followed by rectifier circuit) is supplied to a few cascaded modules.

3.2.3 Fuel Cells

Fuel cells (FCs) are the electrochemical devices due to the direct conversion of the chemical energy in the fuel into the electricity energy and heat, without undergoing combustion. Similar to the primary batteries, the FCs have two electrodes which are separated by an electrolyte, and the cell reaction can produce electrical energy. But different from batteries, an external storage source supplies the reactants to an FC and the FC continues to operate as long as a fuel supply is maintained.

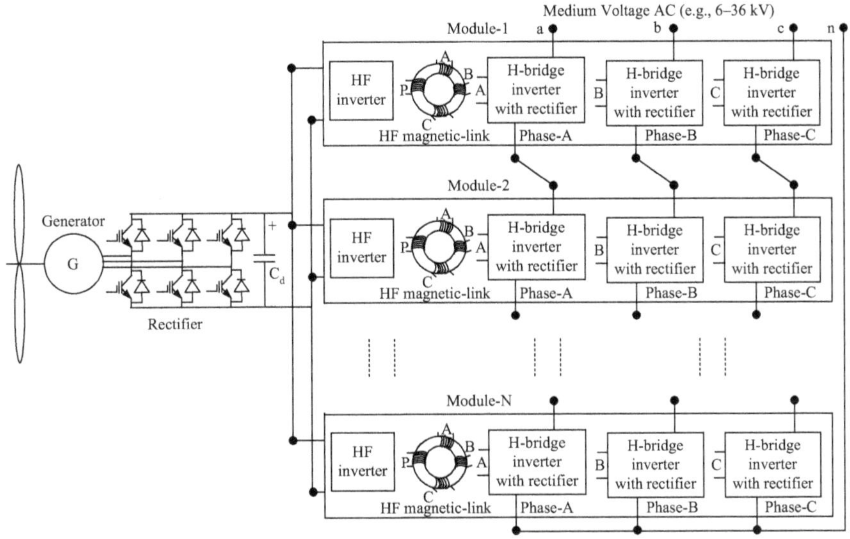

Figure 3.3 Circuit diagram of the modular high-power density wind generation system [22]

FCs have several very appealing advantages, including low operating temperature, higher efficiency than conventional energy conversion processes and good performance even at partial loads. Fuel reformation increases fuel flexibility in the system, but fuel reformer technologies are still in early development stage. Furthermore, the lack of moving parts in the energy converter and modular design make the maintenance easy and improve the system reliability. Efficient and clean energy production processes are very interesting and topical due to the concern about the environment and the earth's limited resources. In addition, FCs are expected to be economically viable once the mass production begins, at least in the fields of distributed power generation and remote systems.

FCs are customarily classified according to the electrolyte used. The most common FC types are as follows [23]:

- proton exchange membrane fuel cells (PEMFCs)
- alkaline fuel cells (AFCs)
- phosphoric acid fuel cells (PAFCs)
- molten carbonate fuel cells (MCFCs)
- solid oxide fuel cells (SOFCs)
- direct methanol fuel cell (DMFCs)
- zinc-air fuel cells (ZAFCs)
- protonic ceramic fuel cells (PCFCs)
- biological fuel cells (BFCs)

Being the electrochemical devices, the FCs involve the thermodynamics, electrochemical, mass transport, charge transport, heat transfer, characterization

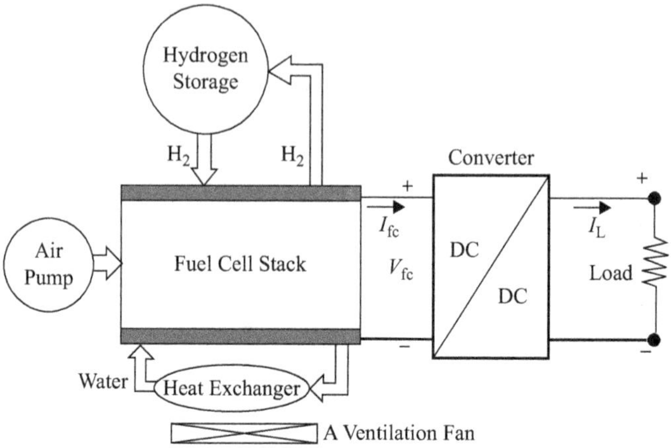

Figure 3.4 A hydrogen–air fuel cell system

technologies, materials and components, failure or degradation mechanisms, operating conditions, power electronics technology, and modeling and control. In practice, according to the application of FC and its efficiency, a FC system can be designed simply or very intricately. The devices to support its operation mainly include a fuel supply pump, an air circulating pump, a coolant circulating pump, a ventilation fan, and an electrical controller as shown in Figure 3.4.

At present, hybrid ICG/EV vehicles made in United States, Japan, and EU such as the GM HydroGen3, Daimler-Chrysler NECAR 5, Toyota FCHV-5, Honda FCX-V5, Audi Q5, and Volkswagen Passat are very successful. However, hybrid electric vehicles (HEVs) still heavily depend on oil. Therefore, fuel cell vehicles (FCVs) will be the next generation of electric-drive vehicles after HEVs, even though the issues of hydrogen refueling infrastructure and high cost of PEMFC have to be solved before commercialization. Before presenting the FCVs architecture, it is necessary to describe hybrid vehicle architectures for a better understanding of the FCVs technology because, generally, FCVs also belong to the HEVs. These hybrid vehicle technologies can increase the efficiency and fuel economy through the application of regenerating energy during braking and storing energy from the ICE coasting [24].

3.2.4 Electric Vehicles

In 1886, one of the most life-changing technologies was invented by German inventor Karl Benz as Benz Patent-Motorwagen which is widely regarded as the first automobile in history. Since then, with more than a century of rapid development, the mobile vehicle industry has promoted a whole new standard throughout the world [25]. However, because of the rapid growth in the number of automobiles in the world, we are now facing more problems of environment and human life, such as oil and diesel consumption or the greenhouse emissions. These issues bring an urgent challenge for researchers to develop new vehicle technologies with cleaner, more efficient and sustainable solutions for transportation [24]. With these requirements, nowadays, many automotive companies are focusing on EVs, HEVs, plug-in hybrid electric vehicles (PHEVs) and fuel-cell vehicles. In the near future, with the rising of EV technology, the world will have a new generation of conventional automobiles.

The first EV was introduced in the second half of last century, and more and more researches have been focusing on this field since 1990s. From that time, the EV industry has experienced an accelerated pace with a large number of manufacturers beginning their own plan for EVs and HEVs technologies. Power electronics is an essential technology for the development of these more environmentally friendly vehicles, providing the means for a clean and efficient transportation system [26]. Most EVs' battery can serve as a backup storage system not only to "valley filling" (store energy by charging during off-peak hours – G2V) but also to "peak shaving" (provide power to the grid by discharging during peak hours – V2G). In this way, the EVs can act as a distributed ESS which is a new way to achieve the smart grid technology.

Besides EVs, HEVs also provides an effective solution to achieve higher fuel economy, better performance and lower emissions, compared with conventional vehicles. The first HEV was Lohner-Porsche Mixte Hybrid produced in 1899 [27]. Since then, a majority of HEV concepts were developed in the last century. The hybrid vehicles can be very different from each other in term of structure, but share the characteristic of having at least two energy source, and one (or more) of these source is electricity. These sources can be a battery, a flywheel, a fuel cell, etc. In a number of cases, one source takes energy from a fuel and converts it into a more qualified form which is either mechanical or electrical. The other source may include energy storage where electrical energy will be generated if some energy has been stored in it. Depending on the vehicle configurations, two or more of these power or energy sources are used [28–30]. For that reason, HEVs combine the advantages of the electric motor drive and an internal combustion engine (ICE) to propel the vehicle. In a conventional vehicle, only 10–15 percent of the energy contained in gasoline is converted to traction but the proportion can be improved to about 30–40 percent in hybrid vehicle. This would reduce the emission and increase the fuel economy.

A typical configuration of a fuel-cell powered HEV is shown in Figure 3.5. This configuration consists of a FC system with DC/DC converter as the primary power source, batteries or supercapacitors with bidirectional DC/DC converter as the peaking power source (PPS), an electric motor drive with DC/AC inverter, and a vehicle controller. With proper control strategy, the motor power (torque) output and the energy flows can be controlled by the vehicle controller, based on the received power/torque command and other signals. When in sharp acceleration where the peak power demand is needed, it will enable both the fuel-cell system and the PSS to supply the propulsion power to the traction motor. During the braking process, the traction motor acts as a generator that provides power back to the PSS. Friction brakes are used when the vehicle must be stopped in a short time or if the PSS is fully charged. If the rated power generation is greater than the load

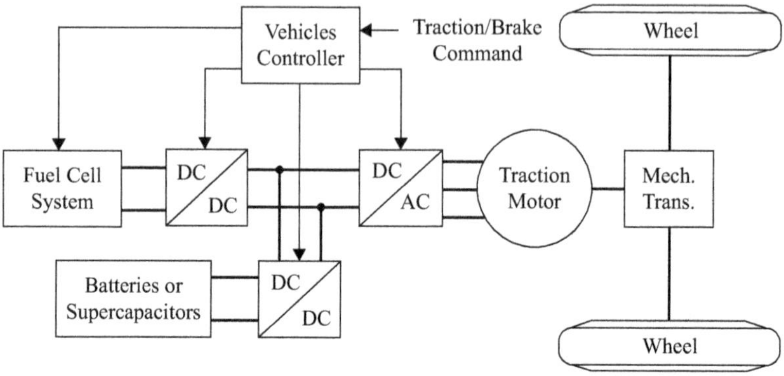

Figure 3.5 Configuration of a typical hybrid fuel cell vehicle

consumption, the excess energy coming from the fuel-cell system can be stored in the PPS.

The HEVs are generally classified as series hybrids and parallel hybrids. In series hybrid vehicles, the engine drives the generator which, in turn, powers the electric motor. In a parallel hybrid vehicle, the engine and the electric motor are coupled to drive the vehicle. The advantage with the series HEV configuration is that the ICE runs mostly at its optimal combination of speed and torque, and therefore, the system will have low fuel consumption and high efficiency. However, because there are two stages of power conversion during the transformation of the energy between the ICE and the wheels, there may be more lost energy in the process. On the other hand, the parallel hybrid vehicle can be driven with the ICE, or the electric motor, or both at the same time. Therefore, it is possible to choose the most desirable combination freely to feed the required amount of torque at any given time. With this technique, the system can achieve the most efficient operating points for the ICE anytime, leading to lower fuel consumption. Energy is also saved due to regenerative braking. For that reason, the parallel HEVs have fewer energy conversion stages so the lesser energy will be lost.

The pure electric vehicle or battery EV relies only on the electricity from the ESS such as batteries. However, at the current stage, the pure EVs or battery electric vehicles (BEVs) are not preferred due to battery technology limitations such as short travel distance and high cost. The other drawback is the lack of fast charging facilities to the battery.

As an intermediate stage, the plug-in HEV may be an excellent choice. Compared to the traditional HEVs, PHEVs can rely on a much greater degree of electrical propulsion and hence possess a few merits such as higher fuel economy, more all-electric ranges, less fuel consumption, and less greenhouse gas emission.

3.2.5 Backup Power Supplies

Besides energy storage units, conventional DGs by combustion systems can be used as backup power supplies when emergency takes place or when fluctuation

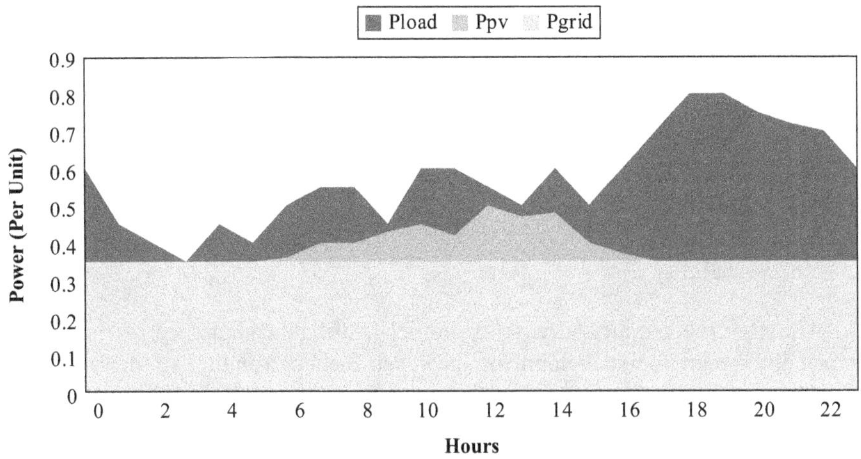

Figure 3.6 Role of backup power generator in renewable source based power grid

demand occurs. The backup generators are used to support the power system in the case of either deficit or excess condition of the distribution grid.

Moreover with the RE penetration, the intermittency of RE sources can be compensated by the support of backup generators [31]. Therefore, the RE sources can operate at the maximum extracted power, meanwhile the grid can serve the base load requirement only. As an addition, the load power demand fluctuation can be handled by the backup generators. Figure 3.6 shows the role of the backup generator where the grid power (Pgrid) provides a constant power meanwhile the RE source from photovoltaic (Ppv) contributes power during the daylight. Thus the gap of the load demand (Pload) can be handled by the generators.

Basically the working principle of the backup power supply generator is identical with the generator used in RE such as hydro or wind turbine. The only difference between them is the prime mover. There are two types of fuel used for the combustion based generators, such as liquid fuel and gaseous fuel. In a microgrid, oil and gas based fuel are preferable as the backup power supplies due to the simplicity of operation as well as maintenance. The DGs are usually synchronous generators and consume diesel as the fuel. Diesel based generators can be ranged from small up to high capacity in the range of kW to MW. Some advantages of DGs in the smart grid applications are:

- physically smaller space required
- simple instalment, operation and maintenance
- providing the dynamic of the power demand
- manageable fuel supply

Furthermore for the environmental concern, the fuel option can be produced from nonfossil sources known as the synthetic or artificial fuel [32]. As for the alternative of fuel choices, the biomass can be utilized as the alternative for the diesel fuel.

Biodiesel is usually produced from vegetable oil through so called trans-esterification process to be converted into the diesel fuel. Using vegetable oil as the fuel has advantages such as [33]:

- portability
- availability
- renewability
- higher heat content
- biodegradability

Although the production cost of biofuel is still the hindrance of this environmentally green fuel development, however the contribution of diesel as the backup power supply is essential in the smart grid network.

3.2.6 MPPT Strategies for PV Based Microgrids

Due to the increasing interest to the PV energy in electrical power applications, it is crucial to increase the efficiency of PV systems by operating them near their MPP. According to the nonlinear nature of PV system the output current and power of the PV array depends on the array terminal operating voltage. On the other hand, the MPP varies with insolation level and temperature. Therefore, the MPPT is a complicated process. To overcome these problems, many tracking strategies have been proposed in the literature [34, 35]. The main requirements for MPPT are simplicity of structure, low cost, quick tracking under changing conditions with small fluctuation of output power. The MPPT methods can be divided into direct and indirect methods. The direct methods measure the real value of PV voltage, current, irradiance and temperature levels, and calculate the MPP using real time data. They don't need a large database or memory and are totally versatile with respect to the load profile. The direct MPPT process is quite independent of the variation of the PV parameters and is not designed for a specific PV array. On the other hand, indirect methods estimate the MPP for any irradiance or temperature levels using measured values of PV voltage, current and irradiance or empirical data along with mathematical expressions and numerical databases. Thus, they are designed for a specific PV array and do not obtain the exact point of maximum power for any irradiance or temperature levels.

3.2.6.1 Direct MPPT Methods

Perturb and Observe (P&O) Method

The perturb and observe method, also known as perturbation method, is the most commonly used MPPT algorithm in commercial PV products [36]. The method is based on a "trial and error" algorithm. In this method the controller increases the reference for inverter output power slightly and then detects the changes on output power. If the output power is increased, it will continue to increase the reference step-by-step until the output power starts to decrease. At this point the controller decreases the reference to avoid collapse of the PV output due to the highly nonlinear PV characteristic. The method is easy to implement although there are some drawbacks due to the slow dynamic response of process.

Incremental Conductance Method

The operation of this method is based on the derivative of PV output power with respect to its output voltage [37]. An incremental conductance algorithm is used to define dP/dV. The maximum PV output power can be achieved when dP/dV approaches zero. In this method the controller calculates dP/dV based on measured values of PV output power and voltage. The method is quite fast in tracking of maximum power however there is a possibility for instability in output power due to the use of derivative algorithm. Another drawback of this method is unsatisfactory results under low levels of insolation.

Parasitic Capacitance Method

This method in general is similar to the incremental conductance method, except that it takes into account the effect of the PV cell's parasitic union capacitance [38].

Forced Oscillations Methods

In this method a small medium-frequency signal is added to the operation voltage of the PV panel. This leads to a ripple on the output power, whose phase and amplitude are dependent on the relative location of the operation point to the MPP. According to the phase shift between oscillations of output power and output voltage, the direction of movement of operating point toward the MPP is defined [39].

3.2.6.2 Indirect MPPT Methods

Curve-Fitting Method

The method is based on the off-line modeling of nonlinear characteristic of PV panel from conventional diode model, using mathematical equations or numerical approximations [40]. The coefficients of the model equation are calculated using several close operating points and the MPP can be found by using resulted coefficients and solving the equation. This process should be repeated every few milliseconds in order to find a fine MPP. The method requires accurate knowledge of the physical parameters relating to the cell material and manufacturing specifications, and the resulted equations are not valid for all climatological conditions. It also requires a large memory and high speed processor for calculations.

Look-Up Table Method

In this case, the measured values of the PV voltage and current are compared with those stored in a look-up table corresponding to the maximum power operating points under concrete climatological conditions. In this method a large capacity of memory is required for storage of the data and the implementation must be adjusted for a specific PV panel. A drawback of this method is the difficulty of recording all possible system conditions [41].

Open-Circuit Voltage and Short-Circuit Current Method

This algorithm is based on the fact that the PV voltage or the current at the MPP is approximately linearly proportional to its open-circuit voltage or short-circuit current respectively. The proportional constant mainly depends on the fabrication process of solar cells, fill factor, and the meteorological conditions [42, 43].

3.3 The Smart Grid and Microgrids

As the electricity demand keeps rising these years, the existing electrical power system is undergoing a major transformation. Due to the limitation of delivery capability and the increase in power demand, it is getting more and more stressed. Generally, the main drawbacks of the existing power systems can be summarized as follows [44]:

1. *Inefficiency:* Almost 8 percent of the total power is lost along transmission lines while only one-fifth of the generation capacity exists to meet the peak demand.
2. *Domino-effect failures:* It is a strictly hierarchical system where power plants at the top of the chain ensure power delivery to customers' loads at the bottom of the chain. In other words, the power flows in only one direction, which will leads to large-scaled blackout triggered by power plants intermittence or even transmission lines problems. The most well-known failure occurred in August 2003, when 50 million customers in the United States and Canada lost power for up to two days due to cascading events.
3. *Instability:* The unprecedented fluctuation of power demand, coupled with increasing penetration level of DGs and lagging investments in the electric power infrastructure, has decreased the system stability.

In order to address the issues of the existing electricity grid mentioned above and at the same time exploit the RE sources effectively, the next generation electricity grid known as the "smart grid" was proposed recently. At the generation level, it is expected to emerge as a well-planned plug-and-play integration of smart microgrids. At the distribution and transmission level, the smart grid integrates small microgrids through dedicated highways for command, data, and power exchanges. Therefore, the efficiency, reliability, and security of the grid can be greatly improved by dynamic optimization of electric-system operations, maintenance, and planning. The smart grid can thus be regarded as a platform to balance the supply and demand sides [45–47].

There are two transmission paths throughout the smart grid, i.e., the digital highway and the physical highway. The information from the smart meters and fault diagnosis monitors is collected by advanced metering infrastructure (AMI) system, and is then sent to the control center through the digital highway. The control center controls the electronic device to optimize the power flow and heal the grid faults in the physical highway. Since the smart grid is an electricity network that can intelligently integrate the actions of many microgrids, the distinct advantage compared to the conventional power grid is that it can efficiently deliver sustainable, economic, and secure electricity supplies. It is noted that the smart grid would and should coexist with the existing electricity grid, and it is required to be self-healing and resilient to system anomalies.

Although much attention has been drawn from all over the world, the development of smart grids is still in the pilot stage. Generally, the key technologies and challenges of smart grids can be summarized as follows:

Power Electronics Control Technology
Control strategies of power electronics with excellent steady-state and dynamic performance are very important in active and reactive power regulation, current, voltage and frequency control. Though some methods have been proposed such as improved droop control, for larger microgrids with numerous inverters, the depth of available literature is less comprehensive and doubts remain as to the scalability of techniques proposed for small-scale systems.

New Semiconductor Device
Silicon based semiconductor device is reaching its physical limits in power handling and switching frequency capability. Seeking new materials technology is very exigent for the smart grid. As the need for RE technologies, energy storage technologies, and the smart grid technologies has grown in recent years, power semiconductor devices with high-voltage, high-frequency, and high temperature operation capability are required.

Information Technology
An increasing number of the smart grid deployments are using the internet technologies, broadband communication, and nondeterministic communication environments. Advanced technique for huge data processing is essential in information collection and energy management within the smart grid.

Security and Privacy Issues
As the grid incorporates smart metering and load management, user and corporate privacy is increasingly becoming an issue. For example, will the power utilities be able to control the customer load without the customer's prior permission? Will the private power consumption information be stolen by cyber-hacker with illegal motivation? Could the electricity consumption patterns lead to the discourse of the customers' information, e.g. how much energy they have used, and when they are at home, at work, or traveling? These are the security and privacy issues needed to be concerned for the construction of the next generation smart grid.

From the power distribution perspective, to mitigate the negative influences on the grid and incorporate effectively the renewable generations, it is a good idea to combine the local utilization, energy storage, and distributed generation to form a grid-friendly distributed generation system, namely microgrid. In addition, considering that most of the power outages and disturbances take place in the distribtution network, the first step toward the smart grid should start at the bottom of the chain, the microgrid system [48].

Based on this background, the next-generation electricity grid is expected to address the major shortcomings of the existing grid. The microgrid is one of such systems in which the distributed power generation is controlled not only to provide power but also to stabilize the fluctuation caused by the RE sources, such as wind power and photovoltaic [49, 50]. Microgrid system is currently a conceptual solution to fulfill the commitment of reliable power delivery for the future power system which can address issues related to operation, control, and stability of the system. Renewable power sources such as wind and hydro offer the best potential for emission free power for future microgrid systems.

A microgrid can operate both in grid-connected and islanded modes which integrates the following features:

- It incorporates DGs capable of meeting local demand as well as feeding the excess energy back to the electricity grid. Such DGs are known as cogenerators and often use renewable sources of energy, such as wind and solar.
- It services a variety of loads, including residential, office, and the whole building.
- It incorporates smart sensors capable of measuring a multitude of consumption parameters (e.g., active power, reactive power, voltage, current, demand, and so on) with acceptable precision and accuracy.

3.3.1 Microgrid Topologies

A microgrid is a cluster of DGs and local loads that can offer many advantages to the current power grid in terms of power autonomy and the ability to incorporate renewable and non-RE sources [51, 52]. In the grid-connected operation, the DGs together with the utility grid supply power to the local loads. If the power generation is greater than the load consumption, the excess can be either stored in the energy storage unit or injected into the grid if there is a need. On the other hand, if the power generation is smaller than the load demand, more power can be imported from the grid. In the islanded operation, the DGs should be able to provide a stable voltage at the point of common coupling (PCC) and pick up the loads automatically. In addition, the microgrid should be reconnected to the utility grid seamlessly when the grid is available.

This new concept is becoming more and more attractive. Compared to a single DG unit, it offers many technical advantages in terms of power quality and reliability. Compared to the whole power system, it presents more control flexibilities. However, due to the discontinuous nature of these DG units, the output power is not stable and may cause negative influence on the quality of electricity or damage the electric appliances. For the spread of RE in demand side, the DG units must be installed with facilities which absorb the fluctuation. In addition, due to the complexity of such a small power system including several kinds of energy sources, ESS and loads, the coordinated control of these DGs to achieve optimal power flow and maintain high power quality becomes a big challenge. From the perspective of the energy sources, the power converters should be controlled to capture the maximum real power and inject the excess into the utility. On the other hand, from the utility/grid perspective, the power electronics interface should also be able to provide reactive power according to the requirement to improve power quality and enhance grid stability [53].

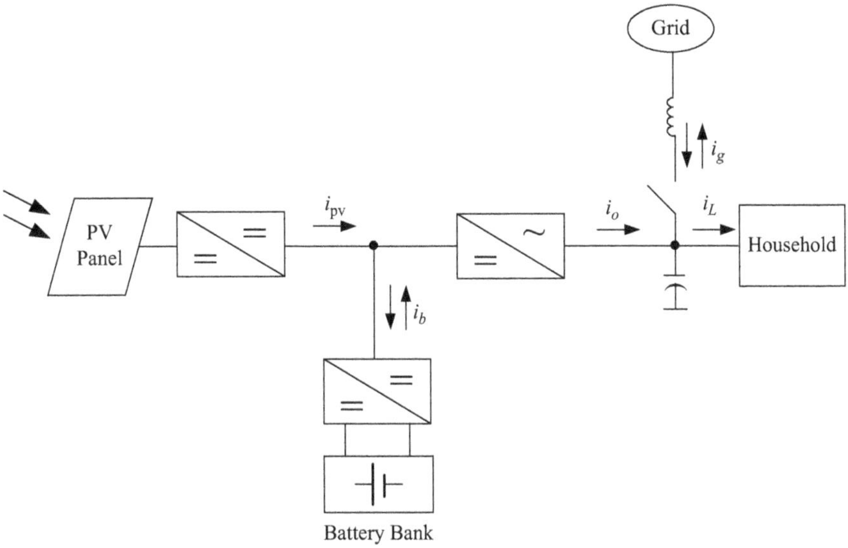

Figure 3.7 Schematic diagram of a simple microgrid for households

Figure 3.7 shows a simple microgrid application for households, where the PV panels are generally installed on the roof of the houses, an ESS is also utilized to absorb the surplus energy when the production is greater than the power consumption.

In such a small power system, the DC output of the PV panel is connected to the common DC bus through a DC/DC converter that is used to achieve MPPT. On the other hand, an energy storage device, such as the battery bank, is used to smooth the gap between the generation and consumption. It is interfaced to the common DC bus through a DC/DC converter which allows the bi-directional power flows for battery charging and discharging. The other side of the common DC bus is connected to a DC/AC converter that is connected to an LC filter before connected to the household electric appliances or the utility grid.

Figure 3.8 shows a typical microgrid system with several DG sources at the Commonwealth Scientific and Industrial Research Organisation (CSIRO), Australia. It consists of three control levels. The first level is the DC microgrid including a small-capacity wind turbine (WT), PV panel, and a battery bank which are connected to a DC PCC. At the second level, single-phase inverters are used to interface PVs to the 240 V AC PCC while a WT and a gas micro-turbine (MT) supply power to this AC common bus through three-phase converters. A load bank controlled in a binary system is utilized to simulate the fluctuated profile of power consumption. There are also wind turbines, PVs, and gas MT with higher capacity at the third level, which is not shown here. In this distributed generation system, the gas MT not only provides a voltage signal to all other inverters for synchronization, but also control the voltage and frequency of AC PCC. The control of the DGs and

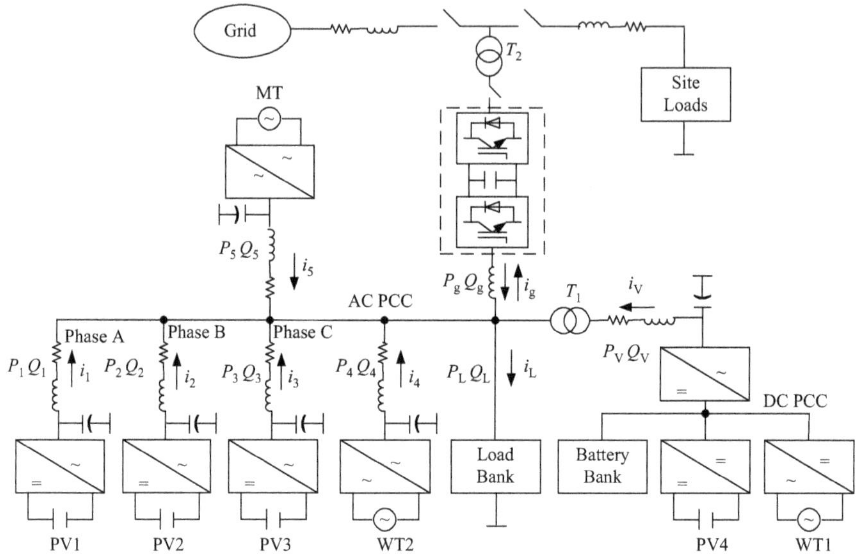

Figure 3.8 A typical microgrid configuration with several DGs

the data acquisition are fulfilled through a supervisory control and data acquisition (SCADA) system.

In such a system, the major part of the electricity produced by DGs is consumed locally, and the excess will be exported to the grid. If the load is greater than local generation, more power can be obtained from the grid. Besides, it can operate in islanded mode in which electric power can continue to be supplied to the load bank. The control center will coordinate/dispatch the generators and energy storage devices according to the local demands. The main objectives of this microgrid are to exploit the RE effectively and efficiently.

Figure 3.9 shows a smart microgrid system built at the University of Technology, Sydney. In this microgrid, there are distributed generation units including doubly fed induction generator (DFIG) based wind power system, photovoltaic (PV), and proton exchange membrane (PEM) fuel cell (FC). In addition, there is a charging/discharging battery based ESS to absorb the fluctuation between power generation and power consumption [54]. The real-time information such as voltage and current of each DG unit and the PCC at common AC bus are monitored by a radio frequency (RF) module which is sent to the control center through wireless transmission. The control center will then determine the future behaviours of the microgrid in order to obtain the optimized power flow and power quality, according to the current real-time information. After that, the control center will deliver the control commands to control the power converters of each DG units.

For the smart operation of a microgrid system, all the DG units should be able to participate in the optimized control mechanism, from the perspectives of both the

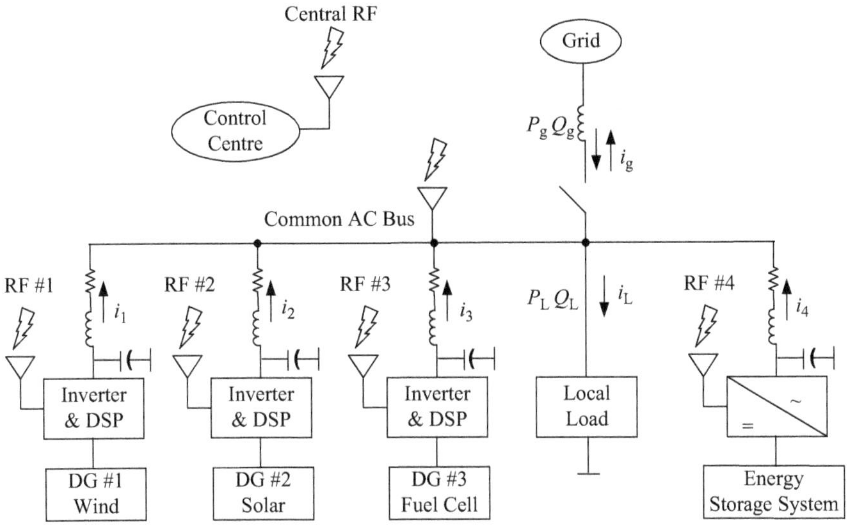

Figure 3.9 A microgrid with multi-DGs and a control center

loads and the utility grid. Generally the basic requirements for a microgrid to achieve smart operation can be described as follows:

1. It provides high quality power for the loads, i.e., establish a stable output voltage regardless of intermittent power generation and the fluctuated power demand.
2. It incorporates RF modules capable of sending the real-time information of power generation and consumption and receiving control demands through wireless transmission, which makes it very useful in the applications where long wire interconnection is impractical, such as rural areas and large high-tech building in cities considering the reduction of cost and space.
3. It can also participate in the voltage support and power quality improvement for the utility grid by compensating reactive power.

Figure 3.10 shows the prospective electricity network of a smart city configuration in the future. Large-scale of renewable power distributed generation units such as wind turbine and PV arrays will supply electrical power to the grid together with the conventional power plants. The intelligent power transmission and distribution system will optimize the power flow and balance the power generation and consumption effectively. At the power demand side, consumers not only can enjoy high-quality power, but also supply the surplus power back to the grid.

3.3.2 Microgrid Control Strategies

With a non-radial system configuration due to the presence of DG units, the power control complexity for a microgrid is substantially increased, and the "plug-and-play" feature is the key to ensure that the installation of additional DG units will

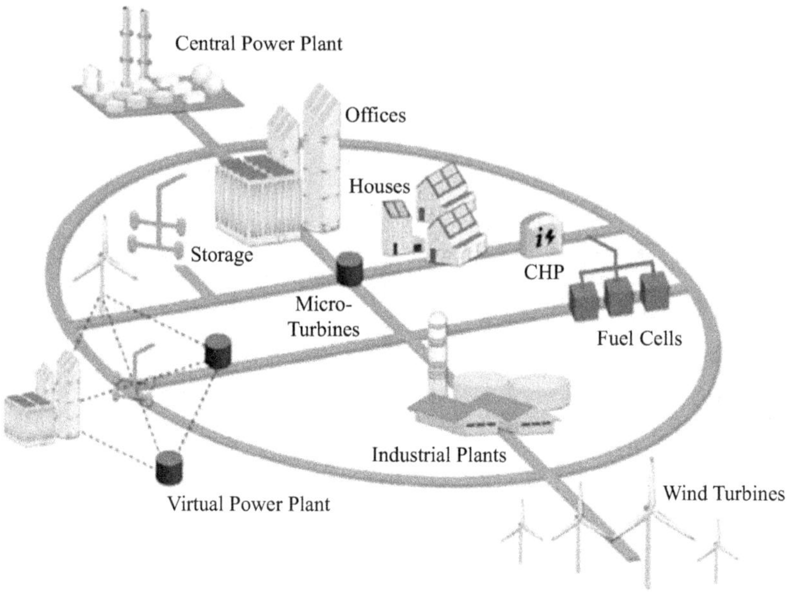

Figure 3.10 Prospective smart microgrid [55]

not change the control strategies of DG units already in the microgrid [47]. To realize this "plug-and-play" characteristic, many control techniques have been developed, which can be categorized into two main groups, namely active load sharing method and droop control method. The first one is derived from control approaches of parallel DC/DC converters and UPS, such as centralized [56], average load sharing [57], and circular chain control [58]. However, these methods need intercommunication lines between DGs to achieve output-voltage regulation and power sharing, which can deteriorate the system stability and reliability.

The second type of control technique of parallel DGs is the droop control without control wire interconnections, which has been commonly used for active and reactive power regulation in microgrids recently [59–64]. In this method, an artificial droop characteristic is introduced, and the active and reactive power supplied by the DGs can be controlled by adjusting the frequency and amplitude of the output voltage, respectively, according to a predefined manner. Therefore, the DGs can meet the new load requirements in a manner determined by its frequency and voltage droop characteristics.

With this technique, the active and reactive power sharing by the inverters is automatic achieved by adjusting the frequency and amplitude of the output voltage. In order to fix the reference voltage generated by the droop controller, generally a multiloop control scheme is implemented, where an inner inductor current feedback loop and outer filter capacitor voltage feedback loop are used.

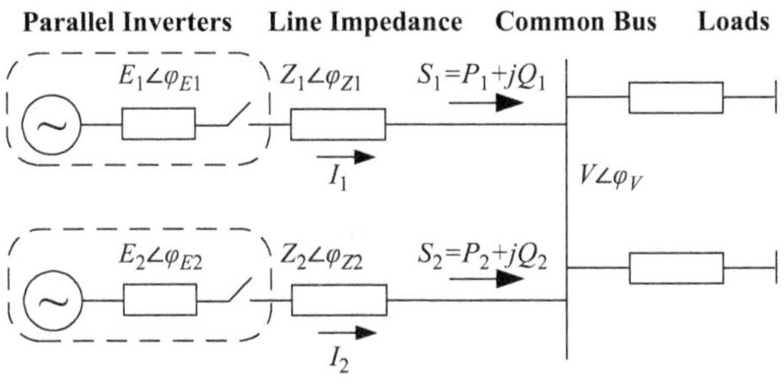

Figure 3.11 Equivalent circuit of parallel-inverters based microgrid

The equivalent circuit of DGs connected to a common AC bus through inverters is presented in Figure 3.11. The active and reactive power flow from the DG to the common AC bus can be expressed as [65]

$$P = \frac{E^2 \cos \varphi_Z - EV \cos(\varphi_E - \varphi_V + \varphi_Z)}{Z} \tag{3.1}$$

$$Q = \frac{E^2 \sin \varphi_Z - EV \sin(\varphi_E - \varphi_V + \varphi_Z)}{Z} \tag{3.2}$$

where E and V are the magnitudes of the inverter output voltage and AC common bus voltage, Z is the line impedance. For a purely inductive line impedance, the line resistance may be neglected, i.e., $\varphi_Z = 90°$, and thus (3.1) and (3.2) become

$$P = \frac{EV \sin(\varphi_E - \varphi_V)}{X} \tag{3.3}$$

$$Q = \frac{E^2 - EV \cos(\varphi_E - \varphi_V)}{X} \tag{3.4}$$

where X is the line reactance.

Further, considering that the phase angle difference $\delta = \varphi_{\mathrm{E}} - \varphi_{\mathrm{V}}$ is typically small, we can assume $\sin(\delta) = \delta$ and $\cos(\delta) = 1$, and consequently,

$$P = \frac{EV\delta}{X} \tag{3.5}$$

$$Q = \frac{E(E - V)}{X} \tag{3.6}$$

As a consequence, the flow of active power is linearly dependent on the phase angle difference (δ) and the flow of reactive power is linearly dependent on the voltage magnitude difference ($E - V$). At this point, similar to the power system theory where a generator connected to the utility drops its frequency when the required power increases, a voltage and frequency droop control method for microgrid can be defined as

$$\omega = \omega^* - m(P - P^*) \tag{3.7}$$

$$E = E^* - n(Q - Q^*) \tag{3.8}$$

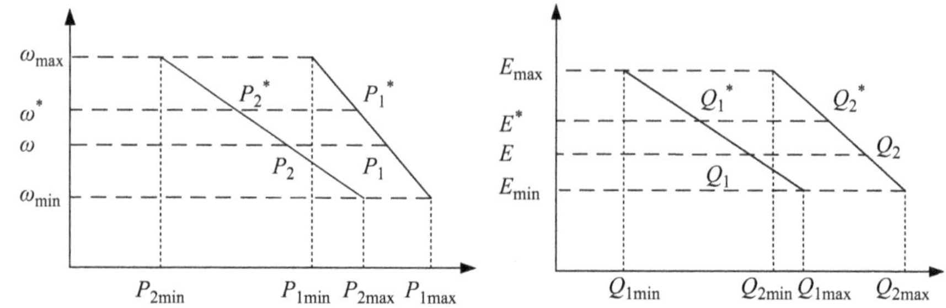

(a) P-É droop characteristics

(b) Q-E droop characteristics

Figure 3.12 Droop characteristics

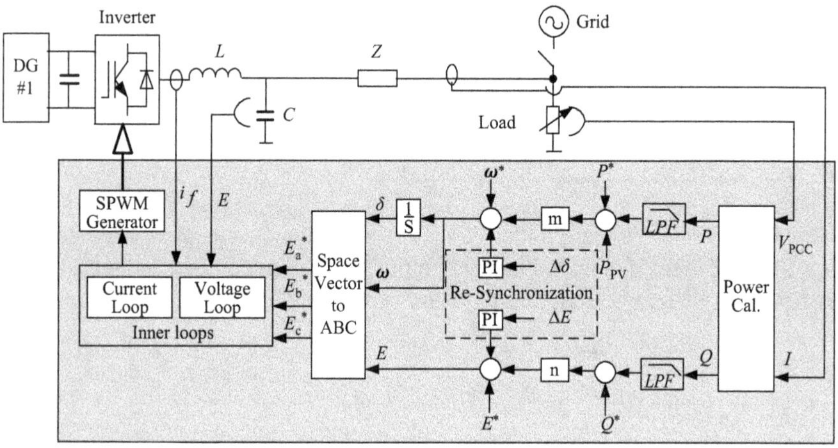

Figure 3.13 Control diagram of the whole microgrid control based on voltage droop method

where P^* and Q^* are the desired active and reactive power, ω^* and E^* the inverter nominal output frequency and voltage amplitude, and m and n the slopes of the droop characteristics. Figure 3.12 depicts the droop characteristics. For example, in P-ω droop control, the active power demand changes in load can be taken up by the DGs in a predetermined manner and the wireless control of parallel inverters is achieved with the utilization of system frequency as a communication link within a microgrid.

Figure 3.13 shows the block diagram of the droop control strategy, consisting of an inner voltage control loop and an outer power control loop. In the outer power control, the instantaneous P and Q are firstly calculated according to the inverter output voltages and output currents. These measured powers together with the power references, $P^*$and Q^*, are then sent to a predefined droop function, in order to obtain the desired voltage, E^*, which will be delivered to the inner voltage control loop. In the voltage control loop, a proportional compensator is used to force the inverter output voltage to track this specified E^*. The outputs of this voltage compensator together with the inner filter inductor currents are then fed to an inner current compensator to produce modulating signals for the sinusoidal PWM (SPWM). This permits each generator to take up changes in the total load in a manner determined by its frequency droop characteristics and amplitude droop characteristics, and essentially utilizes the system frequency as a communication link between the generator control systems.

This type of control method is very suitable for microgrids where several DG units are parallel-connected to a common AC bus. However, this method is subject to two particular problems as follows:

Power coupling: (3.5) and (3.6) are derived based on the assumption that the line impedance is mainly inductive. However, in some cases, line resistance cannot be neglected, especially in a low-voltage microgrid. As a result, the change of δ and E can affect both the active and reactive power flows, which can be explained from (3.1) and (3.2).

Slopes selection: Even the power coupling can be avoided with improved strategies such as virtual power frame transformation or virtual output impedance. The trade-off between the power sharing accuracy and the voltage frequency and magnitude deviation should be taken into account in the selection of m, n.

The microgrid control strategies mainly focus on a two-parallel-DG based system, and droop control method is generally used to control the inverters for autonomous power sharing between DGs. However, it is commonly necessary to combine different kind of DG sources, loads and storage technologies in order to ensure both long-term and short-term energy storage, and thus smooth the gap between power generation and power consumption within the microgrid [66]. Therefore, advanced techniques of microgrids including DGs, load, and ESS for power flow optimization are desired.

3.3.3 Simulation Results and Discussion

A microgrid with two DG units, as shown in Figure 3.14, is used in this simulation. There is one local load connected to each DG unit and one common load connected to common AC bus. The microgrid is connected to utility through a static transfer switch (STS). The voltage droop method is used. The system parameters are listed in Table 3.4.

At the beginning, the microgrid operates in the islanded mode. Figure 3.14 shows the active and reactive power sharing within the microgrid. P_1, P_2, and P_g denote the active power output from DG1 and DG2, and the active power flowing from the utility, respectively. On the other hand, Q_1, Q_2, and Q_g denote the reactive power output from DG1 and DG2, and the reactive power flowing from the utility, respectively. In islanded mode, DG1 and DG2 together supply all the active and reactive power required by the local loads and the common load according to their frequency and voltage droop characteristics. It can be seen that DG1 carries a larger share of active power since it has a stiffer slope while $P_g = 0$ W and $Q_g = 0$ VAR. At 0.1 second, the power consumption of local Load 1 rises to its rated value

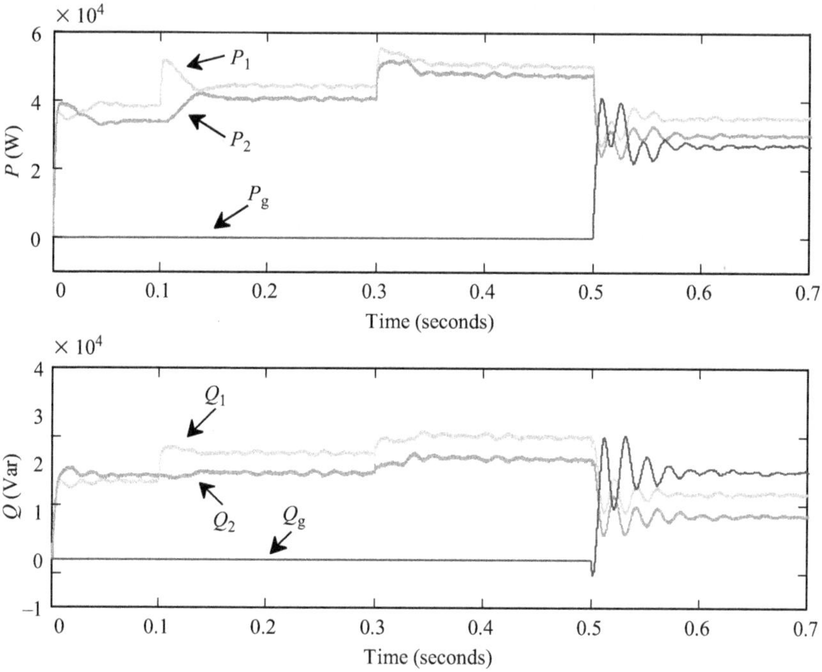

Figure 3.14 Power flows within microgrid

Table 3.4 System parameters of microgrid using voltage droop method

Nominal utility voltage	**380 V**	ω^*	**50 Hz**
Switching frequency	2 kHz	ω_{min}	48.5 Hz
DC source voltage	1,000 V	E^*	310 V
Filter inductance	0.9 mH	E_{min}	290 V
Filter capacitance	250 μF	P_1^*	35 kW
Line impedance	0.05 Ω, 1.6 mH	Q_1^*	15 kVar
Rating of each DG	70 kVA	P_2^*	30 kW
Rating of load 1	30 kW, 10 kVar	Q_2^*	10 kVar
Rating of load 2	20 kW, 6 kVar	m_1, n_1	−3/7e4, −4/8e4
Rating of common load	13 kW, 10 kVar	m_2, n_2	−3/1.1e4, −1/3e3

(30 kW, 10 kVAR), DG1 and DG2 increase the output power automatically to meet the new power requirement.

At 0.3 second, the microgrid starts to synchronize, the increase in both active and reactive output power of DGs can be observed during the process of synchronization. After reconnection (0.5 second) of the microgrid and utility, it is seen that the output powers of DG1 and DG2 drop back to their referenced dispatched values (P_1^*, Q_1^*, P_2^*, and Q_2^*) because part of the power flows from the utility in

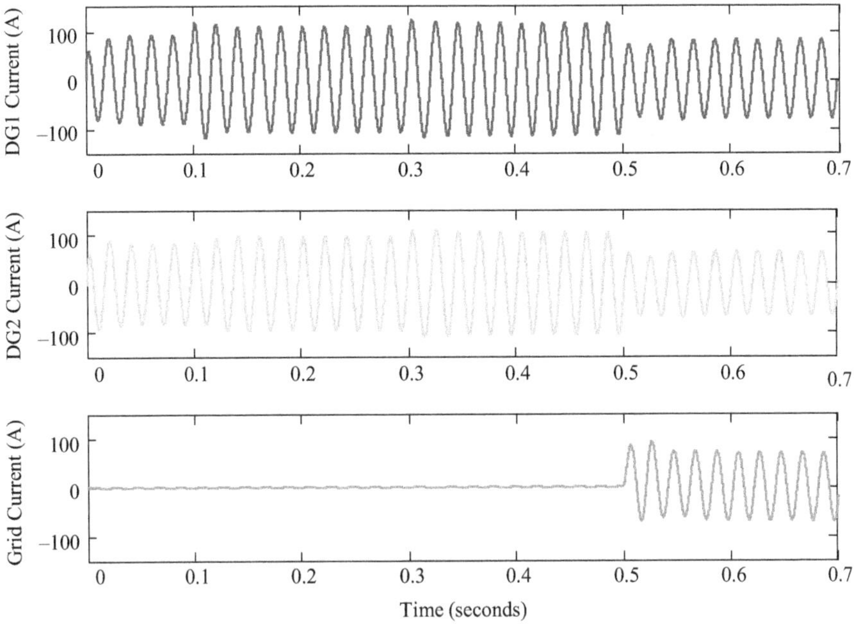

Figure 3.15 Per phase current within the microgrid

grid-connected mode. The DG output current and the grid current are shown in Figure 3.15.

The dynamic response of AC common bus voltage and STS current can be found in Figure 3.16. The AC common bus voltage is very stable during the grid connection process while slight overshoot current can be observed in the STS at 0.5 second. The synchronization process can be observed in Figure 3.17. The AC common bus voltage can track the utility grid voltage in less than 100 milliseconds in terms of frequency, magnitude and phase angle, which guarantees the smooth and fast transition from the islanded mode to the grid connected mode.

3.3.4 Experimental Results and Discussion

The laboratory microgrid structure in the experiment is shown in Figure 3.18. It consists of a solar PV array, a gas micro-turbine, a programmable load bank, and an electric motor. All these units are parallel connected to a common AC bus with nominal line-line voltage of 415 V and nominal frequency of 50 Hz. The microgrid can operate in both modes, connected to the utility grid through the STS or autonomously in islanded mode. The control of the DGs, loads, STS, and the data acquisition are fulfilled through a SCADA system.

3.3.4.1 Grid-Connected Operation

In this experiment, the PVs and the load bank are connected to the common AC bus and the microgrid is then connected to the utility grid through the STS. The total

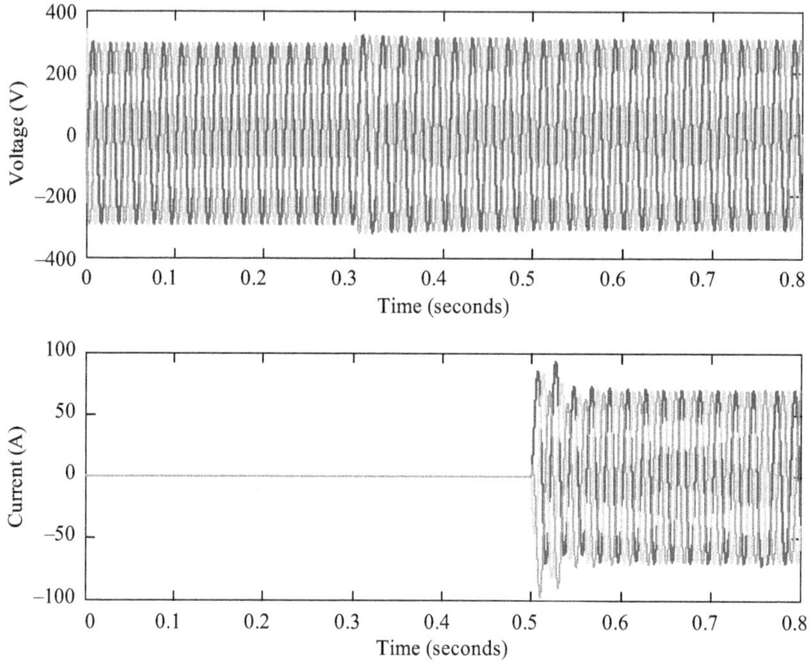

Figure 3.16 Response of voltage and AC common bus and current through STS

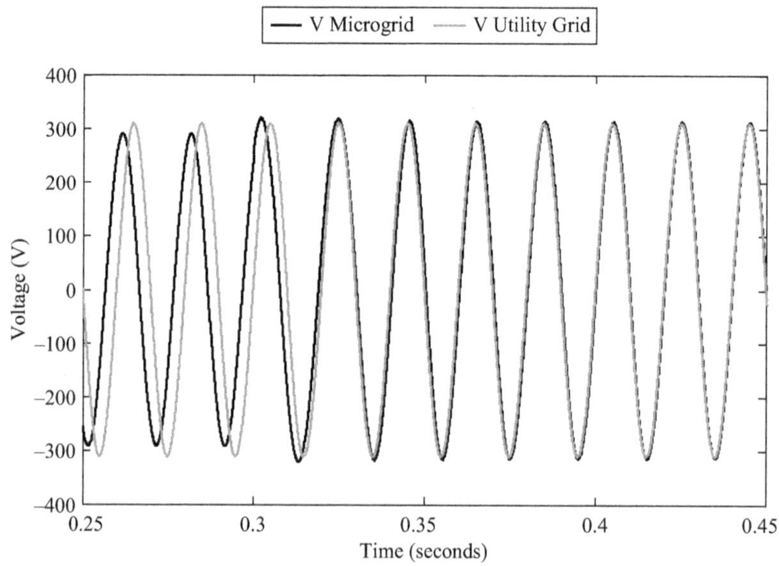

Figure 3.17 Synchronization of microgrid and utility grid

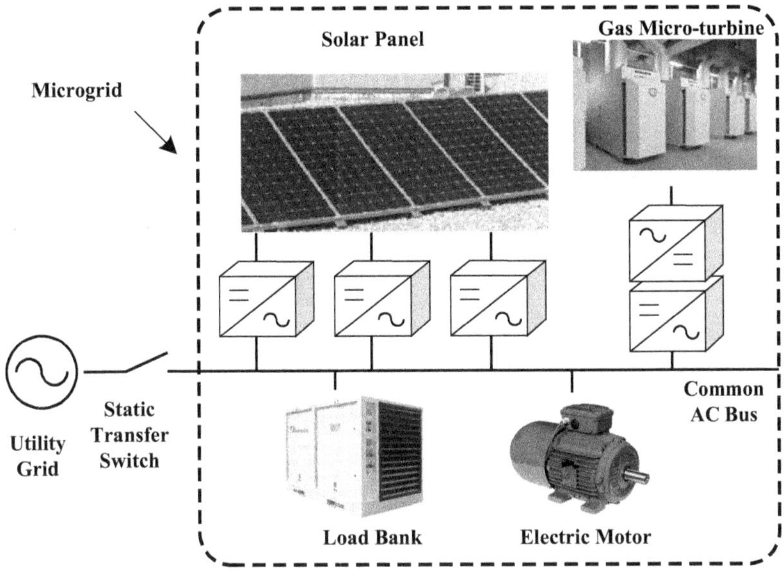

Figure 3.18 Microgrid under study

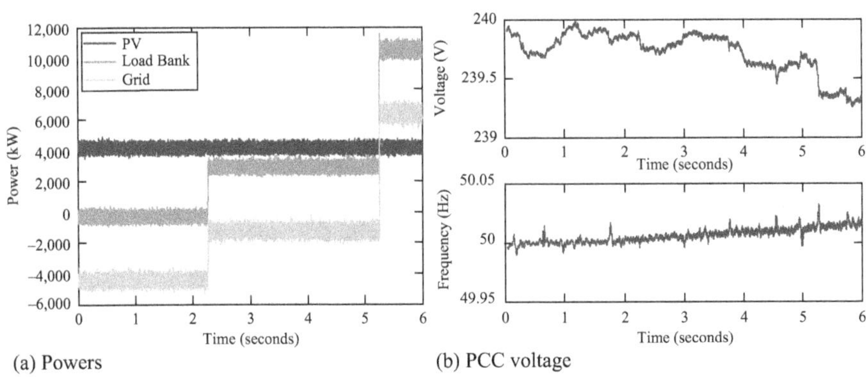

Figure 3.19 Response of powers and PCC voltage in grid-connected mode

power output of the PV systems at the time of this test is about 4.5 kW. The load bank is set to zero at the beginning, then raised up to 3 kW (less than the PV output power) and 10 kW (greater than the PV output power) at 2.2 seconds and 5.3 seconds, respectively. Under such circumstances, the power flow between the microgrid and the utility grid, and the power flow between the PV and load demand, would be interesting to see. The voltage response at the common AC bus in terms of amplitude and frequency will be investigated as well.

Figure 3.19(a) shows the active power sharing between the PV systems and the utility grid. It can be seen that the PV systems produced about 4.5 kW active power

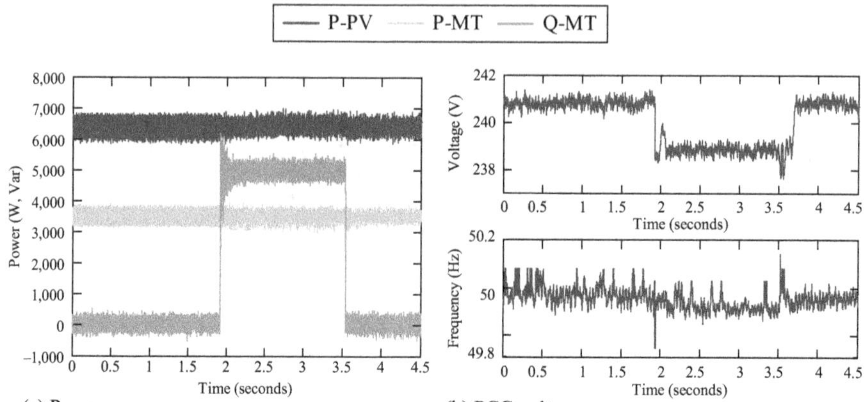

(a) Powers (b) PCC voltage

Figure 3.20 Response of powers and PCC voltage in grid-connected mode

throughout this test. At no-load operation at the beginning, the power flow from the utility is –4.5 kW, which indicates that all the power generated from the PVs was fed back to the utility grid. When the load bank was set to 3 kW at 2.2 seconds, the power flow from utility becomes –1.5 kW. This is as expected since the PVs provided all the load demand, resulting in only 1.5 kW (4.5 kW – 3 kW = 1.5 kW) to be exported to utility. When 10 kW was loaded at 5.3 seconds, it can be observed that extra 5.5 kW (10 kW – 4.5 kW = 5.5 kW) power was imported from the utility because now the load demand is greater than the power generation from the PV systems.

Figure 3.19(b) presents the voltage response at the PCC during the variation of the load demand. It can be seen that the amplitude and frequency of the PCC voltage are quite stable, presenting a voltage variation of only 0.6 V and frequency variation of 0.02 Hz, which is as expected due to the connection to the utility grid.

3.3.4.2 Islanded Operation

When the microgrid was isolated from the utility grid, the micro-turbine system is used as a master in islanded operation. The total output active power of the PV systems at the time of this experiment was about 6.5 kW. The load profile was set in such a pattern that there was a constant 10 kW resistive load throughout this test while a 5 kVAR inductive load was added and removed at 1.9 seconds and 3.5 seconds, respectively. The aim of this test is to demonstrate the feasibility of the microgrid in islanded mode under the fluctuated load profile.

Figure 3.20(a) shows the power sharing between the PV systems and the micro-turbine system in islanded mode. It can be seen that the micro-turbine system initially supplied active power together with PVs to the load bank. At $t = 1.9$ seconds, the micro-turbine system increased its reactive power from 0 to 5 kVAR to meet the new load demand, while the PV systems continued to supply around 6.5 kW

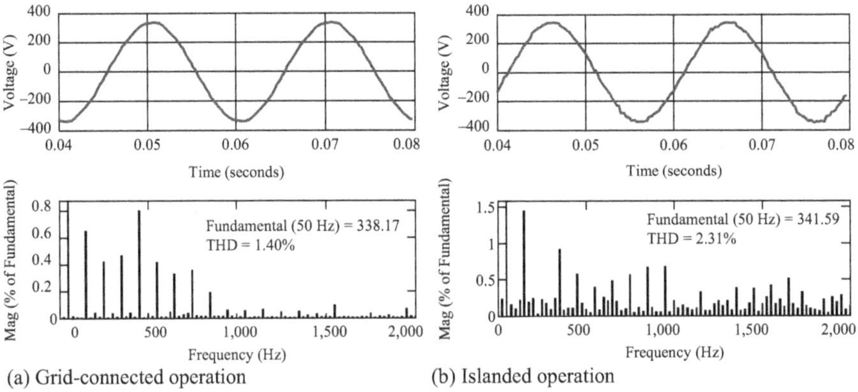

Figure 3.21 Harmonic spectrum of common AC bus voltage

Table 3.5 Power quality comparison

Performance	Voltage distortion	Flicker
Standard limits	5.0	0.8
Grid-connected	1.40	0.12
Islanded	2.31	0.56

active power. At $t = 3.5$ seconds, the inductive load was removed, the micro-turbine system continued to provide 3.5 kW active power while the reactive power dropped back to 0 kVAR.

Once again, the response of the common AC bus voltage is the main concern and it is analyzed during the variation of load demand. Due to the excellent control capability of the micro-turbine system, the common AC bus voltage is stabilized during the whole islanded operation. The magnitude and frequency of PCC voltage during the load change are shown in Figure 3.20(b).

3.3.4.3 Power Quality Comparison

Figure 3.21 compares the harmonic spectrum of the common AC bus voltage at grid-connected and islanded operation. It can be seen that both grid-connected and islanded operation can provides high-quality power, with only 1.40 percent voltage THD and 2.31 percent voltage THD, respectively.

In order to better validate the feasibility of the proposed microgrid operation, the power quality results are summarized at Table 3.5, where the performance of the common AC bus voltage is compared with the IEEE Standard 1547 [67]. The results show that the microgrid at grid-connected mode supplies better quality power compared with islanded mode. Most importantly, the power quality of both grid-connected and islanded operation can meet the standard requirement.

3.4 Conclusions

Because of the rapid growth in the number of automobiles in the world, we are now facing more problems of environment and human life, such as oil and diesel consumption or the greenhouse emissions. These issues bring an urgent challenge for researchers to develop new vehicle technologies with cleaner, more efficient and sustainable solutions for transportation. With these requirements, nowadays, many automotive companies are focusing on EVs, HEVs, and PHEVs driven by clean energy. In the near future, with the rising of EV technology, the world will have a new generation of conventional automobiles. Scientists all around the world are now seeking energy and environment solutions from different renewable resources and till today only wind and solar PV energy sources are found to be suitable for future large-scale generation. By generating electric power from our abundant renewables, such as wind and solar energy sources, we can solve a big portion of energy deficiency. It is expected that more than 80 percent future wind and solar PV power generation systems will be connected with the grids by 2030. Therefore, it is essential for scientists and researchers to find the most effective technologies for the grid integration and control of wind and solar PV power generation systems. The power converter topology, system stability and control of grid-connected DGs have attracted considerable interest in recent years, as the existing technologies are not suitable for large-scale systems. Research is needed to push these technologies to solve two enormous challenges, i.e. energy and environment by replacing conventional power plants with clean power plants.

References

[1] G. N. Sorebo and M. C. Echols. *Smart Grid Security*. USA: CRC Press; 2012.

[2] R. Strezelecki and G. Benysek. *Power Electronics in Smart Electrical Energy Networks*. Germany: Springer; 2008.

[3] P. Chiradeja and R. Ramakumar. "An approach to quantify the technical benefits of distributed generation." *IEEE Transactions on Energy Conversion*. 2004;**19**(4):764–773.

[4] J. Larminie and J. Lowry. *Electric Vehicle Technology Explained*. USA: John Wiley & Sons; 2003.

[5] F. S. Barnes and J. G. Levine. *Large Energy Storage Systems Handbook*. USA: CRC Press; 2011.

[6] M. R. Islam, Y. G. Guo, and J. G. Zhu. "A multilevel medium-voltage inverter for step-up-transformer-less grid connection of photovoltaic power plants." *IEEE Journal of Photovoltaics*. 2014;**4**(3):881–889.

[7] J. L. Santos, F. Antunes, A. Chehab, and C. Cruz "A maximum power point tracker for PV systems using a high performance boost converter." *Solar Energy*. 2006;**80**(7):772–778.

[8] M. R. Islam, Y. G. Guo, J. G. Zhu, and J. Jin "Design and fabrication of a microcontroller based maximum power point tracker for renewable energy

systems." *Journal of Applied Superconductivity and Electromagnetics*. 2011;**2**(1):17–23.

[9] M. R. Islam, Y. G. Guo, J. G. Zhu, and M.G. Rabbani. "Simulation of PV array characteristics and fabrication of microcontroller based MPPT." *6th International Conference on Electrical & Computer Engineering (ICECE)*; Dhaka, Bangladesh, December 2010. pp. 155–158.

[10] M. R. Islam, Y. G. Guo, and J. G. Zhu "A transformer-less compact and light wind turbine generating system for offshore wind farms." *Proceedings of the 2012 IEEE International Conference on Power and Energy*; Kota Kinabalu, Malaysia, December 2012. pp. 605–610.

[11] M. R. Islam, Y. G. Guo, J. G. Zhu, and D. Dorrell. "Design and comparison of 11 kV multilevel voltage source converters for local grid based renewable energy systems." *Proceedings of the 37th Annual Conference on IEEE Industrial Electronics Society*; Melbourne, Australia, November 2011. pp. 3596–3601.

[12] M. R. Islam, Y. G. Guo, and J. G. Zhu. "A medium-frequency transformer with multiple secondary windings for medium-voltage converter based wind turbine generating systems." *Journal of Applied Physics*. 2013;**113**(17): 17A324–17A324-3.

[13] M. R. Islam, Y. G. Guo, and J. G. Zhu. "An amorphous alloy core medium frequency magnetic-link for medium voltage photovoltaic inverters." *Journal of Applied Physics*. 2014;**115**(17):17E710.

[14] M. R. Islam, G. Lei, Y. G. Guo, and J. G. Zhu. "Optimal design of high-frequency magnetic-links for power converters used in grid connected renewable energy systems." *IEEE Transactions on Magnetics*. Vol. 50, No. 11, article 2006204, November 2014. doi:10.1109/TMAG.2014.2329939

[15] D. Sun, B. Ge, F. Z. Peng, A. R. Haitham, D. Bi, and Y. Liu. "A new grid-connected PV system based on cascaded H-bridge quasi-Z source inverter." *Proceedings of the 2012 IEEE International Symposium on Industrial Electronics*; Hangzhou, China, May 2012. pp. 951–956.

[16] H. Choi, W. Zhao, M. Ciobotaru, and V. G. Agelidis. "Large-scale PV system based on the multiphase isolated DC/DC converter." *Proceedings of the 3rd IEEE International Symposium on Power Electronics for Distributed Generation Systems*; Aalborg, Denmark, June 2012. pp. 801–807.

[17] S. Kouro, C. Fuentes, M. Perez, and J. Rodriguez. "Single DC-link cascaded H-bridge multilevel multistring photovoltaic energy conversion system with inherent balanced operation." *Proceedings of the 38th Annual Conference on IEEE Industrial Electronics Society*; Montreal, QC, Canada, October 2012. pp. 4998–5005.

[18] M. R. Islam, Y. G. Guo, and J. G. Zhu. "A high-frequency link multilevel cascaded medium-voltage converter for direct grid integration of renewable energy systems." *IEEE Transactions on Power Electronics*. 2014;**29**(8):4167–4182.

[19] C. H. Ng, M. A. Parker, Li Ran, P. J. Tavner, J. R. Bumby, and E. Spooner. "A multilevel modular converter for a large, light weight wind turbine generator." *IEEE Transactions on Power Electronics*. 2008;**23**(3):1062–1074.

[20] X. Yuan, J. Chai, and Y. Li. "A transformer-less high-power converter for large permanent magnet wind generator systems." *IEEE Transactions on Sustainable Energy*. 2012;**3**(3):318–329.

[21] F. Deng and Z. Chen. "A new structure based on cascaded multilevel converter for variable speed wind turbine." *Proceedings of the 36th Annual Conference on IEEE Industrial Electronics Society*; Glendale, AZ, November 2010. pp. 3167–3172.

[22] M. R. Islam, Y. G. Guo, J. G. Zhu, H. Y. Lu, and J. X. Jin. "High-frequency magnetic-link medium-voltage converter for superconducting generator-based high power density wind generation systems." *IEEE Transactions on Applied Superconductivity*. 2014;**24**(5):5202605.

[23] G. Hoogers. *Fuel Cell Technology Handbook*. New York: CRC Press; 2003.

[24] M. Ehsani, Y. Gao, and A. Emadi. *Modern Electric, Hybrid Electric, and Fuel Cell Vehicles Fundamentals, Theory, and Design*. 2nd edn. Power Electronics and Applications Series. New York: CRC Press; 2010.

[25] C. Alaoui. *Electric Vehicle Energy Management System*. Ph.D. Dissertation, University of Massachusetts Lowell; 2001.

[26] I. Husain. *Electric and Hybrid Vehicles: Design Fundamentals*. USA: CRC Press; 2003.

[27] UltimateCarPage.com. Lohner-Porsche Mixte Voiturette [online]. 2007. Available from http://www.ultimatecarpage.com/car/3456/Lohner-Porsche-Mixte-Voiturette.html. [Accessed 9 Nov 2014].

[28] A. Emadi, M. Ehsani, J. M. Miller. *Vehicular Electric Power Systems: Land, Sea, Air, and Space Vehicles*. Power Engineering Series, no. 22. New York: CRC Press; 2003.

[29] C. C. Chan. "An overview of electric vehicle technology." *Proceedings of the IEEE*. 1993;**81**(9):1202–1213

[30] R. Mauldin. *Performance Evaluation of Fuel Cell-Battery Hybrid Electric Vehicle*. Ph.D. Dissertation, University of Nevada, Las Vegas; 2005.

[31] M. R. Islam, M. R. Islam, and M. R. A. Beg. "Renewable energy resources and technologies practice in Bangladesh." *Renewable and Sustainable Energy Reviews*. 2008;**12**(2):299–343.

[32] M. R. Islam, M. Parveen, M. R. Islam, and H. Haniu. "Technology development for bio-crude oil from pyrolysis of renewable biomass resource." *Proceedings of 1st International Conference on the Developments in Renewable Energy Technology*; Dhaka, Bangladesh, December 2009. pp. 1–4.

[33] A. Demirbas. "Biodiesel fuels from vegetable oils via catalytic and non-catalytic super-critical alcohol transesterifications and other methods: A survey." *Energy Conversion and Management*. 2003;**44**(13):2093–2109.

[34] B. Subudhi, R. Pradhan. "A comparative study on maximum power point tracking techniques for photovoltaic power systems." *IEEE Transaction on Sustainable Energy*. 2013;**4**(1):89–98.

[35] V. Salas, E. Olías, A. Barrado, and A. Lázaro. "Review of the maximum power point tracking algorithms for stand-alone photovoltaic systems." *Solar Energy Materials and Solar Cells*. 2006;**90**(11):1555–1578.

[36] Ch. Hua, J. Lin, and Ch. Shen. "Implementation of a DSP-controlled PV system with peak power tracking." *IEEE Transaction on Industrial Electronics*. 1998;**45**(1):99–107.

[37] K. H. Hussein, I. Muta, T. Hoshino, and M. Osakada. "Maximum photovoltaic power tracking: an algorithm for rapidly changing atmospheric conditions." *IEE Proceedings Generation Transmission Distribution*. 1995;**142**(1):59–64.

[38] A. Branbrilla, M. Gambarara, A. Garutti, and F. Ronchi. "New approach to photovoltaic arrays maximum power point tracking," *30th Annual IEEE Power Electronics Specialists Conference*; Charleston, SC, June 1999. pp. 632–637.

[39] K. K. Tse, H. S. H. Chung, S. Y. R. Hui, and M. T. Ho. "A novel maximum power point tracking technique for PV panels." *32th Annual IEEE Power Electronics Specialists Conference*; Vancouver, BC, June 2001. pp 1970–1975.

[40] M. A. Hamdy. "A new model for the current-voltage output characteristics of photovoltaic modules." *Journal of Power Sources*. 1994;**50**(1–2):11–20.

[41] H. E.-S. A. Ibrahim, F. F. Houssiny, H. M. Z. El-Din, and M.A. El-Shibini. "Microcomputer controlled buck regulator for maximum power point tracker for DC pumping system operates from photovoltaic system." *Proceedings of IEEE International Fuzzy Systems Conference*; Seoul, South Korea, August 1999. pp. 406–411.

[42] M. A. S. Masoum and H. Dehbonei. "Design, construction and testing of a voltage-based maximum power point tracker (VMPPT) for small satellite power supply." Proceedings of 13th Annual AIAA/USU Conference on Small Satellite; 1999. SSC99-XII-7.

[43] T. Noguchi and S. Togashi. "Short-current pulse-based adaptive maximum power point tracking for a photovoltaic power generation system." *Electrical Engineering in Japan*. 2002;**139**(1):65–72.

[44] H. Farhangi. "The path of the smart grid." *IEEE Magazine Power and Energy*. 2010;**8**(2):18–28.

[45] J. Wang, A. Q. Huang, W. Sung, Y. Liu, and B. J. Baliga. "Smart grid technologies." *IEEE Industrial Electronics Magazine*. 2009;**3**(2):16–23.

[46] J. Hu, J. Zhu, and G. Platt. "Smart grid – the next generation electricity grid with power flow optimization and high power quality." *Proceedings of IEEE Int. Conf. on Electrical Machines and Systems*; Beijing, China, 2011. pp. 1–6.

[47] K. Moslehi and R. Kumar. "A reliability perspective of the smart grid." *IEEE Transaction on Smart Grid*. 2010;**1**(1):57–64.

[48] E. J. Coster, J. M. A. Myrzik, B. Kruimer, and W. L. Kling. "Integration issues of distributed generation in distribution grid." *Proceedings of the IEEE*. 2011;**99**(1):28–39.

[49] F. Katiraei and M. R. Iravani. "Power management strategies for a microgrid with multiple distributed generation units." *IEEE Transaction on Power System*. 2006;**21**(4):1821–1831.

[50] B. Zhao, X. Zhang, and J. Chen. "Integrated microgrid laboratory system." *IEEE Transaction on Power System*. 2012;**27**(4):2175–2185.
[51] J. M. Guerrero, F. Blaabjerg, T. Zhelev, K. Hemmes, E. Monmasson, S. Jemei *et al*. "Distributed generation: Toward a new energy paradigm." *IEEE Mag. Ind. Electron*. 2010;**4**(1):52–64.
[52] J. Hu, J. Zhu, D. G. Dorrell, and J. M. Guerrero. "Virtual flux droop method – A new control strategy of inverters in microgrids." *IEEE Transaction on Power Electronics*. 2014;**29**(9):4704–4711.
[53] J. Hu, J. Zhu, and D. G. Dorrell. "Model predictive control of inverters for both islanded and grid-connected operations in renewable power generations." *IET Renewable Power Generation*. 2014;**8**(3):240–248.
[54] J. Momoh. *Smart grid: Fundamentals of design and analysis*. USA: Wiley-IEEE Press; 2012.
[55] J. Potocnik. *European SmartGrids Technology Platform: Vision and Strategy for Europe's Electricity Networks of the Future*. European Commission; 2006.
[56] A. P. Martins, A. S. Carvalho, and A. S. Araújo. "Design and implementation of a current redundant operation system." *Proceedings IEEE IECON*; 1995. pp. 584–589.
[57] X. Sun, Y.-S. Lee, and D. Xu. "Modeling, analysis, and implementation of parallel multi-inverter system with instantaneous average-current-sharing scheme." *IEEE Transaction on Power Electronica*. 2003;**18**(3):844–856.
[58] T. F. Wu, Y.-K. Chen, and Y.-H. Huang. "3C strategy for inverters in parallel operation achieving an equal current distribution." *IEEE Transaction on Industrial Electronics*. 2000;**47**(2):273–281.
[59] M. C. Chandorkar, D. M. Divan, and R. Adapa. "Control of parallel connected inverters in standalone AC supply systems." *IEEE Transaction on Industrial Applications*. 1993;**29**(1):136–143.
[60] J. M. Guerrero, J. Matas, L. G. de Vicuña, J. Matas, M. Castilla, and J. Miret. 'A wireless controller to enhance dynamic performance of parallel inverters in distributed generation systems'. *IEEE Transaction on Power Electronics*. 2004;**19**(5):1205–1213.
[61] Y. Li, D. M. Vilathgamuwa, and P. C. Loh. "Design, analysis, and real-time testing of a controller for multibus microgrid system." *IEEE Transaction on Power Electronics*. 2004;**19**(5):1195–1204.
[62] J. Hu, J. Zhu, and G. Platt. "A droop control strategy of parallel-inverter-based microgrid." *Proceedings. IEEE ASEMD Conf.*; 2011. pp. 188–191.
[63] K. D. Brabandere, B. Bolsens, J. V. Keybus, A. Woyte, J. Driesen, and R. Belmans. "A voltage and frequency droop control method for parallel inverters." *IEEE Transaction on Power Electronics*. 2007;**22**(4):1107–1115.
[64] Y. Li and C. N. Kao. "An accurate power control strategy for power-electronics-interfaced distributed generation units operating in a low-voltage multibus microgrid." *IEEE Transaction on Power Electronics*. 2009;**24**(12):2977–2988.

[65] A. Tuladhar, H. Jin, T. Unger, and K. Mauch. "Control of parallel inverters in distributed AC power system with consideration of line impedance effect." *IEEE Transaction on Industrial Applications*. 2000;**36**(1):131–138.

[66] L. Valverde, C. Bordons, and F. Rosa. "Power management using model predictive control in a hydrogen-based microgrid." *Proceedings IEEE IECON2012*; 2012. pp. 5653–5660.

[67] IEEE. IEEE Standard for Interconnecting Distributed Resources with Electric Power Systems. IEEE Standard 1547. *IEEE*; 2003.

Chapter 4

Power Conversion Technology in the Smart Grid and EV

*M.A. Mahmud**

Abstract

The chapter presents a dynamical model and power conversion technology of electric vehicles (EVs) used in smart grids. The efficient power conversion of EVs in smart grids depends on the operation of bi-directional converters as these EVs need to be either charged or discharged. In this chapter, the mathematical model of a bi-directional converter used in EVs is developed and a nonlinear controller is designed to facilitate the power conversion in the smart grid environments. Since the power conversion of EVs in smart grids requires the communication, a nonlinear partial feedback linearising distributed controller based on the communication with different EVs is proposed to ensure high power quality and system stability.

4.1 Introduction

The recent innovations in renewable energy integration at low or medium voltage networks have led to the concept of smart grids. In smart grids, the connection of conventional synchronous generators is very uncommon and these generators are normally responsible for supplying load demand in conventional power systems. The renewable energy sources (RESs) have intermittent characteristics for which it is essential to store the extra energy into storages devices after fulfilling the demand. For example, effective charging and discharging schedules of PHEVs could support the integration of RESs by storing energy during the off-peak and deliver it back to the grid during the peak. These features of EVs pose several opportunities and challenges in energy management strategies of smart grids [1]. Power conversion in smart grids with EVs plays an important role as frequent charging and discharging of EVs are essential.

The integration of huge number of EVs into smart grids as loads might cause several problems such as transformer or line overloading and voltage stability [2–4].

*School of Software and Electrical Engineering, Swinburne University of Technology, Hawthorn, VIC 3122, Australia

On the other hand, power quality is an important issue when EVs are used to supply the load as power electronic inverters are used as interfacing units. An effective switching scheme is essential for efficient power conversion in smart grids with EVs as well as to maintain the power quality and stability of whole system. By considering these problems, several investigations have been performed in [5, 6] so that EVs could be advantageous for the smart grid operations.

Although a great deal of attention has been paid for investigating the impacts of EVs on smart grids along with optimal charging and discharging of EVs, a very little focus has been made on the development of appropriate power conversion technology [7–10]. The charging of EVs has huge impacts on the operation of smart grids as these vehicles consume a large amount of electrical energy which leads to undesirable peak demands. From the smart grid operators' point of view, power losses are another important concern as a significant amount of power is lost during the charging of EVs. Since the charging of EVs increases the load on smart grids, voltage stability is another key issue [11, 12]. Some efficient charging schemes have been recently developed to minimise the power loss and overloading problems [13].

During the discharging mode of EVs, the batteries of EVs act as distributed energy resources by locally meeting the demand during peak hours which in turn reduces the stress on overloaded smart grids. The amount of power delivered from vehicles to the grid is estimated by the aggregator in which a communication link is used to communicate between vehicle owners and the smart grid service providers (SGSPs) [14]. A sudden discharge of batteries used in smart grids may cause a voltage variation problem and this in turns causes voltage stability problems. Very few power conversion techniques are available in the literature to eliminate these problems. A fuzzy-based frequency controller is proposed in [15] to alleviate frequency and power fluctuations in tie-lines with an application to V2G systems. The approach presented in [15] provides satisfactory results for controlling active power but the reactive power control is uncovered which is a key factor for maintaining voltage stability. The control of power flow has been demonstrated using a fuzzy logic controller in [16] for voltage compensations and peak shavings. However, the main limitations of fuzzy logic controllers are that a fuzzy system cannot fully capture the dynamical model of V2G systems and require more fine-tuning and simulation before making it operational [17]. Therefore, it is essential to consider model-based control approach to enhance the power quality and stability of smart grids with EVs.

A mathematical model of EVs has been proposed in [5] for economic evaluation of these vehicles in smart grids. The model developed in [5] is useful for cot benefit analysis of smart grids but could not be used for technical development such as power conversion. A static model of EVs is used in [13] to control charging in smart grids without any contacts. A robust load frequency control scheme is used in [18] for the penetration of wind energy in smart grids where the controller is designed using H_∞ control scheme for the linearised model. The dynamical model used in [18] is mainly developed by considering the dynamics of the load and wind generators rather than EVs.

Most of the smart grid literature from the control point of view, so far exist in the literature, treats EVs as ideal voltage sources behind the inverters and some of them consider the dynamics of inverters only in order to design proportional integral (PI) and other linear controllers. But the consideration of ideal voltage sources behind the inverters is valid when the dynamics of smart grids are dominated by the generators connected to high voltage transmission levels. However in smart grids with EVs, low voltage power networks are considered and the system dynamics are dictated EVs. Therefore it is essential to consider the dynamics of original dynamics of EVs for an efficient operation of smart grids. Therefore, special attention has been paid in this chapter to develop a meaningful dynamical model of EVs which is suitable for the implementation of model-based controller.

Centralised controllers are normally used for power conversion with power electronic interfaces in either smart grids or microgrids [19–21]. The main limitations of these centralised controllers are the inclusion of very slow control loops and low-bandwidth communication systems. Moreover, the use of centralised controllers may cause the failure of whole system if one of the units fails down. In addition to these, most of the centralised controllers are designed based on the linearised model of smart grids and linear techniques are used for power conversion through inverters connected to RESs.

Distributed power conversion schemes have some features in which the failure of any unit does not shut down the whole system. In recent years, the concept of distributed control scheme has gained much popularity for the control of smart and microgrids where the controller at each unit communicates with others [22]. A distributed control scheme is proposed in [23] where proportional integral controllers are used to achieve the desired control objectives. Another cooperative PI controller is used in [24] in order to maintain an appropriate power balance between the supply and demand. In [25], a distributed controller is designed for DC microgrids where a linearised microgrid model is considered. Recently, a network-based distributed controller is proposed in [26, 27] for linearised model of islanded microgrids. However, these linear controllers can operate only over a fixed set of operating points and have not been implemented in smart grids with EVs.

Nonlinear power conversion or switching control schemes overcome the limitations of linear controllers by converting appropriate power from EVs over a wide range of operating conditions. Feedback linearization technique transforms a nonlinear system into a fully or partially linear system and linear control techniques can be employed to stabilise the whole system [28–31]. In the literature of power systems, the application of feedback linearisation techniques on large-scale interconnected power systems has been considered in a decentralised manner due to the geographical separation of generators [32]. But in the recently growing smart grids, the EVs are connected very close to each other and the performance of one EV affects the dynamic performances of others. Therefore, it is essential to consider the interactions among different EVs which can be done by implementing the feedback linearising controllers in a distributed manner.

This chapter aims to design a nonlinear distributed controller for power conversion in smart grids with EVs by considering the actual dynamical models of EVs. Since the partial feedback linearisation is a model-based approach, a comprehensive mathematical model of EVs, which matches with the control proposed theory, is formulated in this chapter. The regulation of active and reactive power is taken as the control objectives as either any or both these quantities need to be consumed or supplied by EVs. Partial feedback linearisation technique is used to design the controller and the stability of internal dynamics has been ensured before designing the controller. Since the proposed distributed scheme requires the communication among different EVs, graph theory has been used to formulate a generalised communication scheme and incorporated with the feedback linearising controller. Some case studies have been presented to demonstrate the performance of the proposed scheme.

4.2 Dynamical Modelling of EVs

For effective power conversion in smart grids with EVs, it is essential to develop a meaningful system model which has the capability to adapt with various power conversion technologies as overshadowed in the earlier section. In the literature of power conversion of EVs in smart grids, the mathematical model is not well-developed for the implementation of model-based controller as most of the existing literature is mainly based on the intelligent controllers [15, 16]. A model-based investigation has been performed very recently in [33] by considering a static model of EVs. But still now, it is very hard to locate an appropriate dynamical model of EVs in the literature of smart grids. This section aims to develop a new generalised mathematical model of EVs to represent the dynamics of the system which can be used for both charging and discharging mode by adjusting the controllable factors. However before developing the model of EVs, a battery model is briefly reviewed as this is a major part of any smart grid integrated with EVs.

Electrochemical batteries are normally used in EVs which have the capability to store energy based on the requirement. There exists a rich literature on the modelling of batteries which are mainly based on the chemistry of batteries and most of them are very complex for practical electrical engineering applications [34, 35]. The most commonly used battery model is proposed in [36] and the electrical circuit model of this battery is shown in Figure 4.1. The battery model as presented in [36] is

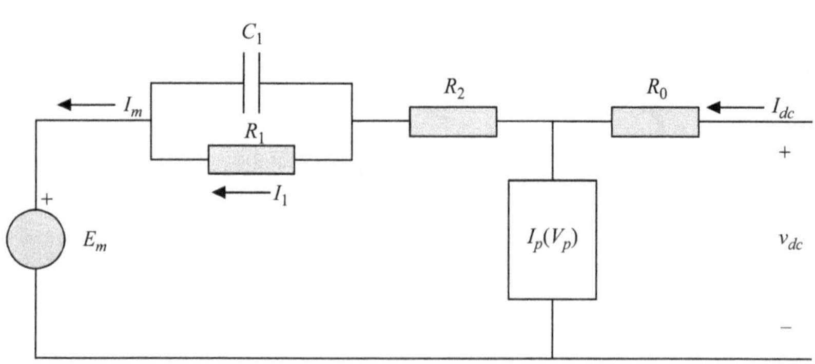

Figure 4.1 Equivalent circuit diagram of an EV battery

basically the analogous representation of an electrochemical model in which electrical components such as electromotive forces, resistors, capacitors and inductors are used to represent the electrochemistry of batteries.

In the model as described in [36] and presented by Figure 4.1, the charge stored in the battery is the integral of only a part I_m of the total current I_{dc} entering into the battery. The detail of the battery elements such as resistors (R_0, R_1 and R_2), capacitor (C_1) and internal voltage (E_m) can be seen in [36]. Since parasitic reactions often present in the battery, nonreversible parasitic branch models (with subscript p in Figure 4.1) draw some current but does not participate in the main, reversible reaction. Therefore, the parasitic branch can be omitted and the model of EVs can be revised accordingly.

Let EVs used in a smart grid, N numbers of EVs are connected either to consume or to supply power and the connection of ith EV at i^{th} node of the smart grid is shown in Figure 4.2 where the subscript 'i' denotes i^{th} PHEV and the smart grid node. The EV model as shown in Figure 4.2 can be connected either to a single-phase or a three-phase smart grid node and can also be modelled both single- and three-phase systems. In this section, first the model is developed for single-phase model nodes and then for three-phase nodes.

4.2.1 Dynamical Modelling of EV Connected to Single-Phase Smart Grid Node

From Figure 4.2, it can be seen that $I_{mi} = I_{dci}$. Now by applying Kirchhoff's current law (KCL) at the node where the resistor R_{1i} and capacitor C_{1i} are connected in parallel, it can be written as

$$I_{dci} = I_{1i} + C_{1i}\frac{dV_{c_{1i}}}{dt} \tag{4.1}$$

where $V_{c_{1i}}$ is the voltage across C_{1i} which is also the voltage across R_{1i} and thus, $V_{c_{1i}} = I_{1i}R_{1i}$. Using this relationship, (4.1) can be simplified as

$$\frac{dI_{1i}}{dt} = \frac{1}{\tau_{1i}}(I_{dci} - I_{1i}) \tag{4.2}$$

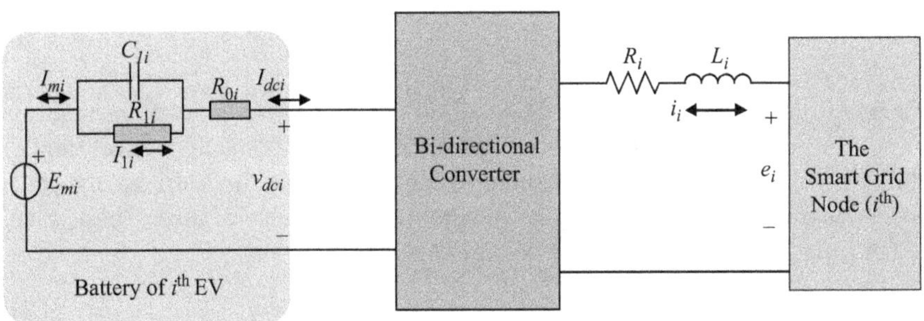

Figure 4.2 EV connected to smart grids

where $\tau_{1i} = R_{1i}C_{1i}$. Now by applying Kirchhoff's voltage law (KVL) at the output-side of the inverter, i.e. at the grid-side, it can be written as

$$\frac{di_i}{dt} = -\frac{R_i}{L_i}i_i + m_i\frac{v_{dci}}{L_i} - e_i \tag{4.3}$$

where m_i represents the switching action of the converter which is a function of modulation index and firing angle, R_i is the resistance of the connecting line, i_i is the output current of the inverter, L_i is the combination of filter and connecting line inductance and e_i is the grid voltage. The modulation index m_i can be adjusted either to charge or discharge EVs.

Equations (4.2) and (4.3) represent, the instantaneous and time-variant model of EVs connected to i^{th} node. But for the purpose of analysis and control, it is essential to transform the model into time-invariant system. To do this, the smart grid-connected EV system can be transformed into *dq*-frame. Let if x is a state variable in the instantaneous form and the corresponding space-phasor is $\vec{x}$, then in the *dq*-frame it can be written as

$$\vec{x} = \left(X_d + jX_q\right)e^{j\omega t} \tag{4.4}$$

where X_d and X_q represent the state variables in d- and q-frame respectively, j is the complex number, t is the time and ω is the angular frequency. The dynamics of the state variable can be written as

$$\dot{\vec{x}} = \left(\dot{X}_d + j\dot{X}_q\right)e^{j\omega t} + j\omega\left(X_d + jX_q\right)e^{j\omega t} \tag{4.5}$$

Using the relations presented by (4.4, 4.5), the grid-connected EV model represented by (4.2, 4.3) can be transformed into the following *dq*-frame:

$$\dot{I}_{1i} = \frac{1}{\tau_{1i}}\left(M_{di}I_{di} + M_{qi}I_{qi} - I_{1i}\right) \tag{4.6}$$

$$\dot{I}_{di} = -\frac{R_i}{L_i}I_{di} + \omega I_{qi} - \frac{E_{di}}{L_i} + M_{di}\frac{v_{dci}}{L_i} \tag{4.7}$$

$$\dot{I}_{qi} = -\omega I_{qi} - \frac{R_i}{L_i}I_{qi} - \frac{E_{qi}}{L_i} + M_{qi}\frac{v_{dci}}{L_i} \tag{4.8}$$

with

$$I_{dci} = mi_i = M_{di}I_{di} + M_{qi}I_{qi} \tag{4.9}$$

where ω is the angular frequency; M_{di} and M_{qi} are the switching functions in d- and q-frame respectively; I_{di} and I_{qi} are the currents in d- and q-frame respectively and E_{di} and E_{qi} are the grid voltages in d- and q-frame respectively. In *dq*-frame, the active power (P_{1pi}) and reactive power (Q_{1pi}) delivered from the vehicle into the single-phase grid or from grid to the vehicle can be written as

$$P_{1pi} = E_{qi}I_{qi} + E_{di}I_{di} \tag{4.10}$$

$$Q_{1pi} = E_{qi}I_{di} - E_{di}I_{qi} \tag{4.11}$$

In dq rotating frame, it can be assumed that $E_{di} = 0$ [31, 37] and in this case, (4.10) and (4.11) can be simplified as

$$P_{1pi} = E_{qi} I_{qi} \tag{4.12}$$

$$Q_{1pi} = E_{qi} I_{di} \tag{4.13}$$

Equations (4.6–4.8) represent the full dynamical model of an EV connected to single-phase node of a smart grid and the three-phase node is developed in the following subsection.

4.2.2 *Dynamical Modelling of EV Connected to Three-Phase Smart Grid Node*

For three-phase smart grid node, the relationship among different currents will be similar to that of single-phase node and but the input current to the inverter can be written as

$$I_{dci} = M_{ai} i_{ai} + M_{bi} i_{bi} + M_{ci} i_{ci} \tag{4.14}$$

Therefore for three-phase system, (4.2) will be modified as

$$\dot{I}_{1i} = \frac{1}{\tau_{1i}} (M_{ai} I_{ai} + M_{bi} I_{bi} + M_{ci} I_{ci} - I_{1i}) \tag{4.15}$$

If the EV is connected to a three-phase node of smart grids, it is essential to use a three-phase six-pulse full bridge converter. In this case, the current flowing through all three phases can be expressed as

$$\dot{i}_{ai} = -\frac{R_i}{L_i} i_{ai} - \frac{e_{ai}}{L_i} + \frac{v_{dci}}{L_i} (2M_{ai} - M_{bi} - M_{ci}) \tag{4.16}$$

$$\dot{i}_{bi} = -\frac{R_i}{L_i} i_{bi} - \frac{e_{bi}}{L_i} + \frac{v_{dci}}{L_i} (-M_{ai} + 2M_{bi} - M_{ci}) \tag{4.17}$$

$$\dot{i}_{ci} = -\frac{R_i}{L_i} i_{ci} - \frac{e_{ci}}{L_i} + \frac{v_{dci}}{L_i} (-M_{ai} - M_{bi} - 2M_{ci}) \tag{4.18}$$

where the subscripts '*a*', '*b*' and '*c*' represent phase-a, phase-b and phase-c respectively and all other variables represent the same as in single-phase smart grid node.

Equations (4.15–4.18) represent the complete model of an EV connected to three-phase smart grid nodes which are instantaneous and time-variant. Like single-phase model, this model also needs to convert into dq-frame so that it can easily be used for power conversion applications in smart grids. For three-phase system, Park's transformation is used to transform from phase components to dq-frame. Using park transformation, (4.15–4.18) can be written as

$$\dot{I}_{1i} = \frac{1}{\tau_{1i}} \left(M_{di} I_{di} + M_{qi} I_{qi} - I_{1i} \right) \tag{4.19}$$

$$\dot{I}_{di} = -\frac{R_i}{L_i} I_{di} + \omega I_{qi} - \frac{E_{di}}{L_i} + M_{di} \frac{v_{dci}}{L_i} \tag{4.20}$$

$$\dot{I}_{qi} = -\omega I_{qi} - \frac{R_i}{L_i} I_{qi} - \frac{E_{qi}}{L_i} + M_{qi} \frac{v_{dci}}{L_i} \tag{4.21}$$

with

$$I_{dci} = M_{ai} i_{ai} + M_{bi} i_{bi} + M_{ci} i_{ci} = M_{di} I_{di} + M_{qi} I_{qi} \tag{4.22}$$

where $I_{dq} = T_{abc}^{dq} i_{abc}$, $E_{dq} = T_{abc}^{dq} e_{abc}$ and $K_{dq} = T_{abc}^{dq} K_{abc}$. Equations (4.19–4.21) represent the full dynamical model of an EV connected to a three-phase node of smart grids. This model is similar to that of a single-phase node. The transformation matrix T_{abc}^{dq} can be written as

$$T_{abc}^{dq} = \frac{2}{3} \begin{bmatrix} \cos \omega t & \cos(\omega t - 120^\circ) & \cos(\omega t + 120^\circ) \\ \sin \omega t & \sin(\omega t - 120^\circ) & \sin(\omega t + 120^\circ) \\ 1/2 & 1/2 & 1/2 \end{bmatrix} \tag{4.23}$$

In dq-frame, the active power (P_{3pi}) and reactive power (Q_{3pi}) delivered from the vehicle into the three-phase grid or from grid to the vehicle can be written as

$$P_{3pi} = \frac{3}{2} \left(E_{qi} I_{qi} + E_{di} I_{di} \right) \tag{4.24}$$

$$Q_{3pi} = \frac{3}{2} \left(E_{qi} I_{di} - E_{di} I_{qi} \right) \tag{4.25}$$

Since it can be assumed that $E_{di} = 0$, (4.24) and (4.25) can be simplified as

$$P_{3pi} = \frac{3}{2} E_{qi} I_{qi} \tag{4.26}$$

$$Q_{3pi} = \frac{3}{2} E_{qi} I_{di} \tag{4.27}$$

The dynamical model of an EV connected to both single-phase and three-phase node of smart grids is same and the only difference is in the power equation. In smart grids, it is essential to control the power in or out from EVs for which it is essential to use a high performance power conversion algorithm. From (4.12, 4.13) and (4.26, 4.27), it can be seen that active power (P) and reactive power (Q) can be controlled just by controlling I_{qi} and I_{di} respectively.

4.3 Power Conversion Problem Formulation in Smart Grids with EVs

Since EVs can be either charged or discharged based on the desired objectives, it is essential to adjust the modulation index and firing angle of the bi-directional converter. The main aim of the proposed controller is to convert either AC power into DC during the charging mode or DC to AC during the discharging mode of EVs. During the discharging model, the EV will supply both active and reactive power into the grid as the load connected to the grid consumes these powers. From the power equations as described by (4.12, 4.13) and (4.26, 4.27), it can be seen that the active power (P) can be regulated by controlling the current I_q and reactive power (Q) regulation can be obtained by controlling I_d as there is nothing to do with the microgrid voltage E_q.

During the charging mode of EVs, reactive power will be zero as the battery can store DC power only. In this case, the modulation index and firing angle need to be adjusted in such a way that $I_d = 0$. The design of feedback linearizing controller depends on the output functions, i.e. the control objectives of the system. To serve the purpose of this chapter, the active and reactive current components I_q and I_d need to be chosen as control objectives.

Feedback linearisation technique may transform the smart grid with EVs into a fully or partially linear one and when the orders of the transformed system is equalled to the orders of original system, the controller design is straight-forward. But if the transformed system is a reduced-order linear system, some additional works need to be done to check the stability of untransformed dynamics (which are also known as internal dynamics) before designing the controller.

In this chapter, distributed partial feedback linearising scheme is used for power conversion in smart grids with EVs. But before designing the proposed distributed control based on partial feedback linearisation, it is essential to know about the feedback linearisation and justify the necessity of this special type of partial feedback linearisation.

4.4 Feedback Linearisation and Feedback Linearisability of Smart Grids with EVs

The EV vehicle model as describe in the Section 4.2 can be represented in the following form of a generalised nonlinear systems:

$$\dot{x}_i = f_i(x_i) + g_{1i}(x_i)u_{1i} + g_{2i}(x_i)u_{2i} \tag{4.28}$$

$$y_{1i} = h_{1i}(x_i) \tag{4.29}$$

$$y_{1i} = h_{2i}(x_i) \tag{4.30}$$

where

$$x_i = \begin{bmatrix} I_{1i} \\ I_{di} \\ I_{qi} \end{bmatrix}, \quad f_i(x_i) = \begin{bmatrix} -\dfrac{1}{\tau_{1i}} I_{1i} \\ -\dfrac{R_i}{L_i} I_{di} + \omega I_{qi} - \dfrac{E_{di}}{L_i} \\ -\omega I_{qi} - \dfrac{R_i}{L_i} I_{qi} - \dfrac{E_{qi}}{L_i} \end{bmatrix}, \quad g_{1i}(x_i) = \begin{bmatrix} \dfrac{1}{\tau_{1i}} I_{di} & \dfrac{1}{\tau_{1i}} I_{qi} \\ \dfrac{v_{dci}}{L_i} & 0 \\ 0 & \dfrac{v_{dci}}{L_i} \end{bmatrix},$$

$$u_i = \begin{bmatrix} u_{1i} \\ u_{2i} \end{bmatrix} = \begin{bmatrix} M_{di} \\ M_{qi} \end{bmatrix}, \quad \text{and} \quad y_i = h_i(x_i) = \begin{bmatrix} h_{1i} \\ h_{2i} \end{bmatrix} = \begin{bmatrix} I_{di} \\ I_{qi} \end{bmatrix}.$$

Based on this mathematical model, an overview of feedback linearising controller design and feedback linearisability of smart grids with EVs are discussed in the following two subsections.

4.4.1 Overview of Feedback Linearisation

The design of feedback linearising controller for efficient power conversion in smart grids with EVs depends on feedback linearisability of the developed model. The feedback linearisability is defined by the relative degree of the system and the relative degree in turns depends on the output functions [38].

The mathematical model of smart grids with EVs as developed in Section 4.2 can be linearised using feedback linearisation when some conditions as described latter are satisfied. Consider the following nonlinear coordinate transformation ($z_i = \varphi_i(x_i)$) for the smart grid with EVs.

$$z_i = \begin{bmatrix} h_{1i} & L_{f_i}h_{1i} & \cdots & L_{f_i}^{r_1-1}h_{1i} & h_{2i} & L_{f_i}h_{2i} & \cdots & L_{f_i}^{r_2-1}h_{2i} \end{bmatrix}^T \tag{4.31}$$

where $r_1 < n$ and $r_2 < n$ are the relative degree corresponding to output functions $h_{1i}(x)$ and $h_{2i}(x)$ respectively, $L_{f_i}h_i(x_i) = \frac{\partial h_i}{\partial x_i}f_i(x_i)$ are the Lie derivative of $h_i(x_i)$ along $f_i(x_i)$ [39]. The definitions of Lie derivative and relative degree are provided in Appendix I. The change of coordinate (4.31) transforms the nonlinear system (4.28–4.30) from x_i to z_i coordinates provided that the following conditions are satisfied for:

$$L_{g_i}L_{f_i}^{k_i}h_i = 0; \quad \text{for either } k_i < r_{1i} - 1 \text{ or } k_i < r_{2i} - 1 \text{ and } h_i = h_{1i} \text{ or } h_i = h_{2i} \tag{4.32a}$$

$$L_{g_i}L_{f_i}^{r_{1i}-1}h_{1i} \neq 0 \quad \text{or} \quad L_{g_i}L_{f_i}^{r_{2i}-1}h_{1i} \neq 0 \tag{4.32b}$$

$$n_i = r_i = (r_{1i} + r_{2i}) \tag{4.32c}$$

where $L_{f_i}h_i(x_i)$ is the Lie derivative of $L_{g_i}L_{f_i}h_i(x_i)$ along $g_i(x_i)$. Since $n_i = r_i$, all states x_i will be converted into new states z_i. In this case, the feedback linearised system is called exactly linearised which can be written as

$$\dot{z}_i = A_i z_i + B_i v_i \tag{4.33}$$

where A_i is the system matrix, B_i is the input matrix and v_i is the new control input for the exactly linearised system.

When $(r_{1i} + r_{2i}) < n_i$, only partial feedback linearisation is possible, i.e. some states are transformed through nonlinear coordinate transformation and some are not. The new states of a partially feedback linearised system can be written as

$$z_i = \varphi_i(x_i) = [\tilde{z}_i \quad \hat{z}_i]^T \tag{4.34}$$

where $\tilde{z}_i$ represents the state vector obtained from nonlinear coordinate transformation of order $r_{1i} + r_{2i}$ and $\hat{z}_i$ the state vector of the nonlinear (remaining) part of order $n_i - (r_{1i} + r_{2i})$. The dynamic of $\hat{z}_i$ is called the internal dynamic of the system which need to be stable in order to design and implement a partial feedback linearising controller for the following partially linearised system which can be written as

$$\dot{\tilde{z}}_i = \tilde{A}_i \tilde{z}_i + \tilde{B}_i \tilde{v}_i \tag{4.35}$$

where $\tilde{A}_i$ is the system matrix, $\tilde{B}_i$ is the input matrix and $\tilde{v}_i$ is the new linear control input for the partially linearised system. The developed smart grid model with EVs could be exactly or partially linearised and the feedback linearisability of this system is shown in the following subsection.

4.4.2 Feedback Linearisability of Smart Grids with EVs

The feedback linearisability of smart grids with EVs represented by (4.28–4.30) can be obtained by calculating the total relative degree (r_i) of the system. The relative degree corresponding to the first output function $h_{1i}(x_i) = I_{di}$ can be calculated as

$$L_{g_i} L_{f_i}^{1-1} h_{1i}(x_i) = L_{g_i} h_{1i}(x_i) = \frac{v_{dci}}{L_i} \neq 0 \tag{4.36}$$

where $r_{1i} = 1$. Similarly, the relative degree corresponding to the other output function $h_{2i}(x_i) = I_{qi}$ can be calculated as follows:

$$L_{g_i} L_{f_i}^{1-1} h_{2i}(x_i) = L_{g_i} h_{2i}(x_i) = \frac{v_{dci}}{L_i} \neq 0 \tag{4.37}$$

which indicates that $r_{2i} = 1$. Therefore, the total relative degree $r_i = r_{1i} + r_{2i} = 2$ and this means that $(r_{1i} + r_{2i}) < n_i$ as $n_i = 3$. From this, it can be said that a smart grid with EVs is partially linearised for the implementation of nonlinear feedback linearising power conversion technique. Therefore partial feedback linearisation approach needs to be used in order to design the controller for this system which is shown in the following subsection.

4.4.3 Partial Feedback Linearisation of Smart Grids with EVs

A nonlinear coordinate transformation can be written as

$$\tilde{z}_i = \tilde{\varphi}_i(x_i) \tag{4.38}$$

where $\tilde{\varphi}_i$ is the function of x_i. For smart grids with EVs, the following coordinate transformation has been considered.

$$\tilde{z}_{1i} = \tilde{\varphi}_{1i}(x_i) = h_{1i}(x_i) = I_{di} \tag{4.39a}$$

and

$$\tilde{z}_{2i} = \tilde{\varphi}_{2i}(x_i) = h_{2i}(x_i) = I_{qi} \tag{4.39b}$$

Using the above transformation, the partially linearised system can be obtained as follows:

$$\dot{z}_{1i} = \frac{\partial h_{1i}(x_i)}{\partial x_i}\dot{x}_i = L_{f_i}h_{1i}(x_i) + L_{g_i}h_{1i}(x_i)u_i = -\frac{R_i}{L_i}I_{di} + \omega I_{qi} - \frac{E_{di}}{L_i} + M_{di}\frac{v_{dci}}{L_i} \tag{4.40a}$$

$$\dot{z}_{2i} = \frac{\partial h_{2i}(x_i)}{\partial x_i}\dot{x}_i = L_{f_i}h_{2i}(x_i) + L_{g_i}h_{2i}(x_i)u_i = -\omega I_{qi} - \frac{R_i}{L_i}I_{qi} - \frac{E_{qi}}{L_i} + M_{qi}\frac{v_{dci}}{L_i} \tag{4.40b}$$

Equation (4.40) can be written in the following form of linearised system:

$$\dot{z}_{1i} = \tilde{v}_{1i} \tag{4.41a}$$

$$\dot{z}_{2i} = \tilde{v}_{2i} \tag{4.41b}$$

where $\tilde{v}_{1i}$ and $\tilde{v}_{2i}$ are linear control inputs which can be obtained by choosing any linear control technique and can be expressed as

$$\tilde{v}_{1i} = -\frac{R_i}{L_i}I_{di} + \omega I_{qi} - \frac{E_{di}}{L_i} + M_{di}\frac{v_{dci}}{L_i} \tag{4.42a}$$

$$\tilde{v}_{2i} = -\omega I_{qi} - \frac{R_i}{L_i}I_{qi} - \frac{E_{qi}}{L_i} + M_{qi}\frac{v_{dci}}{L_i} \tag{4.42b}$$

From (4.41), it can be seen that feedback linearisation techniques decouples the smart grids with EVs. Based on this partial feedback linearisation, the distributed controller can be design for power conversion in smart grids with EVs. However the design of distributed controller in smart grids with EVs requires the information from neighbouring EVs for appropriate power conversion. Since graph theory is considered for communication among different controllers connected to EVs, a brief overview of graph theory is provided first and then the distributed controller is designed in the following section.

4.5 Distributed Controller Design for Smart Grids with EVs

In this section, a distributed controller is designed for the feedback linearised smart grids with EVs. Since feedback linearisation technique decouples the smart grid into several subsystems depending on the number of EVs, the distributed controller can be designed by considering the communication among different subsystems. The communication among different subsystems can represented by using a directed graph (digraph) and in this chapter, the digraph is represented through bi-directional communication among different subsystems.

If a smart grid with EVs is represented by using a digraph G and A_G represents the adjacency matrix, the Laplacian matrix can be defined as

$$L = \Delta - A_G \tag{4.43}$$

where $\Delta = \text{diag}(d_i)$ is the degree matrix with $d_i = \sum_{j \in N_i} a_i$. The adjacency and Laplacian matrices are used to incorporate the communication in the distributed controller among different subsystems.

To achieve the control objectives by incorporating communication among different subsystems, e.g. between different nodes i and j, the control objectives for each subsystem need to be set in a cooperative manner which can be expressed as

$$\mathbf{e}_i = \sum_{j \in N_i} a_{ij}(\mathbf{Y}_i - \mathbf{Y}_j) + \mathcal{G}_i(\mathbf{Y}_i - \mathbf{Y}_{iref}) \tag{4.44}$$

where $\mathbf{e}$ is the tracking error in terms of the local neighbourhood, $\mathcal{G}$ is the pinning gain, $\mathbf{Y}$ represents the output vectors and the subscript i and j represent the neighbouring subsystems. If an EV is connected at the node i, then $\mathbf{Y}_i$ can be written as

$$\mathbf{Y}_i = [I_{di} \quad I_{qi}]^T \tag{4.45}$$

and similarly if another EV is connected at the node j, then

$$\mathbf{Y}_j = [I_{dj} \quad I_{qj}]^T \tag{4.46}$$

Now it is essential to obtain the distributed control law for each subsystem which can be determined by using any linear control technique. But before obtaining the control law, it is essential to analyse the stability of internal dynamics of each subsystem which is shown in Appendix II.

If a PI controller is used to achieve $\mathbf{e} \to \mathbf{0}$, then the linear control law (in the vector form) for i^{th} partially linearised subsystem can be written as

$$\tilde{\mathbf{v}}_i = c\left(\mathbf{K}_p + \frac{\mathbf{K}_i}{s}\right)\mathbf{e}_i \tag{4.47}$$

where $\tilde{\mathbf{v}}_i = [\tilde{\mathbf{v}}_{1i} \quad \tilde{\mathbf{v}}_{2i}]^T$ is the vector of linear control inputs, $\mathbf{K}_p$ is the vector of proportional gain, $\mathbf{K}_i$ is the vector of integral gain and c is the coupling factor. Here proportional and integral gains can be chosen in a similar way as presented in [29]. The coupling factor c is a quantity obtained from the graph theory which can be written as [40]

$$c \geq \frac{1}{\lambda_{\min}} \tag{4.48}$$

where $\lambda_{\min} = \min_{i \in N} \text{Re}(\lambda_i)$ and λ_i is the eigenvalue of $L + G$ with $G = \text{diag}(\mathcal{G}_i)$.

Therefore, from (4.42) partial feedback linearising control law for i^{th} subsystem can be written as

$$M_{di} = \frac{1}{v_{dci}}\left(L\tilde{v}_{1i} + R_i I_{di} - \omega L I_{qi} + E_{di}\right) \tag{4.49a}$$

$$M_{qi} = \frac{1}{v_{dci}}\left(L\tilde{v}_{2i} + \omega L I_{di} + R_i I_{qi} + E_{qi}\right) \tag{4.49b}$$

This distributed controller is used for power conversion in smart grids with EVs. The performance of the designed controller in light of power conversion and stability is discussed in the following section.

4.6 Performance Evaluation

The performance of the designed controller is evaluated on a test smart grid with four EVs as shown in Figure 4.3 in which vehicles are connected to a residential area, i.e. single-phase grid supply point. The corresponding single-line diagram is shown in Figure 4.4. Since the main tasks of EVs are to consume and supply power, a minimum state of charge (SOC) needs to be maintained in order to complete these tasks. In this Chapter, the minimum (SOC) is considered as 40 per cent. The following equation is used to calculate the total available energy of EVs during discharging [16]:

$$S_{discharging} = P_b \times N \times SOC_{min} \tag{4.50}$$

where $S_{discharging}$ is the total available energy for discharging to support the grid, P_b is the kWh of batteries, SOC_{min} is the minimum SOC and N is the total number of EVs connected to the grid.

The desired amount of energy during the charging is stage of EVs can be calculated in a similar manner by considering the maximum SOC which is considered as 100 per cent. The following equation can be used to calculate the required amount of energy for charging each vehicle:

$$S_{charging} = P_b \times SOC_{max} \tag{4.51}$$

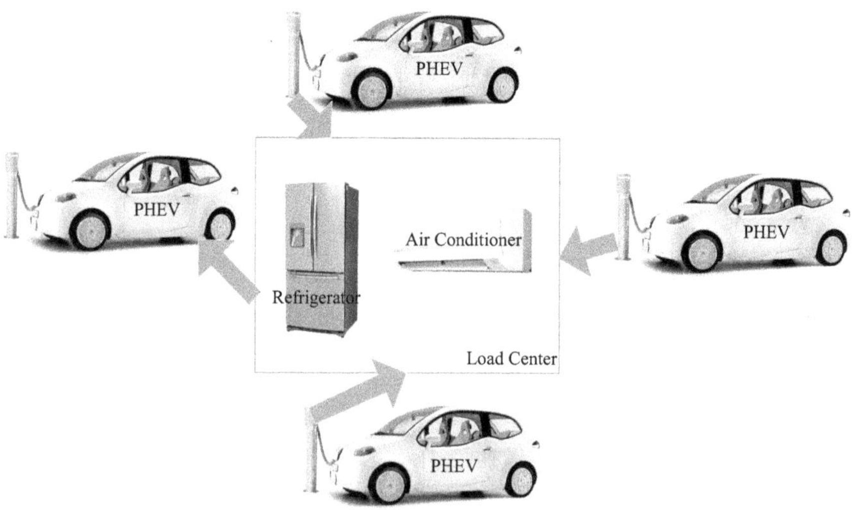

Figure 4.3 Test smart grid with EVs

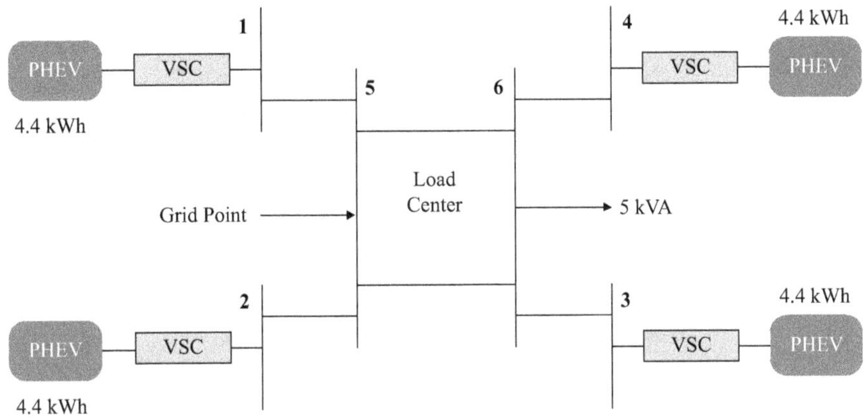

Figure 4.4 Single-line diagram of the smart grid with EVs

where $S_{charging}$ is the total available energy for charging and SOC_{max} is the maximum SOC. In this work, four EVs are connected to the grid and each of them with a battery rating of 4.4 kWh. Therefore the total available energy for discharging is 7.04 kWh. The other parameters of the battery and grid are provided in the Appendix III. During the discharging phase, the batteries of EVs are delivering power to the grid to supply a load of 5 kVA in a residential area and this information is provided by the aggregator. During the charging mode, each EV will consume 4.4 kWh power from the grid.

Since the distributed controller is designed in such a way that the controller designed for each DG unit has the capability to communicate with the controller connected to other DGs, the communication topology for the test microgrid can be represented in a manner as shown in Figure 4.4.

The adjacency and Laplacian matrices are calculated based on the communication topology as shown in Figure 4.5. To evaluate the performance of the designed controller, the pinning gain for each subsystem is considered as 1 from which the corresponding coupling factor is obtained as 0.24. The implementation block diagram of the proposed power conversion scheme is shown in Figure 4.6 for a single-phase system as overshadowed in the beginning of this section.

From Figure 4.6, it can be seen that the single-phase grid voltage and current are transformed into phasor quantity and then from phasor quantity to *dq*-frame for the purpose of implementing the proposed controller with the developed smart grid model with EVs. The controller is a combination of a linear PI controller which includes information from neighbouring EVs and a nonlinear partial feedback linearising controller. The nonlinear terms in partial feedback linearising controller are used to cancel the nonlinearities within the model and make the system ready for achieving desired response with linear controllers. Finally, control inputs are again transformed into phasor from *dq*-frame and then from phasor to sinusoidal. Finally, the sinusoidal control input is used through pulse width modulation (PWM) to the inverter switches where the switching frequency of the inverter is considered as 10 kHz.

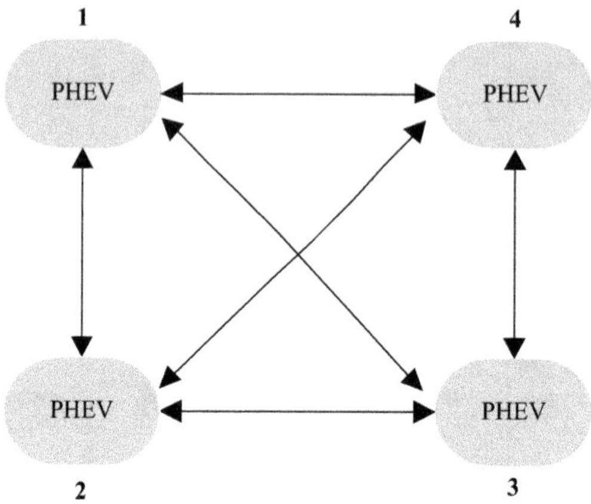

Figure 4.5 Communication among EVs in the smart grid

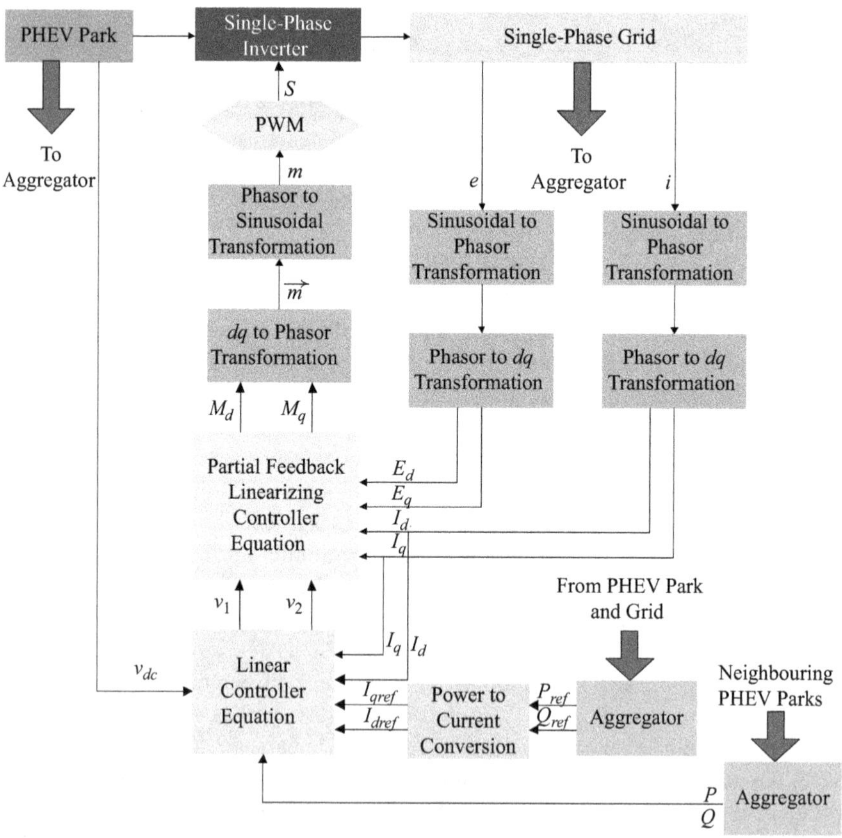

Figure 4.6 Implementation block diagram of distributed power conversion scheme

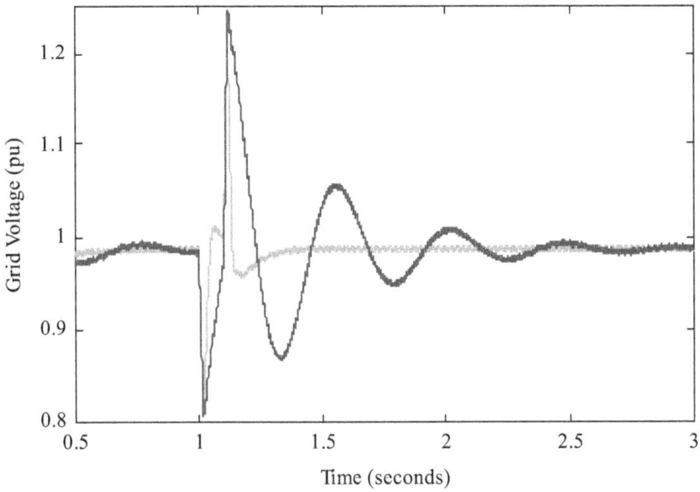

Figure 4.7 Grid voltage during the charging of EVs (Light grey line, proposed distributed controller; dark grey line, decentralised PI controller)

In this chapter, the performance of the developed scheme is evaluated both for charging and discharging phase and compared to that of an existing decentralised Pi controller.

4.6.1 Controller Performance Evaluation During the Charging Mode of EVs

During the charging mode of EVs, the grid voltage will be disturbed due to the addition of new loads and frequency will also be disturbed as the batteries of EVs consume active power. The grid voltage, i.e. voltage at bus-5 is shown in Figure 4.7.

From Figure 4.7, it can be seen that the grid voltage reduce when all EVs are connected at bus-1, bus-2, bus-3 and bus-4 at $t = 1$ second. Each of these EVs consumes 4.4 kWh power to become fully charged. Here the total power consumed by EVs is 17.6 kWh. However, the consumption of this power varies depending on the SOC and charging time. However, a certain amount of power will be consumed by these EVs when they are connected to the grid. The grid voltage should recover to the steady-state as soon as possible in order to maintain the voltage stability. From Figure 4.7, it can also be seen that the voltage recovery is much faster with the proposed controller as compared to the conventional controller. The corresponding frequency response is shown in Figure 4.8.

4.6.2 Controller Performance Evaluation During the Charging Mode of EVs

During the discharging mode, three cases-(i) controller performance at unity power factor operation, (ii) controller performance at a power factor other than unity and (iii) controller performance in case of a line-to-ground fault are discussed in the

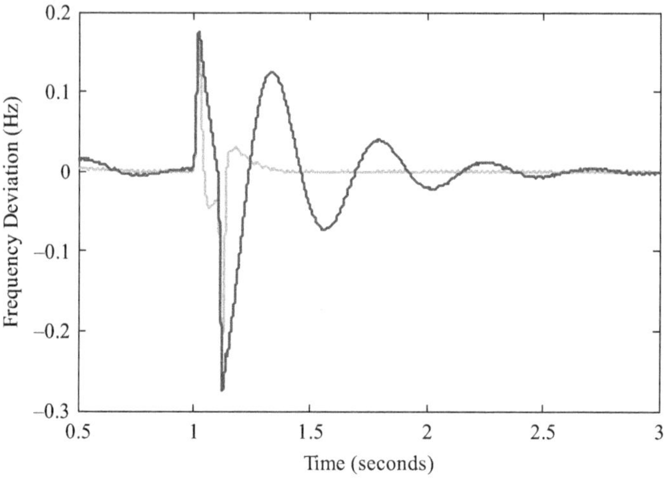

Figure 4.8 Grid frequency response during the charging of EVs (Light grey line, proposed distributed controller; dark grey line, decentralised PI controller)

following have been considered to evaluate the performance of the designed controller.

Case 1: Controller Performance Evaluation at Unity Power Factor

When the power factor of the load is considered as unity, no reactive power will be delivered into the grid as the grid voltage and current will be in phase with each other. The 5 kW power will be delivered into the grid and each vehicle is supplying equal power. At the stage, the current supplying into the load will be 22.72 A which is shown in Figure 4.9. And this current does not contain any harmonic with the designed controller as this is a pure sinusoidal signal which is shown by the light grey line in Figure 4.9. But a conventional proportional integral (PI) controller which is designed for the unity power factor operation of the smart grid with EVs contains some harmonics (dark grey line in Figure 4.9).

Case 2: Controller Performance Evaluation at a Power Factor Other Than Unity

Now if the system needs to operate at a power factor other than unity, the load voltage and current will not be in phase. In this case, some reactive power will be delivered into the load. If the power factor is considered as 0.8, the active power which needs to be delivered into the grid will be 4 kW and that of reactive power will be 3 kVAr. In this case, the designed controller acts in a similar way as considered to the previous case (light grey line in Figure 4.10). But the response of the conventional PI controller will be slower (dark grey line in Figure 4.10) as the reference active and reactive power have been changed.

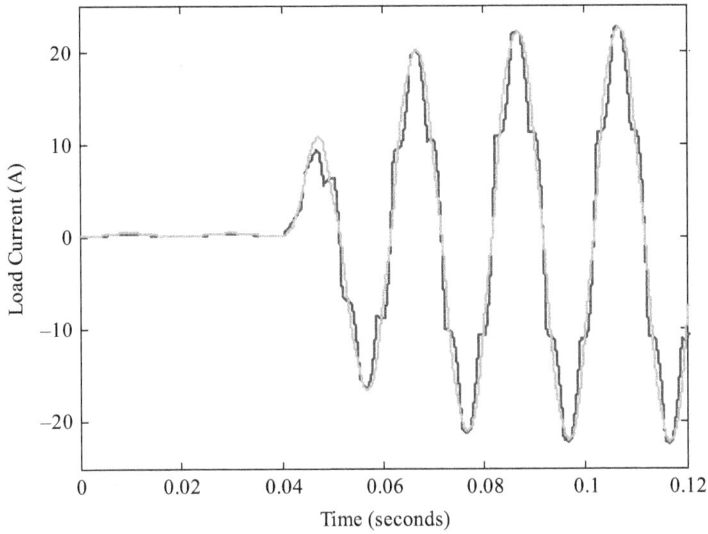

Figure 4.9 Load current at unity power factor during the discharging of EVs (Light grey line, proposed distributed controller; dark grey line, decentralised PI controller)

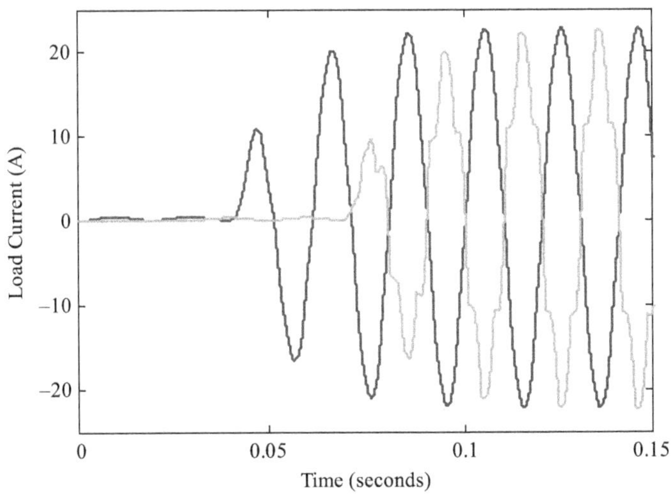

Figure 4.10 Load current at 0.8 power factor during the discharging of EVs (Light grey line, proposed distributed controller; dark grey line, decentralised PI controller)

Case 3: Controller Performance Evaluation in Case of a Line-to-Ground Fault

When a line-to-ground fault is applied at the load bus, the grid voltage becomes zero during the faulted period. After the clearance of fault, the grid voltage should settle down to its original value if the system is stable. If a conventional PI controller is

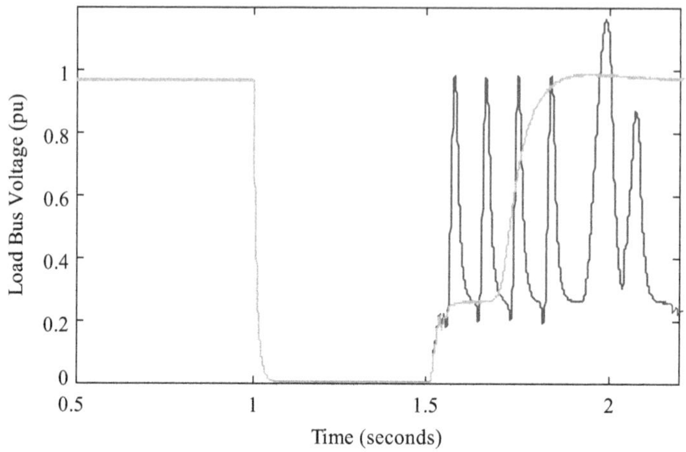

Figure 4.11 Load bus voltage with a line-to-ground fault during the discharging of EVs (Light grey line, proposed distributed controller; dark grey line, decentralised PI controller)

employed, the system is no longer remains stable for a certain period as the applied fault is one of the most severe fault and the PI controller cannot act properly. This scenario is shown in Figure 4.11 through the dark grey line from where it can be seen that the voltage settles down to another value. In this case, the designed controller maintains the stability of the system which is shown in Figure 4.11 (light grey line). From Figure 4.11, it can be clearly seen that the designed controller settles the voltage as soon as the fault is cleared. Thus, the smart grid with the designed control has the ability to maintain stable operation in case of a severe fault.

4.7 Conclusions

A new dynamical model of a smart grid with EVs has been developed and a distributed partial feedback linearising controller has been designed for improving the power quality and stability. The justification, of using the proposed control approach, has also been provided for through the feedback linearizability of the developed model along with the inclusion of the stability of internal dynamics. The designed controller which adopts the flexibilities in communication topology could be useful in practical applications. Simulation results clearly indicate that the proposed approach significantly improves the power quality as compared to the conventional PI controller and enhance the stability of the system as it is independent of operating conditions. The proposed controller acts faster than a PI controller during the changes in operating conditions which saves a huge amount of power. Future works will consider the design and implementation of such controller by considering model uncertainties.

Appendix I: Definition of Lie Derivative and Relative Degree

Lie Derivative: For a given differentiable scalar function $h(x)$ of $x = [x_1 \quad x_2 \quad x_3 \quad \cdots \quad x_n]^T$ and a vector field $f(x) = [f_1 \quad f_2 \quad f_3 \quad \cdots \quad f_n]^T$, the new scalar function, denoted by $L_f h(x)$, is obtained by the following operation:

$$L_f h(x) = \frac{\partial f}{\partial x} f(x) = \sum_{i=1}^{n} \frac{\partial f_i}{\partial x} f_i(x) \tag{A4.1}$$

which is called Lie derivative of function $h(x)$ along the vector field $f(x)$.

Relative Degree: If the Lie derivative of the function $L_f^{r-1} h(x)$ along vector field $g(x)$ is not equal to zero in a neighbourhood, Ω i.e.

$$L_g L_f^i h(x) = 0 \quad i < r - 1 \tag{A4.2}$$

and

$$L_g L_f^{r-1} h(x) \neq 0 \tag{A4.3}$$

then it is said that the system has the relative degree r in Ω.

Appendix II: Stability of Internal Dynamics of Smart Grids with EVs

Partial feedback linearising scheme transforms the third-order system into a second-order linear system. The desired performance of external dynamics can be obtained through the design and implementation of a linear controller. However, to ensure stability, the control law needs to be chosen in such a way that

$$\lim_{t \to \infty} h_i(x_i) \to 0 \tag{A4.4}$$

which implies that the state of a linear system decays to zero as time approached to infinity, i.e. $[\tilde{z}_{1i} \quad \tilde{z}_{2i} \quad \cdots \quad \tilde{z}_{ri}]^T \to 0; t \to \infty$. For each subsystem considered in this work, this means that at steady-state

$$\tilde{z}_{1i} = 0 \tag{A4.5a}$$

$$\tilde{z}_{2i} = 0 \tag{A4.5b}$$

Let the remaining nonlinear state be expressed by the following nonlinear function $\hat{z}_i = \hat{\phi}_i(x_i)$. To ensure stability, this needs to be selected in such a way that it must satisfy the following conditions [41]:

$$L_{g1} \hat{\phi}_i(x_i) = 0 \tag{A4.6a}$$

$$L_{g2} \hat{\phi}_i(x_i) = 0 \tag{A4.6b}$$

For the developed smart grid model with EVs, the above condition will be satisfied if the following nonlinear coordinate transformation is considered.

$$\hat{\phi}_i(x_i) = \hat{z}_i = -\tau_{1i} I_{1i} + \frac{1}{2}\frac{L_i}{v_{dci}} I_{di}^2 + + \frac{1}{2}\frac{L_i}{v_{dci}} I_{qi}^2 \tag{A4.7}$$

Thus, the remaining dynamics of smart grids with EVs can be expressed as follows:

$$\dot{\hat{z}}_i = L_{f_i}\hat{\phi}_i(x_i) = -\tau_{1i} f_{1i} + \frac{L_i}{v_{dci}} I_{di} f_2 + + \frac{L_i}{v_{dci}} I_{qi} f_3 \tag{A4.8}$$

Since $I_{di} = h_{1i}(x_i) = \tilde{z}_{1i}$ and $I_{qi} = h_{2i}(x_i) = \tilde{z}_{2i}$, the above equation can be written as

$$\dot{\hat{z}}_i = -\tau_{1i} f_{1i} + \frac{L_i}{v_{dci}} \tilde{z}_{1i} f_2 + + \frac{L_i}{v_{dci}} \tilde{z}_{2i} f_3 \tag{A4.9}$$

Since $\tilde{z}_{1i} = 0$ and $\tilde{z}_{2i} = 0$, (A4.9) can be written as

$$\dot{\hat{z}}_i = -\tau_{1i} f_{1i} \tag{A4.10}$$

Since $f_{1i} = -\frac{1}{\tau_{1i}} I_{1i}$, (A4.10) can be simplified as

$$\dot{\hat{z}}_i = I_{1i} \tag{A4.11}$$

From (A4.7), 'I' can be calculated as

$$I_{1i} = \frac{1}{\tau_{1i}}\left(\frac{1}{2}\frac{L_i}{v_{dci}} I_{di}^2 + + \frac{1}{2}\frac{L_i}{v_{dci}} I_{qi}^2 - \hat{z}_i\right) \tag{A4.12}$$

If I_{di} and I_{qi} are replaced with $\tilde{z}_{1i} = 0$ and $\tilde{z}_{2i} = 0$ respectively, (A4.12) can be written as

$$I_{1i} = -\frac{1}{\tau_{1i}} \hat{z}_i \tag{A4.13}$$

Equation (A4.13) represents stable internal dynamics of smart grids with EVs and therefore partial feedback linearising controller can be designed for power conversion. It is also clear that the proposed partial feedback linearising scheme divides the dynamics of smart grids with EVs into two parts: one is the external dynamics as described in Section 4.4, and the other is the internal dynamics which needs to be stable to design the controller and here is the proof of stability of internal dynamics.

Appendix III: System Parameters

Battery Parameters: $R_1 = 0.4\,\text{m}\Omega$, $\tau_1 = 7200\,\text{s}$, $R_0 = 2\,\text{m}\Omega$

Grid Parameters: $Voltage\,(rms) = 220\,\text{V}$, $Frequency = 50\,\text{Hz}$, $R = 0.1\,\Omega$ (for each segment) and $L = 10\,\text{mH}$ (for each segment).

References

[1] Galus D. M., Zima M., and Andersson G., 'On integration of plug-in hybrid electric vehicles into existing power system structures'. *Energy Policy.* 2010; 38(11): 6736–6745

[2] Papadopoulos P., Skarvelis-Kazakos S., Grau I., Cipcigan L. M., and Jenkins N., 'Electric vehicles' impact on British distribution networks'. *IET Electrical Systems in Transportation.* 2012; 2(3): 91–102

[3] Fernaández L. P., Romaán, T. G. S., Cossent R., Domingo C. M., and Friías P., 'Assessment of the Impact of Plug-in Electric Vehicles on Distribution Networks'. *IEEE Transactions on Power Systems.* 2011; 26(1): 206–213

[4] Hilshey A. D., Hines P. D. H., Rezaei P., and Dowds J.R., 'Estimating the Impact of Electric Vehicle Smart Charging on Distribution Transformer Aging'. *IEEE Transactions on Smart Grid.* 2013; 4(2): 905–913

[5] Das R., Thirugnanam, K., Kumar P., Lavudiya R., and Singh M., 'Mathematical Modeling for Economic Evaluation of Electric Vehicle to Smart Grid Interaction'. *IEEE Transactions on Smart Grid.* 2014; 5(2): 712–721

[6] Yang H., Chung C. Y., and Zhao J., 'Application of Plug-In Electric Vehicles to Frequency Regulation Based on Distributed Signal Acquisition via Limited Communication'. *IEEE Transactions on Power Systems.* 2013; 28(2): 1017–1026

[7] He Y., Venkatesh B., and Guan L., 'Optimal Scheduling for Charging and Discharging of Electric Vehicles'. *IEEE Transactions on Smart Grid.* 2012; 3(3): 1095–1105

[8] Gunter S. J., Afridi K. K., and Perreault D. J., 'Optimal Design of Grid-Connected PEV Charging Systems with Integrated Distributed Resources'. *IEEE Transactions on Smart Grid.* 2013; 4(2): 956–967

[9] Igualada L., Corchero C., Cruz-Zambrano M., and Heredia F.-J., 'Optimal Energy Management for a Residential Microgrid Including a Vehicle-to-Grid System'. *IEEE Transactions on Smart Grid.* 2014; 5(4): 2163–2172

[10] Shortt A., and O'Malley M., 'Quantifying the Long-Term Impact of Electric Vehicles on the Generation Portfolio'. *IEEE Transactions on Smart Grid.* 2014; 5(1): 71–83

[11] Clement-Nyns K., Haesen E., and Driesen J., 'The Impact of Charging Plug-In Hybrid Electric Vehicles on a Residential Distribution Grid'. *IEEE Transactions on Power Systems.* 2010; 25(1): 371–380

[12] Darabi Z., and Ferdowsi M., 'Aggregated Impact of Plug-in Hybrid Electric Vehicles on Electricity Demand Profile'. *IEEE Transactions on Sustainable Energy.* 2011; 2(4): 501–508

[13] Thirugnana, K., Joy T. P. E. R., Singh M., and Kumar P., 'Modeling and Control of Contactless Based Smart Charging Station in V2G Scenario'. *IEEE Transactions on Smart Grid.* 2014; 5(1): 337–348

[14] Han S., and Sezaki K., 'Development of an Optimal Vehicle-to-Grid Aggregator for Frequency Regulation'. *IEEE Transactions on Smart Grid.* 2010; 1(1): 65–72

[15] Datta M., and Senjyu T., 'Fuzzy Control of Distributed PV Inverters/Energy Storage Systems/Electric Vehicles for Frequency Regulation in a Large Power System'. *IEEE Transactions on Smart Grid*. 2013; 4(1): 479–488

[16] Singhand M., Kumar P., and Kar I., 'Implementation of Vehicle to Grid Infrastructure Using Fuzzy Logic Controller'. *IEEE Transactions on Smart Grid*. 2012; 3(1): 565–577

[17] Khayyam H., Ranjbarzadeh H., and Marano V., 'Intelligent control of vehicle to grid power'. *Journal of Power Sources*. 2012; 201: 1–9

[18] Vachirasricirikul S., and Ngamroo I., 'Robust LFC in a Smart Grid with Wind Power Penetration by Coordinated V2G Control and Frequency Controller'. *IEEE Transactions on Smart Grid*. 2014; 5(1): 371–380

[19] Peas Lopes J. A. P, Moreira C. L., and Madureira A. G., 'Defining control strategies for Microgrids islanded operation'. *IEEE Transactions on Power Systems*. 2006; 21(2): 916–924

[20] Zamora R., and Srivastava A. K., 'Controls for microgrids with storage: Review, challenges, and research needs'. *Renewable and Sustainable Energy Reviews*. 2010; 14(7): 2009–2018

[21] He J., Li Y., and Blaabjerg F., 'An Islanding Microgrid Reactive Power, Imbalance Power, and Harmonic Power Sharing Scheme'. *IEEE Transactions on Power Electronics*. In early access: doi:10.1109/TPEL.2014.2332998

[22] Caldognetto T., Tenti P., Costabeber A., and Mattavelli P., 'Improving Microgrid Performance by Cooperative Control of Distributed Energy Sources'. *IEEE Transactions on Industry Applications*. In early access: doi:10.1109/TIA.2014.2313648

[23] Shafiee Q., Guerrero J. M., and Vasquez J. C., 'Distributed Secondary Control for Islanded Microgrids—A Novel Approach'. *IEEE Transactions on Power Electronics*. 2014; 29(2): 1018–1031

[24] Kim J.-Y., Jeon J.-H., Kim S.-K., Cho C., Park J.-H., Kim H.-M. *et al.*, 'Cooperative Control Strategy of Energy Storage System and Microsources for Stabilizing the Microgrid during Islanded Operation'. *IEEE Transactions on Power Electronics*. 2010; 25(12): 3037–3048

[25] Nasirian V., Moayedi S., Davoudi A., and Lewis F. L., 'Distributed Cooperative Control of DC Microgrids'. *IEEE Transactions on Power Electronics*. In early access: doi:10.1109/TPEL.2014.2324579

[26] Kahrobaeian A., and Mohamed, Y. A.-R. I., 'Networked-Based Hybrid Distributed Power Sharing and Control for Islanded Micro-Grid Systems'. *IEEE Transactions on Power Electronics*. In early access: 10.1109/TPEL.2014.2312425

[27] Shafiee Q., Stefanovic C., Dragicevic T., Popovski P., Vasquez J. C., and Guerrero J. M., 'Robust Networked Control Scheme for Distributed Secondary Control of Islanded Microgrids'. *IEEE Transactions on Industrial Electronics*. 2014; 61(10): 5363–5374

[28] Mahmud M. A., Pota H. R., and Hossain M. J., 'Full-order nonlinear observer-based excitation controller design for interconnected power systems via exact linearization approach,. *International Journal of Electrical Power and Energy Systems*. 2012; 44(1): 54–62

[29] Mahmud M. A., Pota H. R., and Hossain M. J., 'Dynamic stability of three-phase grid-connected photovoltaic system using zero dynamic design approach'. *IEEE Journal of Photovoltaics*. 2012; 2(4): 564–571

[30] Mahmud M. A., Pota H. R., Hossain M. J., and Roy N. K., 'Robust partial feedback linearizing stabilization scheme for three-phase grid-connected photovoltaic systems'. *IEEE Journal of Photovoltaics*. 2014; 4(1): 423–431

[31] Mahmud M. A., Pota H. R., and Hossain M. J., 'Nonlinear DSTATCOM controller design for distribution network with distributed generation to enhance voltage stability'. *International Journal of Electrical Power and Energy Systems*. 2013; 53: 974–979

[32] Mahmud M. A., Pota H. R., Aldeen M., and Hossain M. J., 'Partial Feedback Linearizing Excitation Controller for Multimachine Power Systems to Improve Transient Stability'. *IEEE Transactions on Power Systems*. 2014; 29(2): 561–571

[33] Guenther C., Schott B., Hennings W., Waldowski P., and Danzera M. A., 'Model-based investigation of electric vehicle battery aging by means of vehicle-to-grid scenario simulations'. *Journal of Power Sources*. 2013; 239: 604–610

[34] Hu T., and Hung J., 'Simple algorithms for determining parameters of circuit models for charging/discharging batteries'. *Journal of Power Sources*. 2013; 233: 14–22

[35] Legranda N., Rael S., Knosp B., Hinaje M., Desprez P., and Lapicque F., 'Including double-layer capacitance in lithium-ion battery mathematical models'. *Journal of Power Sources*. 2014; 251: 370–378

[36] Ceraolo M., 'New Dynamical Models of Lead-Acid Batteries'. *IEEE Transactions on Power Systems*. 2000; 15(4): 1184–1190

[37] Kim I.-S., 'Sliding mode controller for the single-phase grid-connected photovoltaic system'. *Applied Energy*. 2006; 83(10): 1101–1115

[38] Mahmud M. A., Hossain M. J., and Pota H. R. (eds.). 'Selection of output function in nonlinear feedback linearizing excitation control for power systems'. Proceedings of Australian Control Conference; Melbourne, Australia, Nov 2011. New Jersey: IEEE; 2011. pp. 458–463

[39] Mahmud M. A., Pota H. R., and Hossain M. J., 'Nonlinear Current Control Scheme for a Single-Phase Grid-Connected Photovoltaic System'. *IEEE Transactions on Sustainable Energy*. 2014; 5(1): 218–227

[40] Bidram A., Davoudi A., Lewis F. L., and Guerrero J. M., 'Distributed cooperative secondary control of microgrids using feedback linearization'. *IEEE Transactions on Power Systems*. 2013; 28(3): 3462–3470

[41] Mahmud M. A., Hossain M. J., Pota H. R., and Roy N. K., 'Robust Nonlinear Controller Design for Three-Phase Grid-Connected Photovoltaic Systems Under Structured Uncertainties'. *IEEE Transactions on Power Delivery*. 2014; 29(3): 1221–1230

Chapter 5

Power Control and Monitoring of the Smart Grid with EVs

M.S. Rahman, F.H.M. Rafi*, M.J. Hossain* and J. Lu**

Abstract

The usage of electric vehicles (EVs) and plug-in hybrid electric vehicles (PHEVs) has increased significantly in recent years due to environmental concerns and hike in the fossil fuel price. These vehicles possess dual dynamic characteristics. They act as a load while in G2V (grid-to-vehicle) mode and as a generator while in V2G (vehicle-to-grid) mode. V2G concept can improve utility grid performance with regard to efficiency, stability and reliability by offering reactive power management and active power control, tracking intermittent renewable energy sources, load balancing and shifting via valley filling support, peak load shaving and current harmonics filtering in the output. On the other hand, G2V includes conventional and fast battery charging that can stress the grid distribution network due to high-power rating of EV batteries and by injecting harmonics. Sophisticated active and reactive power regulation as well as state-of-the-art monitoring system is required to overcome the impacts and to implement successful interfacing. This chapter discusses the impacts of G2V/V2G concepts on the smart grid active and reactive power profile and their optimum control strategies. The importance of and the technologies needed for smart monitoring system in charging/discharging mode are also reviewed. Simulation results show that controlled implementation of V2G can significantly contribute to enhancing dynamic performance and stability of the microgrid under different operating conditions.

5.1 Introduction

It is expected that electric vehicles (EVs) will be a significant part of the connected load in the future smart grid [1]. Nearly one million hybrid electric vehicles (HEVs), plug-in hybrid electric vehicles (PHEVs) and battery electric vehicles (BEVs) will be on the roads of the United States by the year 2015 [2]. The dual

*Queensland Molecular and Nanotechnology Centre, Griffith University, Queensland, Australia

characteristic of EVs – V2G and G2V modes – makes them suitable for demand response program and dynamical electrical storage resources through V2G transaction [3]. They are generally equipped with large batteries ranging from 7.8 to 27.6 kWh, which are connected to the grid to be fully charged [4]. These batteries can be charged from distributed generators in the microgrid, e.g. wind turbines and solar panels during the off-peak time, and supply power back to the grid during peak time. Robust coordinated controls are needed for charging and discharging of EVs along with photovoltaics (PVs) and wind generators. With uncoordinated strategy and power monitoring, this charging can have the following detrimental effects [5]:

- It may augment transformer losses in distribution sides.
- It can bring down transformer life by increasing thermal loading.
- It can increase voltage fluctuation.
- It can enhance harmonic distortions in output current.
- Additional infrastructure and investment will be required to tackle the peak demand.

Several research studies have been conducted for finding the optimum penetration level of EVs into the grid in dual mode. Monitoring power flow variation and designing plausible control methods have become a priority in recent years. Various literatures have proposed different methods of controlling and monitoring power flow during V2G mode (i.e. EV discharging mode) and G2V mode (i.e. EV charging mode) and have tried to quantify the effects by making realistic assumptions. In this chapter, literatures based on the impacts of EV penetration on grid power profile and potential ancillary services that EVs can offer are primarily discussed. Later on, a general discussion on recent trends of EV control strategies as the smart grid support is presented. In the end, an optimized power monitor and controller is presented with real-life case studies.

5.2 Impacts of EV Penetration on Grid Power Profile and Requirements of Its Control and Monitoring

Recently, considerable efforts have been made to investigate the impact of a large number of EVs in distribution grids. In several literatures, the impact of EV penetration has been analysed either by simulation or by doing realistic case studies. In V2G concept, both grid operators and car owners benefit among themselves, which is a reason behind recent rapid EV development. EVs can behave either as loads or as generators. Their behaviour pattern depends on several factors such as the type of charger connections (unidirectional/bidirectional), the number of EVs being charged in the given vicinity, voltage and current levels, battery status and capacity, charging duration and geographical location [6, 7]. Fast charging of these EVs can have a huge impact on distribution grid as a typical EV load is almost double an average household load [8]. Adding up harmonics resulting in low-power factor can be a serious problem if the charger does not utilize state-of-the-art conversion [9]. EVs, with an adequate intelligent connection to grid, interactive hardwires and software control, can serve as direct resources or as spinning reserves during power outage [10–12]. V2G concepts with appropriated control can improve utility grid performance with respect to efficiency, stability and reliability by offering reactive power support [13] and active power regulation, tracking and supporting efficient integration of variable intermittent renewable energy sources (RESs), load balancing and shifting via valley filling, peak load shaving and current harmonic filtering [8, 14, 15]. According to Yilmaz and Krein [16], for maintaining grid reliability and balancing the supply and demand, V2G concepts can provide higher quality ancillary services like quick frequency and voltage regulation, load levelling and peak power management and effective spinning reserve. Most of the existing works have used linear control techniques which provide optimum performance for the nominal working conditions. However, these techniques could not ensure stability under large disturbance when operating points move far away from the nominal operating point. This chapter presents the utilization of controlled EV penetration with PV and wind turbine and includes uncertainty in the system model for enhancing robustness.

5.2.1 Voltage and Frequency Regulation

In order to facilitate proper function of the grid, it is required to unremittingly balance production and consumption of both active and reactive powers, so that the frequency and voltage amplitude can be closer to their nominal values [17]. Active power can be controlled by controlling frequency, and reactive power control relies on voltage control of the system [18]. Thus, frequency and voltage are important factors in determining the power quality of utility grids. Implementing V2G in overall voltage and frequency regulation will pave a path towards active and reactive power quality improvement. Currently, frequency regulation is achieved by the expensive process of cycling large generators. EVs can be used in this case as they can act as medium-sized generators. In [19] an aggregator-based revenue model has been proposed. This model is able to provide regulation up to a certain desired level by organizing EVs. An energy management system (EMS) placed at grid side dispatches appropriate regulation signal within the contracted boundary of the aggregator. Idle EVs respond to the regulation signal based on the developed algorithm. A proper logic of charging and discharging of EVs is implanted in the battery charger with a voltage control to compensate reactive power, which selects the current phase angle to operate in inductive or capacitive mode of charging [20]. Although penetration of a large number of EVs for charging batteries from the grid could be a reason for line overloading and voltage instability at a low voltage network, it can regulate the reactive power within local vicinity [21, 22]. According to the Union for the Coordination of Transmission of Electricity (UCTE) the control of frequency stability in the distribution system can be categorized into three types: primary, secondary and tertiary [23, 24]. This chapter focuses on designing a coordinated control for RESs with EV in the microgrid.

5.2.2 Supporting and Balancing of Intermittent RES Powered Smart Grid

Due to global awareness regarding energy security, fuel diversity and climate change, the use of RESs for power generation is increasing rapidly. Integration of RESs into grid has become a necessity, but unpredictability of the availability of sources has become a prime concern. Currently, the most prevalent RESs are wind power and solar energy. Solar irradiation is not constant throughout the day, and solar power cannot be generated in the night. On the other hand, wind power is more complex due to the unpredictable variations in wind speed. These variations make it strongly intermittent, which leads to overall system imbalance. A number of studies have been done to combine EVs with RESs to serve different purposes such as battery storage, reactive power system and active power control. In [25] an investigation on the possibility of using V2G concept to overcome intermittency in wind power is conducted. Guille and Gross [39] proposed a structure using model predicated control to analyse the positive effect of EVs on wind generator. In [26] an autonomous distributed V2G control has been proposed for providing distributed spinning reserve during intermittency. In [27] different strategies of integrating EVs into wind thermal power plant have been investigated. The authors in [28] proposed a combination of demand response with wind integration to balance the power. V2G concept can provide flexibility in power saving. In an RES integrated smart grid if power generated by renewable sources is high, then without curtailment this energy can be stored in EV batteries with designed charging model. This energy can be further used in residential areas or industrial areas; if not, it can be supplied to the grid later on, thus creating a flexible power supply.

5.2.3 Controlling and Monitoring via 'Smart Charging/Discharging'

It has been mentioned previously that PHEVs or EVs can be used as a dynamic storage and they can act as both generators and loads. V2G can contribute to even out peak energy load by discharging during peak time and charging during low demand (i.e. overnight or off-peak hours). Several literatures proposed different methods for smart vehicle charging systems using various algorithms to level the curve. Takagi *et al.* [40] proposed an algorithm that can be utilized for load levelling by using dynamic electricity pricing curve. This curve can optimize an ideal minimum charge when EV users derogate their electric bill. EVs should be charged late night due to low demand level during that time. A case study of a California grid with four million EVs showed that charging load can be met by the present power system without requiring additional power sources. Mets *et al.* [41] showed that smart charging/discharging can reduce peak load and level the load curve and described different local/global smart charging/discharging and their coordinated control strategies. Chakraborty *et al.* [42] observed that for New York City where 100,000 vehicles were simulated using level 2 charging, a benefit of $110 million per year can be achieved by considering 50% simultaneous EV

Table 5.1 Charging power level for EVs

Criteria	Description	Power and current capacity
Level 1	Opportunity charger (any available outlet)	1.4 kW (12 A) 1.9 kW (20 A)
Level 2	Primary dedicated charger	4 kW (17 A) 19.2 kW (80 A)
Level 3	Commercial fast charger	Up to 100 kW

penetration level and contributing approximately 10% of peaking capacity safely where up to 87.5% penetration of EV charging does not pose any protection issue. For high penetration, load shifting of EV fleets can reduce the overall impact on the grid. This can be achieved by coordinated charging of EVs. The target of a controlled battery charger is to scale down peak load by shifting energy demand [43]. This also gives opportunities for smart charging. A study in 13 U.S. regions without smart charging considering high penetration of EVs showed that additional 160 power plants will be required if every EV is charged in the early evening [44]. In some literatures like [45, 46], it is shown that there is little financial incentive with increased PEV penetration when V2G is used for peak load reduction. According to [44], peak power control has been proven an economically feasible solution in Japan.

Charging power levels for EVs and a compiled comparison between uncoordinated and smart charging are presented in Tables 5.1 and 5.2 respectively [16].

5.3 Hybrid EV Powertrain Architectures

The main challenge of designing an HEV is the control of power flow from various sources to the load with high efficiency. Power flow control is actually dependent upon drive cycles. Usually typical HEVs include more electrical instrumentations than conventional internal combustion engine (ICE) drive vehicles. These include apparatus like electric machines, power electronics (IGBTs/MOSFETs), continuous variable transmissions, embedded powertrain controllers, modern energy storage systems (fuels cells, ultra-capacitors, and flow cells) together with DC/DC and DC/AC converters. The simplest powertrain configuration is composed of ICE, generator, pulse width modulation (PWM) converter, PWM voltage source inverter (VSI), battery, induction motor (IM), transmission linked with wheels. Based on the placement and arrangement of the components, HEV powertrain configuration can be classified into three types [48]:

1. Series HEVs
2. Parallel HEV
3. Combination of series and parallel HEVs

Table 5.2 Smart vehicle charging/discharging system vs. uncoordinated vehicle charging/discharging system

System	Power level	Charging or discharging time	Methods, algorithms and approaches	Requirements and costs	Impact on power distributions system
Uncoordinated charging/ discharging	Most likely at level 1	Continuous (PEV can charge any time depending on users' necessity) Any time full charging or disconnection	Immediately when plugged (decision to charge PEV is independent of price and peak hours) Start after use-adjustable fixed delay Off-peak uncoordinated (during the night when the overall electricity demand is low and generation is mostly base load) Single tariff Dual tariff	No specific requirements, coordination and aggregation Charging PEV is cheap	Reduces the reliability and cost effectiveness Increases the load at peak hours Voltage deviations Extra power losses Overload distribution transformers and cables Increases the monthly electric bill
Coordinated smart charging/ discharging	Most likely at levels 2 and 3	Charging during low demand (off-peak hours, overnight) Discharging daily peaks	Articulated objective functions to peak demand, minimize losses, voltage deviations, load variance, charging cost and maximize load factor, profits for the vehicle owners Quadratic programming (QP, *optimize a quadratic function of several variables in order to determine the optimal continuous charging profile*), dynamic programming (DP, *discrete charge profile, slower than quadratic*) and linear	Sensors and on-board/off-board intelligent meters (real-time monitoring and considerable exchange information, price, battery capacity and SOC) Communication connection and control (it must be bidirectional to report battery status and receive commands) Energy management (measurement/billing) Software (real-time monitoring, management and control) Identification	Optimizes power demand and time Slight effect on peak capacity, shift load and avoid peaks, exploits the grid load factor Lessens voltage deviations, everyday electric costs, line current and transformer load surges Balances the daily load pattern and voltage profile Circumvents incremental grid investments and high energy losses

programming (LP, *to find optimal solution*) Load shifting, interruptions or adjusting the charging rate Multi-agent (MAS) distributed solution (demand side management and hierarchical approach) Computational intelligence-based optimization algorithms Valley filling algorithms (utility-centric assumptions) Particle swarm optimization (PSO) distributed method (*solve the unit commitment – UC*) Real-time nonlinear electricity pricing algorithm (each vehicle can be contracted independently or as a part of an aggregator) Forecast of future electricity prices Customer usage model, power management and financial model Centralized method (directly controls the charging/discharging; obtain global optimum); the final decision to begin or end the charging process is made by a central control entity	Energy efficiency (PFC, filtering and quality analyser) Charging/discharging infrastructure	No substantial impact on transformer Maximizes exploitation of renewable sources and makes efficient use of existing generation capacity Make the most of consumer convenience through use of available infrastructure Increases operating competence

(Continues)

Table 5.2 (Continued)

System	Power level	Charging or discharging time	Methods, algorithms and approaches	Requirements and costs	Impact on power distributions system
			Decentralized method (price signal broadcast, collect and store the trip history; obtain global optimum) Time-of-use (TOU) price with empirical method (retail TOU rates, dual-use programme, time-differentiated tariffs) Incentive-based charging (time-varying price) Real-time smart load management (RT-SLM) control Price signal-based, load signal-based, renewable energy-based charging (preferred operating point (POP) algorithms) Market-based multi-agent approach Local energy control approaches (determine the charging times and rates based on the predicted local base load) and global energy control strategy (determine a charging schedule based on global load information)		

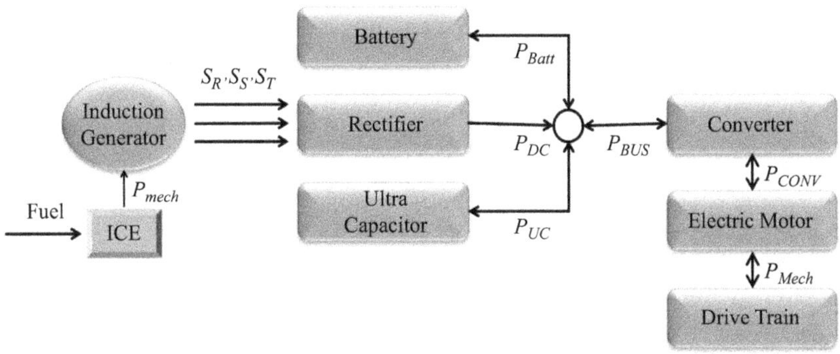

Figure 5.1 Series HEVs powertrain

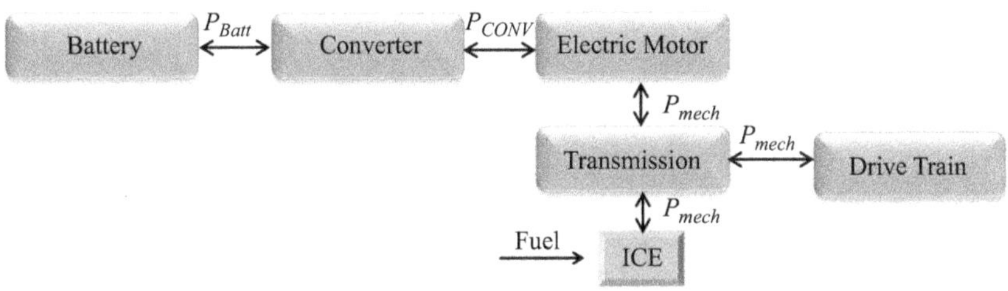

Figure 5.2 Parallel HEVs powertrain

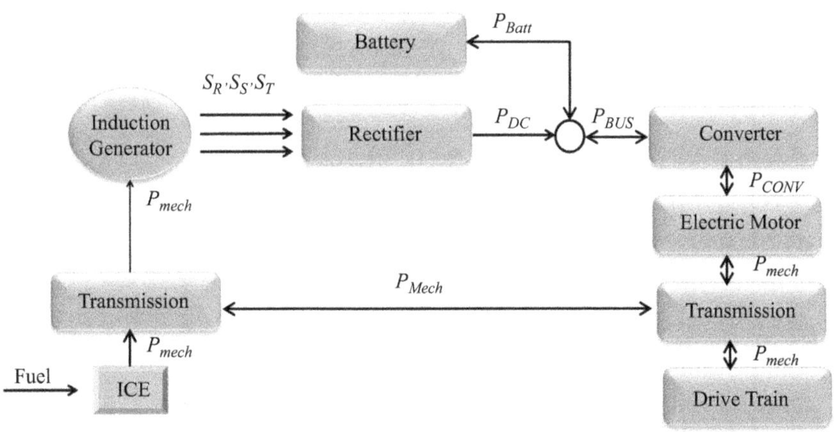

Figure 5.3 Combination of series and parallel HEVs powertrain

These three powertrain configuration and their overall power flow are illustrated in Figures 5.1, 5.2 and 5.3 respectively [49].

A comparative feature analysis of these three types of powertrain is provided in Table 5.3 [50].

Table 5.3 Features of different powertrain configurations

	Series hybrid configuration	**Parallel hybrid configuration**	**Combination of parallel and series HEVs configuration**
Features	Optimized effective power plant Fast 'black box' facility exchange possible Segmental power plant possibilities Long effective life Optimized efficient traction drive line Mature well-proven technology Engine downscaling Outstanding transient response Space packing advantages Zero emission process possible	Financial gain at high cost Retarder option but at difficulty risk Zero emanation process possible	Zero emission procedure possible
Disadvantages	Bigger traction drive system Watchful design algorithms a prerequisite Multiple energy translations	Costly system Control complexity Careful design algorithms a requirement High voltages needed for effectiveness Complex space packing Hino HIMR	Very expensive system Control intricacy Careful strategy algorithms a prerequisite Complex space packaging Outdated by coupled hybrid topology
Vehicle systems/ applications	TEMSA Avenue Hybrid Tesla ultra-light rail Orion VII Conventional light rail Wrightbus electricity New Tesla buses	Bus/heavy truck market	Fiat experimented Nissan experimented

5.4 Industrially Used Control, Monitoring and Management Strategies

To extract optimal benefits from EVs as a smart grid support, a number of control strategies can be used in a drive train for vehicles with different mission requirements. The control objectives of the HEVs are as follows [51]:

- Providing driver power demand
- Operating each component of the vehicle with optimal efficiency
- Extracting regenerative braking energy with high efficiency
- Maintaining state-of-charge (SOC) of EV batteries within desired level

The IM drive used in EV and HEV is supplied by a DC source (battery, fuel cell, ultra-capacitor or flow cell). These DC sources hold a constant terminal voltage. A latched DC/AC inverter along with DC source will provide a variable frequency and variable voltage as per requirement. The DC/AC inverter consists of power electronic switches and power diodes. For DC motor control, PWM technique is widely used, whereas for controlling IM, FOC (field-oriented control) and DTC (direct torque control) are popular approaches. These control strategies use traditional proportional-integral-derivative (PID) controllers but control algorithms like adaptive control, fuzzy control and neural network are also used industrially [51–53].

5.4.1 EV Induction Motor (IM) Control

5.4.1.1 Voltage and Frequency (Volts/Hz) Control of IM

The equation of IM speed, i.e. $n = \frac{60 f_s}{p}(1 - s)$, indicates that the speed of an IM (n) can be controlled by varying the supply frequency (f_s). PWM inverters can easily provide variable frequency supply with good quality output wave shape. Hence, the open loop volts/Hz control has become a widely used technique for speed control of IM drives. But it is not suitable for high accuracy control. The frequency control requires control in applied voltage as implies the stator flux equation, i.e. $\Psi_s = U_s/\omega_s$ (resistance drop is neglected, which remains constant). Otherwise, if only frequency is controlled, the flux would change accordingly.

From the stator flux equation we can see that stator flux is inversely proportional to supply frequency, which means an increase in frequency causes the flux to decrease, thereby decreasing the torque developed by the motor. Conversely if frequency is decreased, the flux will increase, which may lead to magnetic saturation. Since in PWM inverters the voltage and frequency can be controlled independently, these drives are fed from a PWM inverter. The control scheme is simple, with motor being supplied by three-phase supply dc-link and PWM inverter.

This drive does not require any feedback so precise controlling is not possible. As a result it is mostly used in low performance applications. Based on the required speed, the frequency command is applied to the inverter. The phase voltage command is directly generated from the frequency command by a gain factor that controls the input DC voltage of inverter. By utilizing this method, the speed of the motor cannot be controlled precisely because the previously mentioned frequency control only controls the synchronous speed [54, 55]. Depending on different loading conditions, there might be a small fluctuation in the motor speed. With high speed, this fluctuation will be negligible. At low speed the frequency is also low, so if the voltage is reduced then because of the large drop in resistance across the stator will impact the overall performance of the motor. The relationship between voltage and frequency at the time of low speed is given by $U_s = U_0 - kf_s$ where U_0 is the voltage drop in the stator resistance, U_s the voltage across stator, f_s the supply current frequency and k a constant.

5.4.1.2 Speed Vector Control of IM

The main purpose of a controller designed for IM is to ensure high performance with minimum loss. In a separately excited DC machine, developed torque can be controlled by controlling armature current. By doing so and by maintaining the flux constant, fast transient response can be achieved.

Field-oriented control (FOC) or the vector control of AC machines has a similar control technique as applicable for separately excited DC machines. Similar to DC machines, the torque for AC IMs is developed by mutual interaction between armature current and flux.

But the difference in the AC machines is that the power is fed to the stator only. The main challenge is to isolate individual current components responsible for the production of torque and flux. The objective of an FOC is to isolate individual stator current components that are responsible for the production of flux and torque. FOC is obtained by magnitude, phase and frequency control of the stator current by tuning and controlling inverter output. Because of the controlling parameters, i.e. magnitude, phase and frequency, FOC is often referred to as 'vector control'.

In order to obtain independent control of flux and torque in IM, the stator (or rotor) flux linkages phasor is maintained fixed over time in its magnitude. Also its phase is maintained stationary with respect to current phasor. The FOC structure can be classified into two different structures:

1. Direct control structure: where flux position is checked and regulated by flux sensors
2. Indirect control structure: where flux position is estimated with the measurement of rotor speed

In indirect control structure, IM is represented in synchronously rotating reference frame, i.e. *d-q* frame. To obtain this control strategy, controller equations are derived in *d-q* reference model. This model is also referred to as 'dynamic model of IM' and can be found in literatures. The block diagram of the FOC used to control a VSI of an IM is presented in Figure 5.4 [51, 55, 56].

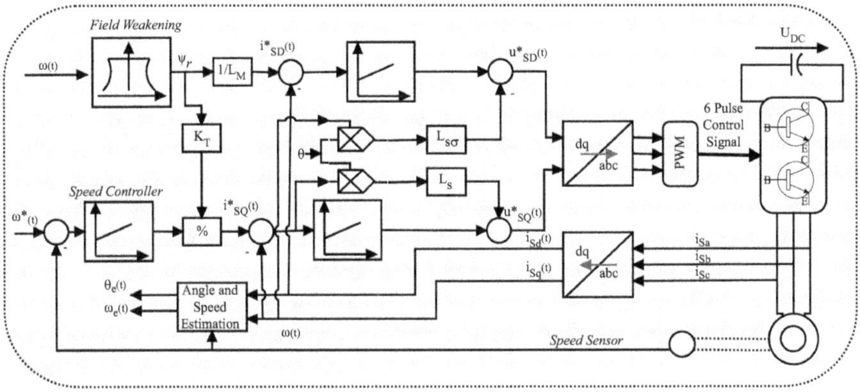

Figure 5.4 Block diagram of VSI induction motor FOC

Generally, a closed loop vector control scheme results in a complex control structure as it consists of the following components:

1. Two PID controllers: one for flux and torque control and another for speed control
2. Decoupling network for voltage and current
3. Complex coordinate transformation
4. *d-q* to *a-b-c* transformation
5. Voltage or current modulator
6. Flux and torque estimator

5.4.2 Requirement of EV Battery Management and Monitoring

Battery management is important for Li-ion batteries to ensure energy availability, lifetime and safety of the energy storage system. The major inputs of an electronic battery management system (BMS) are battery current, voltage and temperature over time. Battery management is in charge of battery protection and SOC, state-of-health (SOH) and state-of-function (SOF) estimation. Additional tasks include controlling the heating/cooling subsystem and the main power switch, ensuring the isolation of high voltage from the vehicle and implementing isolated communication with the in-vehicle network.

From an electronic point of view, main challenges are the accurate and synchronous measurement of the battery current and the voltage of the pack cells, along with data communication over a number of voltage domains and fulfilment of Automotive Safety Integrity Level-C (ASIL-C) safety requirements. Typical accuracy targets are 0.5–1% for currents up to 450 A and 1–2 mV and 0.1% for voltages at cell and pack level respectively. This rigorous demand for voltage accuracy is mainly driven by $LiFePO_4$ compound which is one of the preferred technologies for automotive applications, as it features a good trade-off among energy density, costs, safety properties and lifetime with cycle resistance. A very flat characteristic for the open circuit voltage (V_{OC}) versus SOC stands for this battery technology. This makes accurate SOC estimation from voltage measurement very difficult, especially when the SOC is in the range of 20% to 80%. Other automotive Li-ion chemistries, such as Li-Titanate (which has even better cycle resistance, fast charge properties but lower energy density) or Li-Mangan, are less demanding in terms of voltage measurement accuracy than $LiFePO_4$ cells. From a semiconductor component point of view, it is essential to design for integrated error compensation techniques and accordingly to the ISO26262 design flow. Moreover, having high-precision production test equipment in place and going for comprehensive product qualification is required for an accurate assessment of the product parameters. Considering a linear and offset free measurement system, the main error sources in shunt current measurement are the variation of amplifier gain, shunt resistance and analog to digital converter (ADC) reference over temperature and time. The error sources in voltage measurement are ADC reference variation over temperature and time, common mode voltage variation along the battery string and the tolerances affecting the resistive voltage divider. In BMSs, no re-calibration is foreseen and full accuracy must be maintained for all the vehicle life. However, there are physical reasons for long-term drift of the measurement system. Most significant ones are related to the threshold shift of metal–oxide–semiconductor (MOS) transistors when biased – mechanical stresses that may induce MOS parameter shift and long-term relaxation effects. Comprehensive qualification tests with the help of highly accurate automatic test equipment resources provide the basis for good accuracy predictions over lifetime. Pre-aging of the electronics in biased high-temperature operating environment needs to be considered to minimize long-term accuracy degradation and to achieve the highest accuracy [57].

5.4.3 Techniques for EV Battery Management

Architectural choices for implementing a BMS can be influenced by the physical characteristics of the battery and its targeted application. Ten to over one hundred of high-capacity elementary cells are series-connected to build up the required battery voltage in high-power applications. The overall cell string is usually segmented into modules consisting of 4 to 14 series-connected cells. It means that a battery can be considered to be made of three nested layers: namely the elementary cell, the module and the pack (i.e. the series of modules). Each layer can serve as an intelligent platform for the effective implementation of a subset of the BMS monitoring and management functions. This approach leads to a hierarchical architecture of the BMS. In that hierarchy the most inner layer is the cell monitoring unit (CMU), one for each cell of the string. The middle layer consists of the module management units (MMUs), one for each module in which the string is partitioned. The MMU uses the basic monitoring functions implemented by the inner CMUs and provides higher-level services to the pack management unit (PMU), which supervises all the battery strings. A dedicated and custom bus can be used to connect each CMU to the relevant MMU. A shared galvanic-isolated controller area network (CAN) bus is the preferred choice to implement the communication between the MMUs and the PMUs. The latter also embeds the interface

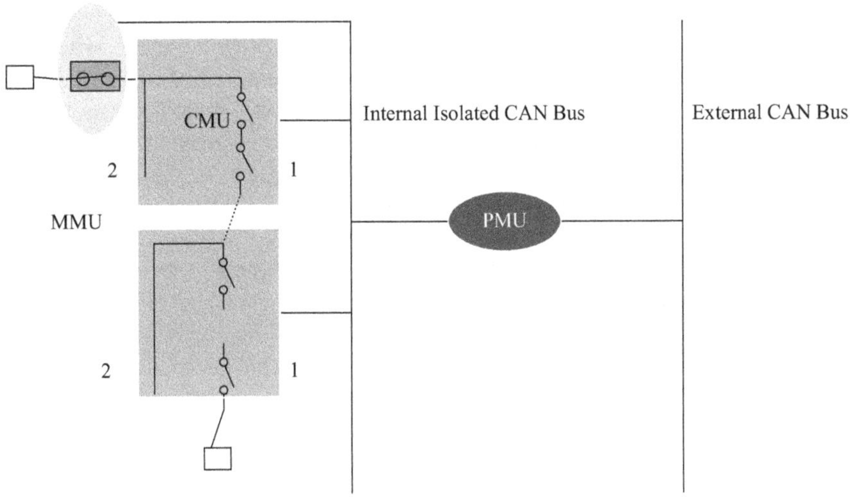

Figure 5.5 Hierarchical model of BMS

between the battery and the other control systems of the targeted application. In an EV, the PMU is linked to the vehicle management system (VMS) through the external CAN bus or other higher-speed fault-tolerant busses like FlexRay [58]. A high-level hierarchical BMS model is shown in Figure 5.5.

The hierarchical model shown in Figure 5.5 is very flexible, as the BMS functions can freely be distributed and replicated (in case of redundancy requirement) over all the three layers of the platform. An example of the hierarchical platform is presented later on this section. In that platform, only the two outer layers are exploited, as the designed BMS embeds an MMU and a PMU for each battery module. This is a common choice, mainly driven by the fact that providing each cell with a dedicated CMU can be relatively expensive and can increase the overall self-discharge rate of the battery in a non-negligible way. However, the current trend is to build up the battery by series-connecting very high-capacity cells, instead of groups of parallel-connected cells with smaller capacity. This way the cost and power consumption of the CMU may become affordable, if compared to the cost (roughly 400 €/kWh) and self-discharge rate (typical value is 3% a month) of a very high-capacity cell. The use of the cell layer can bring some advantages in implementing the BMS monitoring tasks. In fact, the CMU can easily act as a gauge measuring the voltage and the temperature of the related cell. This provides the basic support for adding redundancy to this key BMS function in an effective way. In addition, as the CMU can be embedded in the cell case, it allows some valuable information, such as the serial number, the lifetime, the number of cycles to be evaluated and stored into the cell itself. This enables an easy tracking of the cell history and the potential use in a second market application, such as the smart grid [59], when the progressive degradation of the usable capacity of the cell makes it no longer suitable for an EV.

Protection of battery against overcharge, deep discharge and over-temperature is usually achieved breaking the battery current flow through the main switch unit (MSU) contactor. This method can be combined with the possibility of excluding a segment of the string from the current path by the two-way switch present in each module illustrated in Figure 5.5. This way the bypassed segmented cells are protected, while the battery is still capable of delivering power to the load. Furthermore, the insertion of the two-way switches enables the dynamic reconfiguration of the string.

Mechanism described above also provides a direct method for charge equalization. If a segment reaches the full charge, it can be disconnected from the current path, so that the charging of the remainder cells of the string can continue. Depending on granularity of the switch, this method can be applied to a string segment made up of a single cell [60], allowing charge balancing at cell level. However, this approach requires a pair of bidirectional switches for each cell of the battery, which is impracticable due to complexity. Thus, this method seems to be more attractive if applied to the segment level instead of the cell one. This is because only one high-current two-way switch is needed per module. Different approaches have been proposed for charge equalization. A good review of different balancing techniques is provided in [61]; readers are suggested to have a look on that literature. A comparison is shown in Table 5.4, where the above-described method is reported as *cell/module bypass*. It falls in the category of active balancing techniques. The extra energy stored in the more charged cells is not dissipated as heat through a bleeding resistor, as it happens in passive equalization, but instead is transferred to the less charged cells. Different active balancing techniques are possible, depending on the way the energy is redistributed among the cells. However, a really good compromise between circuital complexity of the active balancing method and achievable efficiency has to be found to make active balancing competitive against passive equalization. Passive balancing implementation is really straightforward, because it requires just one controlled switch and resistor per cell. The most promising approaches seem to be the *module-to-cell* and the *shared cell-to-cell* techniques [62], in which a single DC/DC converter is used to equalize the charge among the cells of a module. In fact, a single converter can be designed with a very high efficiency and also the number of the bulky passive components (inductors and capacitors) is minimized when compared to the *distributed cell-to-cell* technique. An interesting implementation of the *module-to-cell* employing a flyback DC/DC converter is reported in the next section.

MMU contains the hardware implementation of the charge equalizer circuit, while the overall balancing procedure is usually supervised by the PMU, which controls the amount of charge stored in each cell of the string. To achieve this purpose, the PMU needs to estimate the SOC of each cell of the battery string, that is, the charge present in the cell divided by its nominal capacity. The latter corresponds to the amount of charge that can be stored in the cell when it is fully charged. As stated above, this is true only at the beginning of the battery life, as the actual cell capacity gradually decreases with time. Thus, an accurate modelling of

Table 5.4 Features of different battery management techniques

<table>
<tr><th colspan="3">Methods</th><th>Advantages</th><th>Disadvantages</th></tr>
<tr><td>Passive</td><td colspan="2">Cell-to-heat (one exploiting resistor and switch per cell)</td><td>• Very simple
• Very inexpensive</td><td>• 0% efficiency
• Slow</td></tr>
<tr><td rowspan="4">Active</td><td colspan="2">Module-to-cell (charge allocation from a battery unit to a single cell by means of a galvanic-insulated DC/DC converter)</td><td>• Comparatively simple
• Good efficiency
• Fast</td><td>• Switch network
• High isolation voltage of the DC/DC</td></tr>
<tr><td rowspan="2">Cell-to-cell</td><td>Distributed (charge transfer from adjacent cells)</td><td>• Moderate efficiency
• Moderately fast</td><td>• Bulky
• Complex control</td></tr>
<tr><td>Shared (charge transfer from cell A to tank, then from tank to cell B)</td><td>• High efficiency
• Fast</td><td>• Switch network</td></tr>
<tr><td colspan="2">Cell/module bypass (a cell/module disconnection from the current path)</td><td>• High harmonizing efficiency
• Very fast and flexible</td><td>• High current switches
• Complex to implement
• Decrease battery competence during normal operation</td></tr>
</table>

the aging effect is very important for a precise estimation of the remaining charge stored in a cell. Assuming that the initial value is known, SOC can be evaluated by integrating the battery current over time. This method is called Coulomb counting and is commonly used in low-power applications, as portable consumer devices. However, Coulomb counting is very sensitive to measurement errors, particularly those caused by offset temperature drifting of the current sensor, which may lead to large SOC errors over time, because of the current integration. Another simple method to estimate the SOC is to make use of the relationship between SOC and open-circuit voltage (OCV). However, OCV dependence on SOC in Li-ion batteries is very small, so the measurement of OCV must be very accurate. This requires the battery being in steady state, a condition that is reached only after a long time (often several minutes or even hours) with no load connected. Thus, this approach is unsuitable for real-time SOC estimation in EVs, where the battery is continuously charged or discharged with high currents. Among other methods reported in the literature [63], such as discharge tests, internal resistance measurement, or neural networks, model-based algorithms (such as Kalman filters [64] or mix algorithms [65]) seem to be the most attractive for online SOC estimation in EVs. This is because they use onboard noisy voltage and current measurements and they do not require long tuning times as artificial neural networks do. However, model-based methods require an accurate modelling of the cell behaviour for all the cell lifetime and in all the operating conditions. This implies that temperature effects on the cell behaviour must also be taken into account [66]. These techniques for battery capacity monitoring are elaborately discussed in later consecutive sections. A comparative analysis of different battery management techniques is given in Table 5.4.

5.4.4 Techniques for EV Battery Capacity Monitoring

EV users need to know their cars' battery status. For example, they need to know how much charge is left in their batteries in terms of SOC compared to the fuel meter present in conventional ICE-driven vehicles. They also need to know when to recharge the battery or when they need to replace their battery to have optimum service. They actually need to have a reliable battery capacity monitor to serve this purpose. The primary objective of a battery system designer is to ensure reliability of the system by monitoring, collecting and analysing relevant data. To keep off unexpected charge draining and to avoid sudden system collapse, a better monitoring system is a prerequisite. Apart from this, another objective is to optimize recharging to maximize battery life cycle performance as battery life cycle is partially dependent upon the charging duration and patter. To guarantee better performance and to ensure reliability of the system, a battery monitoring system should be capable of detecting the available energy and effective capacity as well as should have the capability to predict the solution. For these reasons, effective EV battery parameters' monitoring is a prime focus for optimal penetration into the grid.

Mainly three types of battery capacity monitoring techniques are used in industrial sectors:

1. Voltage profiling
2. Coulomb counting
3. Ohmic capacity measurement

5.4.4.1 Voltage Profiling

The most commonly used technique for battery monitoring is measuring battery voltage to determine how much energy is stored/left in the device. The main challenge is that voltage profiling requires specific testing arrangements to observe the relation between battery terminal voltage and its capacity under temperature and loading variance. Predefined and modelled table of capacity values for extensive range of voltage current and temperature is stored in memory chip. This technique is suitable for smaller battery applications such as laptops, mobile phones and tablets.

The main advantage of this approach is that it requires less cost to implement. However, the disadvantages are many.

- Restricted application of the capacity table for specific type of batteries.
- The correlation between battery voltage and capacity can only be developed when the relaxation time of the battery has passed a certain period of charge/discharge.
- The capacity table loses its validity as battery ages because voltage–capacity relationship of the battery changes over time.
- Only tested values for a certain level of temperature and load variance can be precisely monitored and programmed in memory.
- For proper system design, details of the battery parameters, duty application, user behaviour and user environment need to be realized.

Due to these shortcomings, voltage profiling technique is not an ideal solution for EV batteries where multiple variables including stochastic parameters have to be considered Moreover, voltage profiling cannot provide information regarding battery aging or battery life cycle [67].

5.4.4.2 Coulomb Counting

The basic principle of Coulomb-counting technique is to find out the starting point of the battery status then track current flow into and out of the battery over time. This will further ensure how much energy is stored in or drained from the battery. Similar to the previous method, this technique is also suitable for smaller battery applications such as portable devices. This technique is also used in some other larger applications such as recreational vehicles (RV) and marine industries. This technique can often be identified as a current shunt of Hall-effect coil added to the negative terminal of the battery to measure accurate current flow.

Coulomb counting has two main advantages: First, it does not require expensive technology, as a result it is efficient to implement, especially in smaller battery applications. Second, it provides a precise measurement of battery current flow. Moreover, Coulomb counting overcomes a key limitation of the voltage profile approach: it is not vulnerable to the errors originated from the voltage-to-capacity relationship in an environment with dynamic parameters. It means that this technique can be used for EV batteries.

Nevertheless, important shortcomings remain in this technique:

- Core chemical reactions can cause self-discharge that cannot be measured but can be a significant factor over time.
- Internal charge and discharge incompetence at different capacity rates is difficult to precisely compensate for, especially when batteries age.
- Internal charge and discharge incompetence with temperature variance, battery age and cycling history is difficult to exactly compensate for.
- Surface/shallow discharges (common in many applications) can lead to an incorrect supposition about total capacity of the battery.

- This technique updates the total capacity value and corresponding parameters only if there is a full charge or discharge, which is not practical and is difficult to detect and measure.
- To avoid accumulation error, which increases with battery age, recurrent calibration is required.

In order to reimburse the shortcomings, a designer needs to understand the specific battery parameters and the system response over time with variable range of conditions and loads the battery will be dealing with. Battery response on programme error compensation algorithms such as Peukert's exponent should be properly studied before implementation. For these reasons Coulomb-counting technique is suitable at the early stage of the battery when cycling performance is unchanged and error compensation algorithms are introduced [67].

5.4.4.3 Ohmic Capacity Measurement

Researchers have developed a number of Ohmic methods to test battery over almost a decade. The main principle is to find out the internal resistance/conductance/impedance of the battery which is dependent upon the available capacity of the battery. It means that if internal resistance of the battery increases, the battery will deliver less energy or it will lose the capability of storing energy. This relation between impedance/conductance/internal resistances and capacity is much studied and accepted in industries. It should be mentioned that this relationship between impedance and battery stored energy is exact for batteries with similar chemical or constructional characteristics. This correlation will vary for different batteries. The limitations mentioned for earlier two methods can be tackled by Ohmic capacity measurement because it can precisely measure battery characteristics for different charge and discharge patterns. The most interesting thing about this method is that it can overcome almost all the shortcomings of previously mentioned methods, such as aging effects, life cycle and temperature variance. Primarily, this method tests the battery response towards variable signal, load or charging current. Then by applying Ohm's law the internal impedance is measured. Though the basic principle of this method is somewhat simple but choosing a proper type of load or charge applied as an input is a major challenge.

There are three major types of Ohmic measurement. This chapter will discuss briefly about those types.

1. AC conductance method
2. DC load–voltage recovery method
3. Large magnitude pulse resistance (LMPR) and stacked large pulse resistance (SLPR) method

AC Conductance Method: This method, which is also known as dynamic conductance method, uses a low-amplitude AC signal that varies within a nominal OCV measured for battery. This signal is applied across battery terminals, and the voltage response of the battery is measured. With that response, the internal impedance value is determined. This determined internal impedance is then transformed into capacity value by using functional table. This method is well suited for static batteries, but batteries that operate online or provide support to the load cannot use this method. These batteries have two types of erratic effects. First, noise generated from usual alternator and other loads associated with it extends the frequency and amplitude of the applied signal, which actually obscures the overall voltage response. Second, due to dynamic loading, the nominal battery voltage gets lifted up or down, causing distorted response for marginal periods. So the main limitation of AC conductance method is that it is not suitable for EV application because of its dynamic and unpredictable behaviour. It is also not a suggested method for larger batteries. Over many years researchers are trying to somewhat lessen these shortcomings when measured with a system where noise and dynamic loading coexist. However, this method is suitable for offline automotive batteries with large capacity.

The DC Load Method: This method includes retrieval of voltage by applying some known DC load and observes the output battery voltage to be stabilized at some low level. After that the DC load is removed and the voltage is retrieved. The internal impedance then is calculated by dividing the voltage change with load current. This method has all the advantages that AC conductance method has and in addition it is really inexpensive compared to AC conductance method. The most significant attribute of this method is that it can tackle the noise sensitivity problem of the previous method. In online system, inherited AC noise can be detected whether the DC load is applied or removed, and the advantage is that the voltage response is not vulnerable to the noise with frequency higher than usual AC ripple. These outstanding characteristics make DC load voltage retrieval method appropriate for batteries used in dynamic environments such as telecommunications, UPS and automotive storage applications. These batteries are exposed to electronic noise but by nature not vulnerable to dynamic loading. These batteries spend almost 99% of their life cycle fully charged and as a backup. The limitation of voltage retrieval method is evident when there is a requirement of dynamic loading. The applied DC load and measured output period are comparatively long (i.e. a second or more than that). In dynamic application it is expected that momentous change in duty loading or charging rate will significantly affect the output. Thus, affected output voltage response can significantly affect the internal impedance measurement because of the residue error. As a result this method is not an advisable solution for battery capacity monitoring where dynamic loading exists. This method can be used in telecommunication applications but cannot be used while they are contributing active duty loading like foreseeing available runtime. An alternative approach combining AC conductance method and DC load method has been established for online backup batteries that are tested in float voltage. This hybrid method uses signals with low-amplitude and high-frequency (approx. 215 Hz) pulses to produce equivalent voltage response, which is above the VOC of the battery, and then internal impedance is measured. This method undoubtedly serves better performance than AC conductance and DC load methods, but limitations regarding real-time measurement with extremely high dynamic duty loading cannot be overcome. Apart from this limitation, this method can improve noise-related problems that AC conductance method and voltage retrieval DC load method cannot address.

Large Magnitude Pulse Resistance (LMPR) and Stacked Large Pulse Resistance (SLPR) Method: Another method that is introduced in this chapter is the LMPR method. This method provides notable performance over previously mentioned methods. The basic principle is to apply comparatively large (1 C to 5 C) load pulses for a minuscule timing (1 ms) and simultaneously measuring the voltage slope. The test current is then measured through known impedance by dividing with the voltage drop. The internal impedance of the battery is then determined by dividing the voltage slope across the battery by the test current. The short duration and large pulse application make this method suitable for dynamic application such as EV battery management. This method does not rely on the load state of the battery during the short test. The comparative change in battery voltage due to large pulse and transient determines the internal impedance. It does not matter whether the voltage is low due to large series-connected load or the voltage is stepped up due to the connection of chargers, the only thing that matters is the change in voltage across batteries in that short duration. This method can overcome the limitations of the AC conductance method and can also tackle the dynamic duty loading issue of the voltage retrieval method. Thus, the LMPR method is by far the most suitable method for determining internal impedance of dynamically loaded battery system. This method is capable of providing fuel gauge indication of the battery capacity. Moreover, this method has inherent capability to pay off the battery aging and life cycle capacity loss. Real-time measurement of the battery capacity and life cycle can be determined by this method by comparing for multiple capacity cycle. The LMPR method can provide radical improvement in case of accurate and stable fuel gauge indication. In this method, the primary battery voltage can be ignored due to the large pulse size and small step duration and therefore voltage slope at the time of active pulse is significant. Nonetheless, in exceptional cases a large charger or large capacitive load connected to the battery deviates the values in the LMPR method. The battery voltage will be stepped up the nominal voltage if a charger is connected with momentous charge (e.g. 0.5 C). During active pulsation, the total voltage drop (V_0) is composed of V_1 and V_2 where V_1 is the voltage of the charger and V_2 is the voltage drop or the battery response to pulsation. V_2 is responsible for measuring the internal impedance. Usually $V_2 \gg V_1$, so V_1 can be ignored and only V_2 can be used for further calculations. But if the system requirement is accuracy, then V_1 cannot be ignored. So this assumption entirely depends on the system requirement. This error can also be generated by large capacitive induction across the battery. It means if a large capacitive load is connected across the battery, it can discharge into the load and this absorption of voltage will change the battery response, which is not desirable. This problem can be solved by efficient measuring circuitry that can neglect this fringe voltage response due to capacitance or chargers and measure only the battery voltage, or by increasing the size of the pulse at the time of low charger/capacitance impact. Nevertheless, this is not a practical solution for larger systems. An optimum solution can be the SLPR method. In this method, two pulses are stacked together.

The first pulse completely saturates the voltage response of the connected loads. It also generates a new reference voltage. The second pulse detriments the battery attribution with high certainty. The second pulse is also called the measurement pulse. The SLPR method has been confirmed to be better in terms of accuracy and stability in extreme dynamic storage application in testing. This method can determine in and out current flow of a battery with high accuracy if the small errors are ignored. This can be done by the measurement of the internal impedance of the battery over certain time intervals. The calculated current (in or out) can be determined as the change in restrained capacity (Ah) linked with the two measured internal impedances divided by the total timer interval. In both LMPR and SLPR methods, the battery power can be utilized as the main energy source for monitoring. Though it varies with the size of the batteries, the energy consumption is approximately 40 mAh/day or less for usual automobile battery if sampled over 60 seconds. This range is usually lower than self-discharging rate. An optimum design will include structured logical firmware that will automatically level the frequency based on battery activity. It means that whenever the battery is not actively used the frequency will be scaled back by one in every 600 seconds. The frequency will be promptly increased up by one in every 10 seconds when there is fast battery charging.

A lot of other battery monitoring methods are available and used industrially. Each of these methods has its advantages and disadvantages and suitability of precise usage. If an application designer requires high dynamic duty loading with noise tolerant characteristics and accurate under loading, then LMPR or SLPR method can be a reliable choice. But if we consider inexpensive implementation, AC conductance or DC loading method would be a preferable choice for EV battery management [67].

5.5 V2G Communication System

For implementing viable V2G framework, the grid operators, aggregators and corresponding groups of EV need constant communication among themselves and their respective components. Existing communication infrastructure can be used for contracting and command signals that ancillary service providers usually use, which are mostly based on fibre optic and broadband communications. The requirements for V2G communications can be identified based on five key factors: bandwidth, latency, reliability, security and mobility. These can be achieved by a variety of communication technologies such as Power-line Communications and HomePlug, Wireless Personal-Area Networking and ZigBee, Interference Management and Cognitive Radio. A brief introduction of some of these technologies is presented below, and clustered V2G structure and communication between aggregators are illustrated in Figures 5.6 and 5.7 respectively. In Figure 5.8 overall control strategies are illustrated.

Power-Line Communication and HomePlug: In PLC high-frequency data signals are superimposed on top of the distribution voltage for data transmission

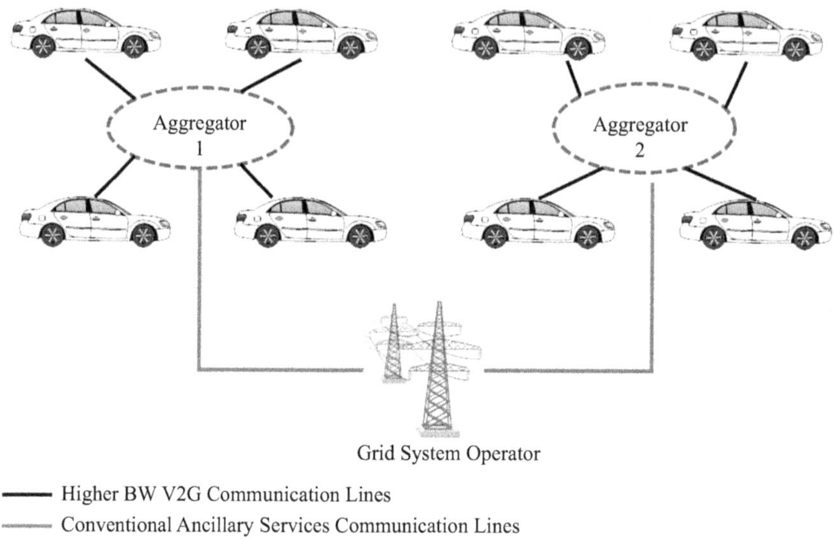

Figure 5.6 Distributed aggregator controller

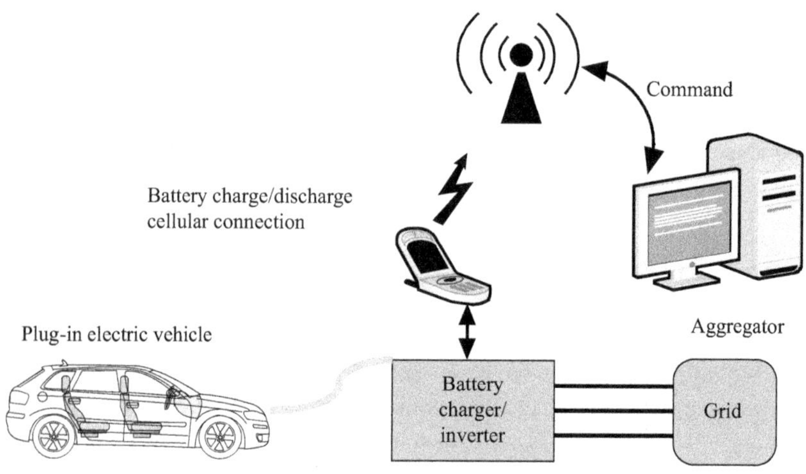

Figure 5.7 Interaction of an aggregator and an EV over cellular network

through existing power-line conductors. Typically, transformers prevent PLC signal propagating over high-voltage lines, so this technique is suitable for medium voltage distribution level. Recent implementation of PLC technology has mainly limited the communication data transfer over residential-side power lines that lead up to neighbourhood. This has successfully reduced damaging antenna effect from medium voltage power lines. With this technology command and price signal

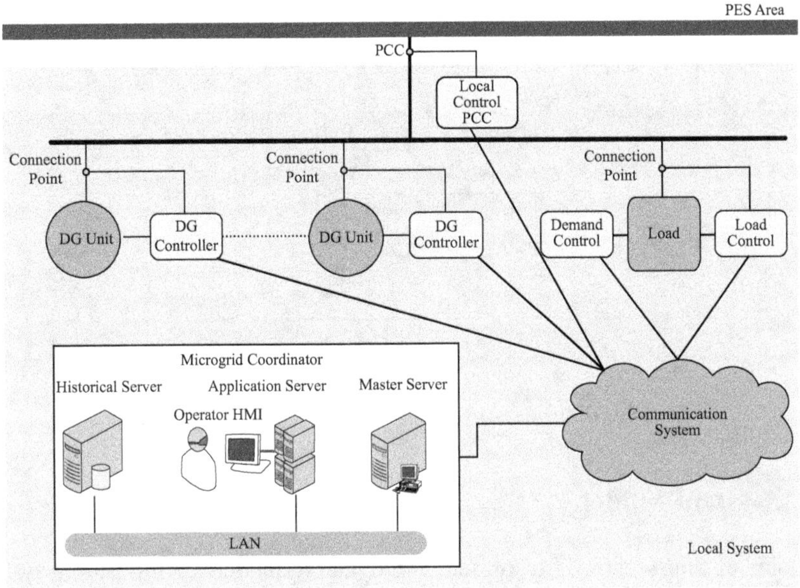

Figure 5.8 Overall smart grid and EV control strategy

can be sent by aggregators to PLC receivers which is actually called HomePlug and is being deployed in some regions. PLC can also contribute in home energy management [68, 69].

Wireless Personal-Area Networking and ZigBee: ZigBee refers to a combination of high-level wireless communication technologies and protocols using small, inexpensive and low-power digital radios based on IEEE 802.15.4-2006 wireless personal-area network standard. ZigBee operates in dual frequency band, i.e. 2.4 and 900 GHz, which serves the flexibility to choose most proper frequency band in noisy radio environment. ZigBee is mainly designed for sensing and automation application like Home Energy Management. ZigBee has fast transmission rate capacity (approx. 20–250 kbps), which can update data of voltage and frequency regulation services in about one in every second. Because of its large range – approximately 400 m – it can be used to cover EV parking lot with minimum number of transceivers [69].

Interference Management and Cognitive Radio: Since several V2G communication technologies utilize same frequency band, there could be wireless interference and congestion in densely populated regions. Therefore, the coexistence strategies (i.e. dynamic and distributed channel allocation) need to be designed and implemented cautiously. Moreover, cognitive radio techniques can be used for spectrum sharing across different V2G/V2H and home area network technologies [69].

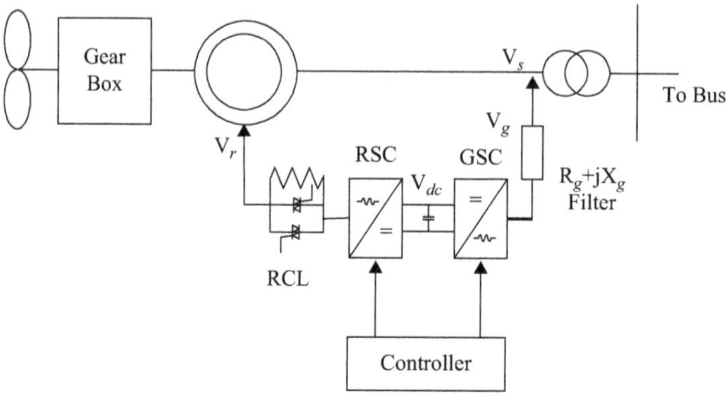

Figure 5.9 Schematic diagram of a DFIG

5.6 System Model

The aim of this section is to present the dynamic models of the power system devices under consideration, which are described in the following sections.

5.6.1 Wind Turbine Modelling

The most important components of a wind turbine, as shown in Figure 5.9, are rotor, drive train and the generator. The wind turbine rotor of radius R_i converts the wind energy supplied to the rotor shaft that rotates at the speed ω_m. The power from the wind relies on the wind speed, V_w, the air density, ρ, and the swept area, A_{wt}. From the available power in the swept area, the power on the rotor is given based on the power coefficient $c_p(\lambda, \theta)$, which depends on the pitch angle of the blade, θ, and the ratio between the speed of the blade tip and the wind speed, denoted tip–speed ratio, $\lambda_r = \frac{\omega_m R_i}{V_w}$. The aerodynamic torque applied to the rotor by the effective wind passing through the rotor is given as [29]

$$T_{ae} = \frac{\rho}{2\omega_m} A_{wt} c_p(\lambda_r, \theta) V_w^3 \tag{5.1}$$

where c_p is approximated by the following relation [30]:

$$c_p = (0.44 - 0.0167\theta)\sin\left[\frac{\pi(\lambda - 3)}{15 - 0.3\theta} - 0.00184(\lambda - 3)\theta\right]$$

A two-mass drive train model of a wind turbine generator system (WTGS) is used in this chapter. The drive train attached to the wind turbine converts the aerodynamic torque, T_{ae}, on the rotor into the torque on the low-speed shaft, which is scaled down through the gearbox to the torque on the high-speed shaft. The first mass term stands for the blades, hub and low-speed shaft, while the second mass term stands for the high-speed shaft, having inertia constants, H_m and H_g.

The shafts are interconnected by a gear ratio, N_g, combined with torsion stiffness, K_s, and torsion damping, D_m and D_g, resulting in torsion angle γ. The normal grid frequency is f. The dynamics of the shaft are represented as follows [29]:

$$\dot{\omega}_m = \frac{1}{2H_m}[T_{ae} - K_s\gamma - D_m\omega_m] \tag{5.2}$$

$$\dot{\omega}_g = \frac{1}{2H_g}[K_s\gamma - T_e - D_g\omega_g] \tag{5.3}$$

$$\dot{\gamma} = 2\pi f\left(\omega_m - \frac{1}{N_g}\omega_g\right) \tag{5.4}$$

The generator gets the mechanical power from the gearbox through the stiff shaft. The relationship between the mechanical torque and the torsional angle is given by

$$T_m = K_s\gamma \tag{5.5}$$

A transient model of a squirrel cage induction generator (IG) is given by the following algebraic–differential equations [29, 30]:

$$\dot{s} = \frac{1}{2H_g}[T_m - T_e] \tag{5.6}$$

$$\dot{E}'_{dr} = -\frac{1}{T'_0}[E'_{dr} + (X - X')i_{qs}] + s\omega_s E'_{qr} \tag{5.7}$$

$$\dot{E}'_{qr} = -\frac{1}{T'_0}[E'_{qr} - (X - X')i_{ds}] - s\omega_s E'_{dr} \tag{5.8}$$

$$V_{ds} = R_s i_{ds} - X' i_{qs} + E'_{dr} \tag{5.9}$$

$$V_{qs} = R_s i_{ds} + X' i_{qs} + E'_{qr} \tag{5.10}$$

$$v_t = \sqrt{V_{ds}^2 + V_{qs}^2} \tag{5.11}$$

where

$X' = X_s + X_m X_r/(X_m + X_r)$ is the transient reactance
$X = X_s + X_m$ is the rotor open-circuit reactance
$T'_o = (L_r + L_m)/R_r$ is the transient open-circuit time constant
v_t is the terminal voltage of the IG
s is the slip
E'_{dr} is the direct-axis transient voltage
E'_{qr} is the quadrature-axis transient voltage

V_{ds} is the d-axis stator voltage
V_{qs} is the q-axis stator voltage
T_m is the mechanical torque
$T_e = E_{dr}i_{ds} + E_{qr}i_{qs}$ is the electrical torque
X_s is the stator reactance
X_r is the rotor reactance
X_m is the magnetizing reactance
R_s is the stator resistance
R_r is the rotor resistance
H_g is the inertia constant of the IG
i_{ds} and i_{qs} are d- and q-axis components of the stator current respectively
ω_s is the synchronous speed

5.6.2 PV System Modelling

The PV system, as shown in Figure 5.10, has mainly two portions: (a) solar energy conversion and (b) electrical interface with the electrical network (power electronic inverter).

Different maximum power point tracking (MPPT) systems, such as perturb and observe (P&O), incremental conductance (IC) and hill climbing, are available to extract maximum power from the PV array. The MPPT controller generates the maximum power DC voltage reference for the PV array. The PV array can be regulated to work on maximum power point (MPP) with or without DC–DC converter depending on the controller design. The PV array current–voltage relationship can be written as [31–33]

$$i_{pvi} = N_{pi}I_{Li} - N_{pi}I_{si}\left[\exp\left[\alpha_{pi}\left(\frac{V_{pi}}{N_{si}} + \frac{R_{si}i_{pvi}}{N_{pi}}\right)\right] - 1\right] - \frac{N_{pi}}{R_{shi}}\left(\frac{V_{pvi}}{N_{si}} + \frac{R_{si}i_{pvi}}{N_{pi}}\right) \tag{5.12}$$

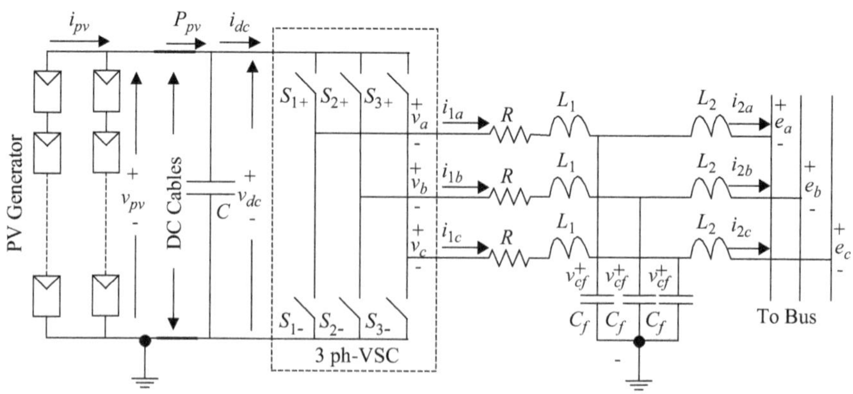

Figure 5.10 PV system connected to the grid

where

I_{Li} is the current corresponding to light
I_{si} is the reverse saturation current chosen as 9×10^{-11} A
N_{si} is the number of cells in series
N_{pi} is the number of modules in parallel
R_{si} and R_{shi} are series and shunt resistances of the array
I_{pvi} is the current flowing through the array
V_{pvi} is the output voltage of the array
The constant $\alpha_{pi} = \frac{q_i}{A_i k_i T_{ri}}$

where

$k_i = 1.3807 \times 10^{-23}$ J/K is the Boltzmann constant
$q_i = 1.6022 \times 10^{-19}$ C is the charge of an electron
A_i is the p-n junction ideality factor with a value between 1 and 5
T_{ri} is the cell reference temperature

The PV generated power is transferred to the loads and the remaining power to the grid via power electronics inverter. Generally, inverter can be classified on the basis of operation as follows:

(i) Voltage source inverter (VSI)
(ii) Current source inverter (CSI)

Although, the CSI inverter has inherently filtering capability over VSI inverters, the latter one is commonly used because of control flexibility in grid tied and islanded operations.

The inverter control can be classified on the basis of controlled parameter as follows:

(i) Voltage controlled
(ii) Power controlled
(iii) Current controlled

The PV system exhibits stochastic characteristics because of nonlinear sun irradiance dependency, and thus implementing linear controller will not describe the proper stability analysis of the system. Two control loops are designed to make efficient control operation of the VSI. The outer control loop regulates the DC voltage reference and the inner control current loop makes the VSI to follow the active and reactive power reference independently.

The DC voltage regulation control equation can be written as

$$\dot{V}_{pvi} = \frac{1}{C_i}(i_{pvi} - i_{dci}) \tag{5.13}$$

The DC current of the PV system and inverted output current can be related with the inverter switching signals as

$$i_{dci} = i_{ai}K_{ai} + I_{bi}K_{bi} + i_{ci}K_{ci} \tag{5.14}$$

Equation (5.13) can be written as

$$\dot{V}_{pvi} = \frac{1}{C_i} i_{pvi} - \frac{1}{C_i}(i_{ai}K_{ai} + I_{bi}K_{bi} + i_{ci}K_{ci}) \tag{5.15}$$

The nonlinear model of the grid tied inverter with LCL filter can be written as [34]

$$\begin{aligned}
\dot{i}_{1ai} &= -\frac{R_i}{L_{1i}} i_{1ai} - \frac{1}{L} e_{ai} + \frac{V_{pvi}}{3_{L1i}}(2K_{ai} - K_{bi} - K_{ci}) \\
\dot{i}_{1bi} &= -\frac{R_i}{L_{1i}} i_{1bi} - \frac{1}{L} e_{bi} + \frac{V_{pvi}}{3_{L1i}}(-K_{ai} + 2K_{bi} - K_{ci}) \\
\dot{i}_{1ci} &= -\frac{R_i}{L_{1i}} i_{1ci} - \frac{1}{L} e_{ci} + \frac{V_{pvi}}{3_{L1i}}(-K_{ai} - 2K_{bi} + 2K_{ci}) \\
\dot{V}_{cfai} &= \frac{1}{C_{fi}}(i_{1ai} - i_{2ai}) \quad \dot{i}_{2ai} = \frac{1}{L_{2i}}(V_{cfai} - e_{ai}) \\
\dot{V}_{cfbi} &= \frac{1}{C_{fi}}(i_{1bi} - i_{2bi}) \quad \dot{i}_{2bi} = \frac{1}{L_{2i}}(V_{cfbi} - e_{bi}) \\
\dot{V}_{cfci} &= \frac{1}{C_{fi}}(i_{1ci} - i_{2ci}) \quad \dot{i}_{2ci} = \frac{1}{L_{2i}}(V_{cfci} - e_{ci})
\end{aligned} \tag{5.16}$$

where K_{ai}, K_{bi} and K_{ci} are the binary input switching signals for a grid tied system. This system uses phase-locked loop (PLL) to synchronize with the distribution system, but a stand-alone system uses a separate frequency regulation control loop instead of PLL.

The system dynamics from (5.15) and (5.16) can be transferred to *dq* frame using the angular frequency ω_i of the grid as [35]

$$\begin{aligned}
C_i\dot{V}_{pvi} &= i_{pvi} - i_{1di}K_{di} - i_{1qi}K_{qi} \\
L_{1i}\dot{i}_{1di} &= -R_i i_{1di} + \omega_i L_{1i} i_{1qi} - V_{cfdi} + K_{di}\, V_{pvi} \\
L_{1i}\dot{i}_{1qi} &= -R_i i_{1qi} - \omega_i L_{1i} i_{1di} - V_{cfqi} + K_{qi}\, V_{pvi} \\
L_{2i}\dot{i}_{2di} &= +\omega_i L_{1i} i_{1qi} + V_{cfdi} - E_{di} \\
L_{2i}\dot{i}_{2qi} &= -\omega_i L_{2i} i_{2di} + V_{cfqi} - E_{qi} \\
C_{fi}\dot{V}_{cfdi} &= \omega_i C_{fi} V_{cfqi}\text{-}C_{fi}(i_{1di} - i_{2di}) \\
C_{fi}\dot{V}_{cfqi} &= -\omega_i C_{fi} V_{cfdi}\text{-}C_{fi}(i_{1qi} - i_{2qi})
\end{aligned} \tag{5.17}$$

The PLL synchronization angle is utilized in the *abc-dq0* transformation and chosen such that the *q*-axis of the *dq* frame is aligned with the grid voltage vector, resulting in $E_{qi} = 0$, and the instantaneous real and reactive power delivered to the grid can be written as

$$P_i = \frac{3}{2} E_{di} I_{di} \quad \text{and} \quad Q_i = -\frac{3}{2} E_{di} I_{qi}$$

5.6.3 EV Energy Storage System (EV-ESS) Modelling

Several battery models are available, such as simple battery model, modified battery model, Thevenin equivalent battery model, resistive Thevenin battery model, modified Thevenin battery model and dynamic model. Among these models, the dynamic model proposed by the author in [36] offers flexible electrical characteristics analysis for the battery integrated power system analysis. The electrochemical battery models only represent general electrical equivalent circuits, making a trade-off between complexity and simulation quality, whereas the dynamic model represents the correlated relation among battery voltage, SOC, depth of discharge (DOC), battery temperature and battery current with user convenient circuitry model for proper power system analysis.

Batteries are mainly used for

(i) Uninterruptable power supplies (UPS)
(ii) Battery energy storage plants (BESP)
(iii) Electrical vehicles

Battery models are basically bipolar, which can be represented by electromotive force and a series of internal impedances such as a Laplace function variable 's' which is regarded as a linear battery model shown in Figure 5.11.

However, as the battery behaviour is far from being linear and E and Z are the dependent functions of electrolyte temperature (θ) and SOC, during the charge storing mechanism, only a portion of the battery current (I_m) stores in and the remaining portion passes through another branch known as parasitic branch [36]. The parasitic branch draws some current without affecting the main battery operation, and converts the entered energy in its electromotive branch (E_p) into another form of energy. For example, the power loss in the real and especially in the parasitic impedances takes part in increasing the overall battery temperature making it hot. This parasitic branch with the linear battery model is shown Figure 5.12.

This battery temperature has significant impact on overall battery performance, reliability as well as on its lifetime. In conventional general batteries, the charge efficiency is considered very near to unity. As the parasitic branch takes no part in

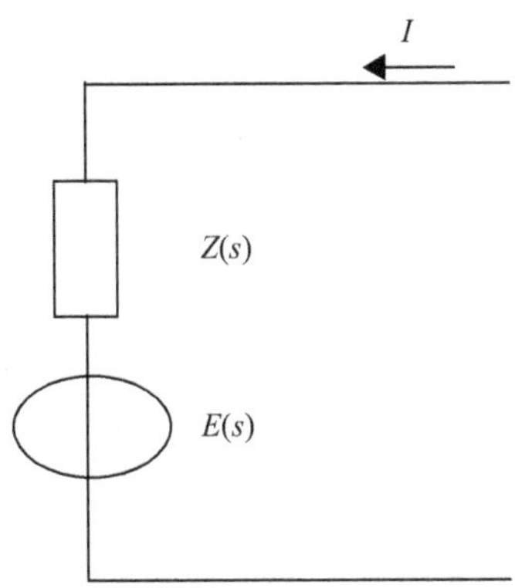

Figure 5.11 Linear battery model

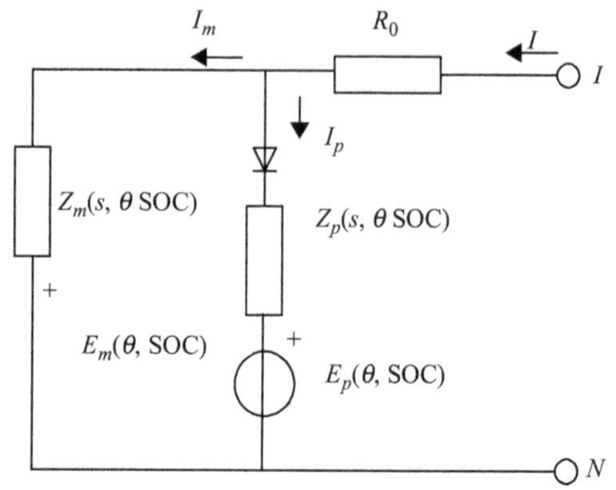

Figure 5.12 Simple battery equivalent circuit with parasitic branch [36]

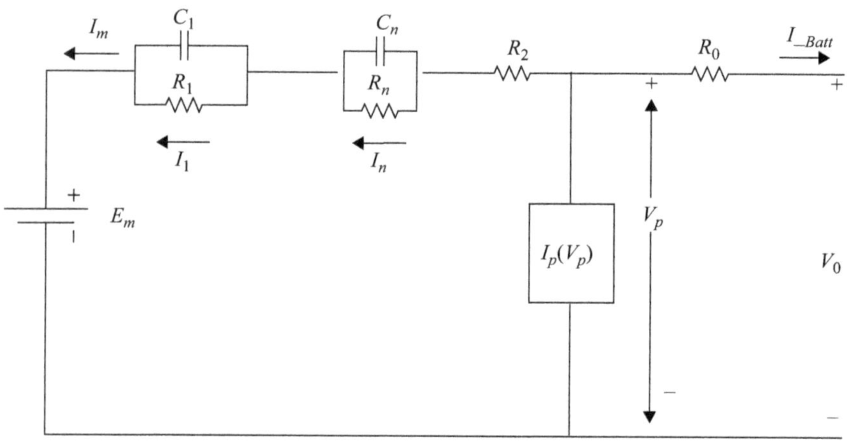

Figure 5.13 Third-order battery aggregate model

the main reversible reaction, during analysis this branch is considered as an inactive portion for analytical simplicity.

The charging and discharging mechanism is electrically represented by several *R-C* blocks as shown in Figure 5.13. The number of *R-C* blocks for the battery model depends on the application of the battery in different sections. However, to make the model accurate and simple, it is optimized accordingly with the frequent current/voltage shapes, considering one or two *R-C* blocks in the battery model [36]. During the nearly fully charged period, the impedance of the main branch increases instantly, which causes the parasitic branch terminal voltage and consequently the current, I_p, to rise. This phenomenon is represented by considering one of the *R-C* block parameters such as $C_n = 0$ and $R_n = R_n(\mathrm{SOC})$, which results in R_n approaching infinity keeping pace with the battery full-state approach. The battery SOC calculation takes into account the battery capacity as a function of electrolyte temperature, discharge current and voltage at discharge point. The charge relation with electrolyte

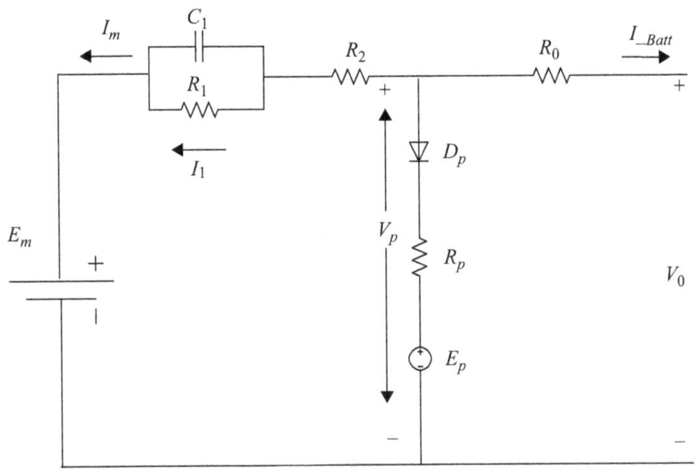

Figure 5.14 Third-order battery equivalent electrical network

temperature and discharge current can be simplified as having inverse characteristics. Higher temperature and lower current will draw more charge than constant temperature and constant discharge current.

Simplified third-order battery model is presented in Figure 5.14. The dependency of the battery parameters on the electrical circuit network is shown in the following sections [36, 37].

5.6.3.1 Main Branch

The battery capacity dependency can be expressed as

$$C(I,\ \theta)_{I,\theta=constant} = C_0(I)\left(1+\frac{\theta}{-\theta_f}\right)^{\varepsilon} \quad \theta > \theta_f \tag{5.18}$$

where

θ_f is the electrolyte freezing temperature = −40°C

$C_0\,(I)$ is the battery capacity at 0°C and an empirical function of discharge current, which is expressed as

$$C_0(I) = \frac{K_c C_0^*}{1+(K_c-1)\left(\frac{I}{I^*}\right)^{\delta}} \tag{5.19}$$

where

$C_0^* = C_0\,(I^*)$

I^* is the reference current for the battery (nominal battery current)

K_c and δ are empirical constant coefficients that depend on the battery reference current I^*

Battery SOC represents how full a battery is with reference to the maximum capacity and how able it is to deliver at the given temperature θ. And, DOC represents how full the battery is with reference to the actual discharge regime. Both SOC and DOC are dependent on the extracted charge as shown below:

$$\mathrm{SOC} = 1 - \frac{Q_e}{C(0,\theta)} = 1 - \frac{Q_e}{K_c C(I^*)} \tag{5.20}$$

$$\mathrm{DOC} = 1 - \frac{Q_e}{C(I_{avg},\,\theta)} \tag{5.21}$$

where

$$Q_e(t) = Q_{einit} + \int_0^t -I_m(\tau)d\tau \tag{5.22}$$

$Q_e(t)$ is the extracted charge from the battery in a fully charged condition, and when $t = 0$ the battery is considered full

Q_{einit} is the initial value of the extracted charge required for the simulation simplicity and perfectness

I_{avg} is postulated to be a filtered value of real battery current $I(t)$ during transient phenomena

I_{avg} can be represented as

$$\begin{aligned} I_{avg} &= I_k \\ \frac{dI_1}{dt} &= \frac{1}{\tau_1}(I_m - I_1) \end{aligned} \tag{5.23}$$

R and C from R-C block can be related with the simulation time constant as

$$\tau_k = R_k C_k \tag{5.24}$$

I_{avg} can be considered equal to I_1 in this battery model [37].

Main branch resistances can be shown with the dependency on SOC and DOC as

$$R_0 = R_{00}[1 + A_0(1 - \mathrm{SOC})] \tag{5.25}$$

$$R_1 = -R_{10}\ln(\mathrm{DOC}) \tag{5.26}$$

However, the R_2 value depends on the battery SOC and increases exponentially during charging while becomes insignificant during discharging operation and is represented as follows:

$$R_2 = R_{20}\frac{\exp[A_{21}(1 - \mathrm{SOC})]}{1 + \exp\left(\frac{A_{22}I_m}{I^*}\right)} \tag{5.27}$$

5.6.3.2 Thermal Model

In general, the electrical equivalent of electrochemical of each electrolyte temperature should be considered while modelling a battery, as each electrolyte has its own distinctive temperature. To avoid further complexity, uniform temperature over the electrolyte in the battery is considered and can be represented as

$$
\begin{aligned}
C_\theta \frac{d\theta}{dt} &= \frac{\theta - \theta_a}{R_\theta} + P_s \\
\theta(t) &= \theta_{init} + \int_0^t \frac{P_s - \frac{\theta - \theta_a}{R_\theta}}{C_\theta} dt
\end{aligned}
\tag{5.28}
$$

where

C_θ is the battery thermal capacitance
θ is the electrolyte temperature
θ_{init} is the initial electrolyte temperature
R_θ is the thermal resistance between the battery and its environment
θ_a is the ambient temperature (temperature of air surrounding the battery)
P_s is the source thermal power (heat generated internally in the battery)
S is the Laplace variable

Now, the main branch voltage can be accurately represented as

$$E_m = E_{m0} - K_E\,(273 + \theta)(1 - \mathrm{SOC}) \tag{5.29}$$

where

E_{m0} is the OCV at full charge and is constant
K_E is constant

5.6.3.3 Parasitic Branch

The voltage in parasitic branch can be written as

$$V_{pn} = I_m R_2 + I_1 R_1 + E_m \tag{5.30}$$

The heat generated power P_s can be represented by the losses in branch resistance R_0 and R_2 [38]:

$$P_s = I_m^2 R_2 + (I_m - I_1)^2 R_0 = R_p I_p^2 \tag{5.31}$$

The nonlinear relation between I_p and V_p following Tafel gassing–current relationship is shown for the parasitic branch as

$$I_p = V_{pn} G_{p0} \exp\left(\frac{V_{pn}}{V_{p0}} + A_p\left(1 - \frac{\theta}{\theta_f}\right)\right) \tag{5.32}$$

The above relation for I_p is valid up to the recombination limit, and it should not be overcome to make sure that the battery works in reversible way. The parasitic branch helps to understand the battery behaviour at the end of a charge and the resistance for this can be simplified as

$$R_p = \frac{V_{pn} - E_p}{I_p}$$

The constant parameters and their corresponding values for a lead acid battery are shown in Table 5.5 [37, 38]. Using the above-mentioned nonlinear models, a control strategy is developed in the next section.

Table 5.5 Third-order battery model constant parameters

Constant parameters	Value
Battery capacity:	
C_0^*	261.6 Ah
K_c	1.18
θ_f	−40°C
ε	1.29
δ	1.40
I^*	49 A
Main branch circuit:	
T_1	5000 s
K_E	0.580e^{-3} V/°C
R_{00}	2.0 mΩ
A_0	−0.30
A_{21}	−8.0
E_{m0}	2.135 V
R_{10}	0.7 mΩ
R_{20}	15 mΩ
A_{22}	−8.45
Parasitic reaction branch:	
E_p	1.95 V
G_{p0}	2 pS
V_{p0}	2 mΩ
A_p	2
Thermal model:	
C_θ	15 Wh°C
R_θ	0.2 °C/W

5.7 Problem Formulation and Control Strategy

In this section a coordinated control strategy is proposed for controlling DFIG, PV and V2G (which is modelled as a battery). Both active and reactive powers of DFIG are controlled in order to control the frequency and voltage. The control strategy is shown in Figure 5.15. In general P_w^{ref} will be the predicted MPP based on wind speed. P_w^{ref} and Q_w^{ref} will be controlled using PI controllers shown in Figure 5.15, which are designed centrally. The desired active and reactive powers are realized by controlling i_{qr} and i_{dr}. To achieve the reference reactive power, the checker block checks the available capacity for i_{qr} with respect to i_{dr} for a converter with a 25% rating of the DFIG. It is unlikely that the set power references and the actual load will balance in a practical system. Droop controllers are proposed in this chapter for the energy storage systems to balance the load and generation. In short, the control strategy is to have one centralized controller and multiple droop controllers [47, 70] to achieve stable operation in both grid-connected and islanded modes and three-phase fault mode.

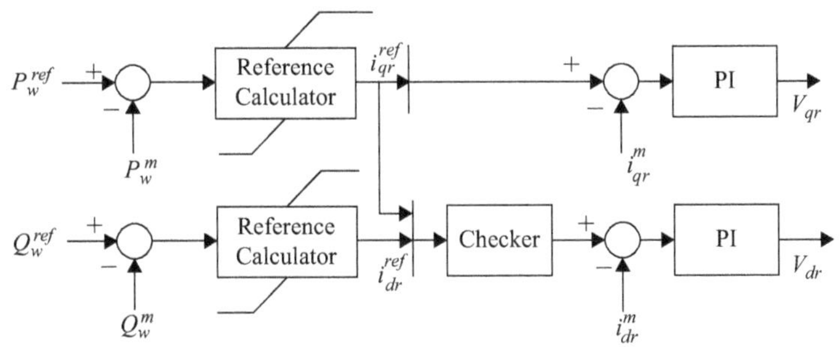

Figure 5.15 Control for DFIG

A PV unit with a battery (represented as a current source) is shown in Figure 5.16. In general P_p^{ref} will be the predicted MPP based on solar radiation and P_p^{ref} and Q_p^{ref} will be controlled using PI controllers. In the voltage control mode of PV units, P_p and Q_p are controlled by the amplitude of the voltage source converter (VSC) terminal voltage. The error signals $P_p^{ref} - P_p$ and $Q_p^{ref} - Q_p$ when fed to the controller produces *d*- and *q*-axis components of the VSC current at their respective reference values, which are again processed to get reference voltage components. These two signals are then divided by v_{pv} to generate K_d and K_q for the PWM.

EV battery can be charged/discharged in a coordinated way by utilizing the proposed controller in Figure 5.17. Depending on the SOC, user requirement, load variation and grid status, the reference power for the battery controller will be generated. The controller will track the reference dynamically and control charging and discharging of the battery. P_{Batt}^{ref} can be increased or decreased to any desired value while PI controller will stabilize the system dynamically. In applications where constant current charging and discharging occur consecutively, power disruption is caused. This disruption can be smoothened by changing the

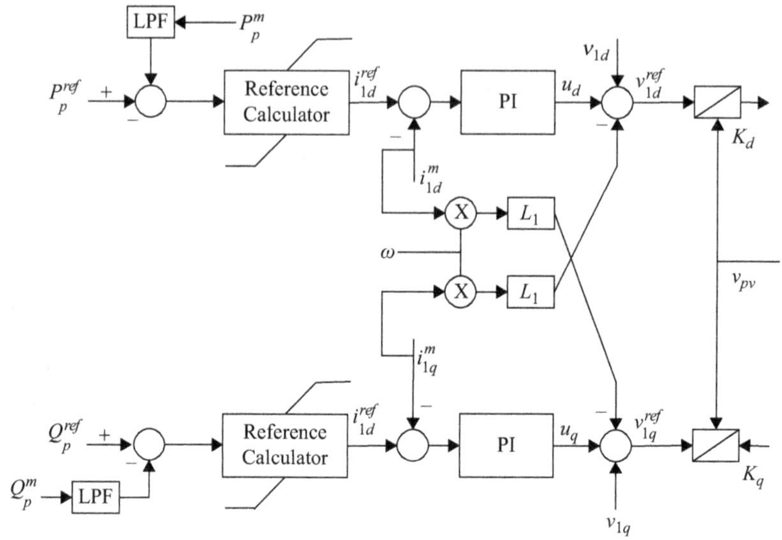

Figure 5.16 Control for PV unit

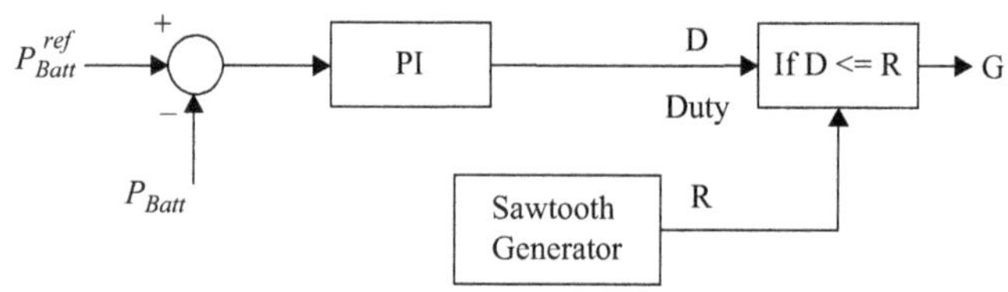

Figure 5.17 EV battery charging/discharging controller

power accordingly. When battery is charged in constant current mode, the voltage of the battery will gradually increase causing ramp-up power rate. Conversely, when the battery is charged in constant voltage mode, the battery current will gradually decrease causing ramp-down power rating. By utilizing these ramp-up and ramp-down rates both constant current and constant voltage modes can be attained [71].

To maintain a stable DC bus voltage and to interlink between DC bus and AC bus (i.e. point of common coupling (PCC)), a neutral clamped VSC with controller is used (Figure 5.18). Measured three-phase voltage and currents are converted into *dq* frame. V_{dq} and I_{dq} are fed to a current regulator where a reference I_{dq}^{ref} is also fed. I_{dq}^{ref} is generated from the voltage regulator. The main task of the voltage regulator is to stabilize the DC bus voltage. A predefined reference DC bus voltage is fed to the voltage regulator. Measured DC bus voltage is compared with the reference DC bus voltage and then controlled by a PI controller.

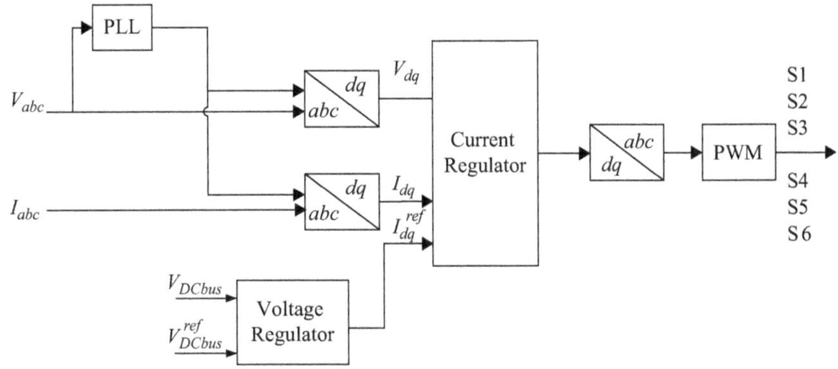

Figure 5.18 VSC controller

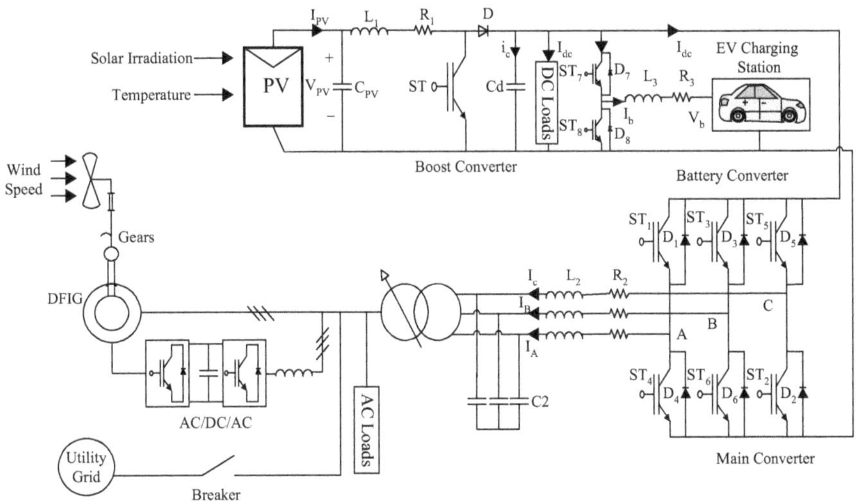

Figure 5.19 Test microgrid

5.8 Simulation Results

To understand the contribution of coordinated V2G integration, a simplistic test microgrid is considered for simulation purpose. Reference [72] is followed for the test microgrid configuration. The system consists of a PV panel of 100 kW and a 9 MW DFIG wind farm connected with utility grid. A 200 kW DC load is connected at the DC bus terminal and a 2 MW, 10 kVar AC load is connected at the AC bus terminal. The PV unit is connected to the system via the VSC. The stator of the DFIG is connected directly to the grid and the rotor via a VSC. The PV module utilizes MPPT algorithm to extract maximum power. Here IC method is used for MPPT purpose. To represent V2G system, a battery bank with output voltage 500 V is connected at the DC bus terminal. The test microgrid is illustrated in Figure 5.19. In the following sections, simulations for three-phase fault at PCC (AC bus) and intentional islanding are carried out and the contribution of V2G during these phenomena is observed.

5.8.1 Severe Three-Phase Fault

The objective of this case study is to demonstrate that coordination of EV can improve stability by maintaining frequency and voltage within statutory limits and by sharing load between PV and DFIG WT during a severe three-phase fault in a grid-connected mode. A three-phase fault is applied on PCC at 3 seconds and cleared after 0.5 second. Figures 5.20 and 5.21 show pre- and post-effect of

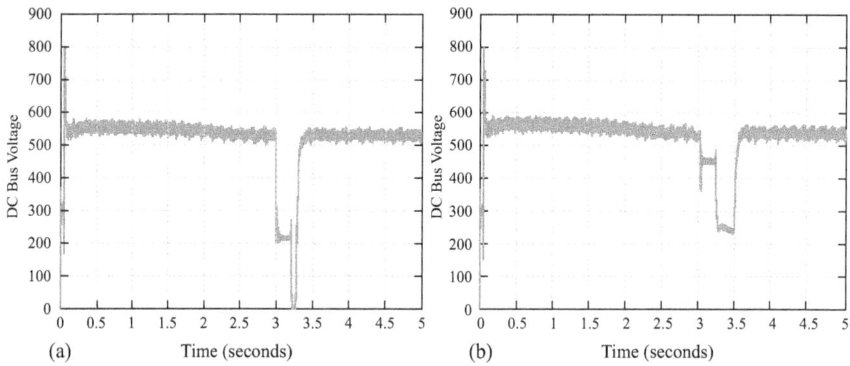

Figure 5.20 DC bus voltage before (a) and after (b) EV ESS introduction (severe three-phase fault)

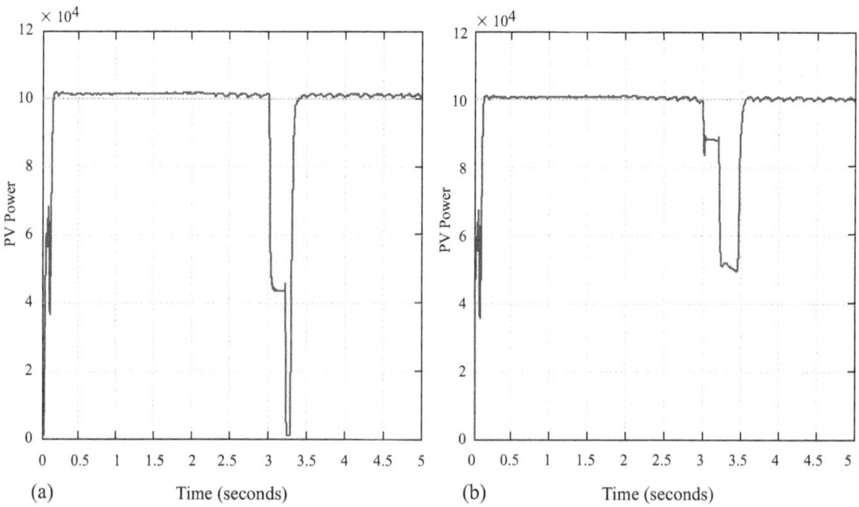

Figure 5.21 Supplied power from PV before (a) and after (b) EV ESS introduction (severe three-phase fault)

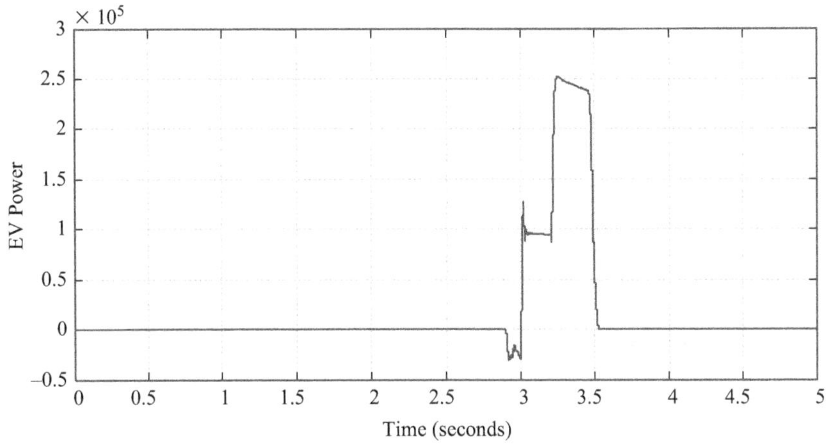

Figure 5.22 Power supplied by EV ESS at the time of three-phase fault

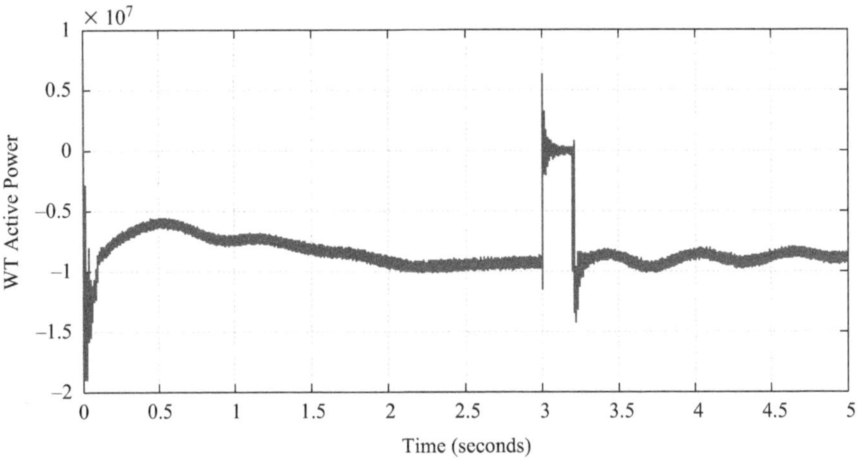

Figure 5.23 Active power of wind generator at the time of three-phase fault

coordinated EV ESS on DC bus voltage and supplied PV power during the fault. Supplied power by EV ESS, DFIG wind turbine active and reactive power profile during severe three phase fault are shown in Figures 5.22, 5.23 and 5.24 respectively. It's evident from the results that EV can effectively control and improve the impact of the fault.

5.8.2 Intentional Islanding

In this case study, an intentional islanding is demonstrated. Before islanding, the microgrid was working in a grid-connected mode, PV and WT were supplying

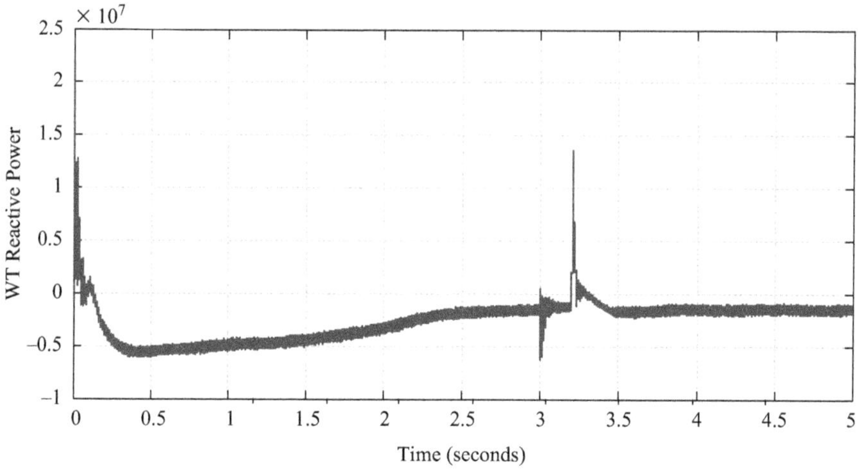

Figure 5.24 Reactive power of wind generator at the time of three-phase fault

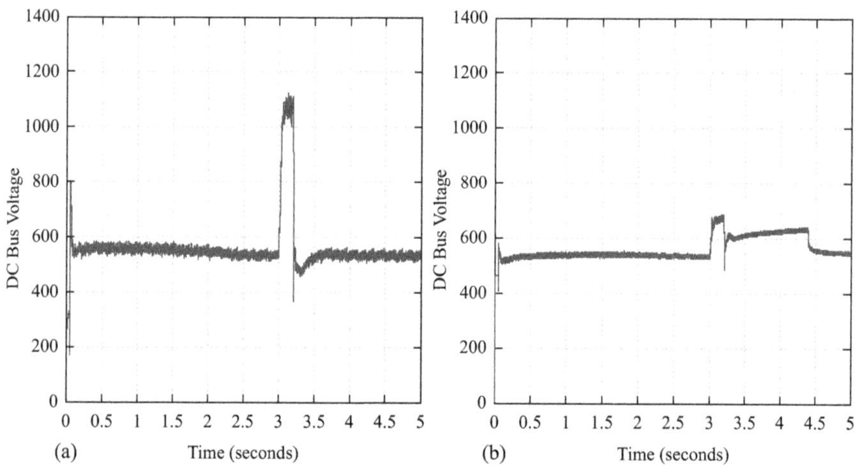

Figure 5.25 DC bus voltage before (a) and after (b) EV ESS introduction (intentional islanding)

100 kW and 9 MW respectively and the remaining power was imported from the grid. At $t = 3$ seconds, an intentional islanded command is applied to the PCC with 0.2 second duration. Figures 5.25 and 5.26 show pre- and post-effect of coordinated EV ESS on DC bus voltage and supplied PV power during islanding. Consumed power by EV ESS, DFIG wind turbine active and reactive power profile during intentional islanding are shown in Figures 5.27, 5.28 and 5.29 respectively. It is evident from the results that EV can effectively control and improve the islanding effect.

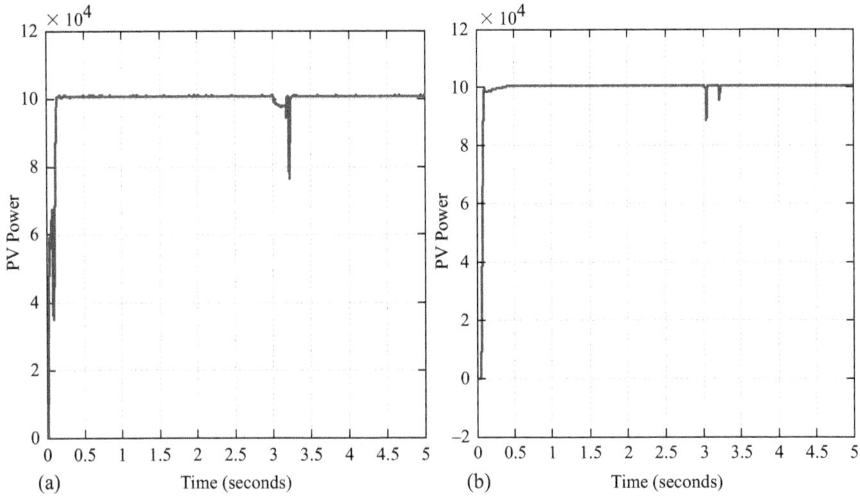

Figure 5.26 Supplied power from PV before (a) and after (b) EV ESS introduction (intentional islanding)

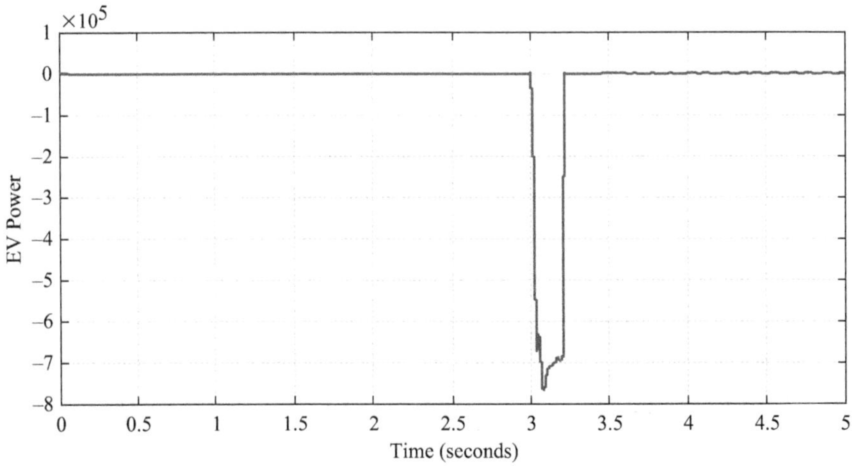

Figure 5.27 Power consumed by EV ESS at the time of intentional islanding

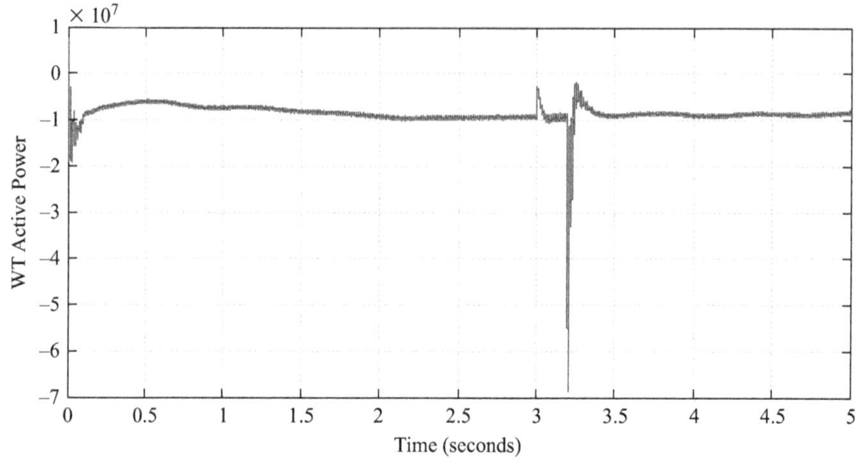

Figure 5.28 Active power of wind generator at the time of intentional islanding

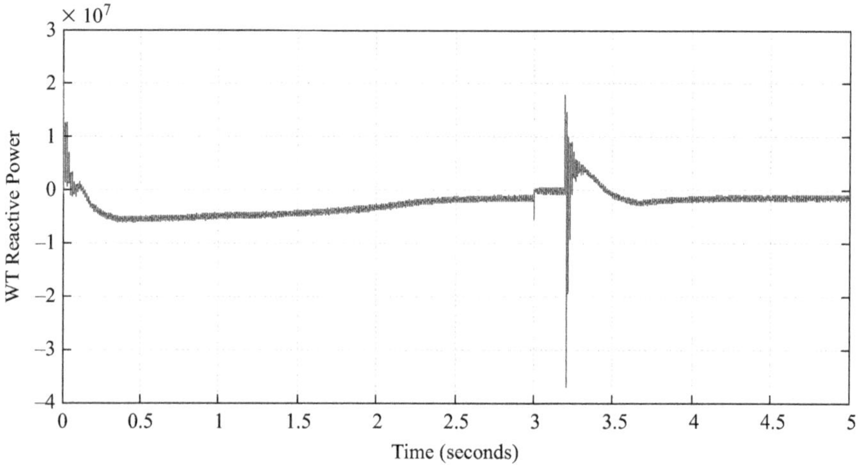

Figure 5.29 Reactive power of wind generator at the time of intentional islanding

5.9 Conclusions

In this chapter, ongoing smart grid power control and monitoring schemes in the presence of EVs have been discussed. Different EV battery monitoring and management techniques along with control strategies for optimum V2G implementation have been depicted. Furthermore, a coordinated control of PV, DFIG and EV aggregator is proposed and simulated in test microgrid under severe environment. The simulation results have validated the viability of optimum V2G implementation in the smart grid power control and monitoring and have proven that coordinated EV penetration in the smart grid with V2G concept can be an outstanding footstep towards efficient power delivery.

References

[1] Bordeau K., Parker G., Vosters G., Weaver W., Kelly J., Wilson D. G. *et al.* 'Distributed control of plug-in hybrid electric vehicles on a smart grid'. *International Symposium on Power Electronics, Electrical Drives, Automation and Motion*; Sorrento, Italy, June 2012, pp. 1412–1416

[2] Saber A. Y., Venayagamoorthy G. K. 'One million plug-in electric vehicles on the road by 2015'. *12th International IEEE Conference on Intelligent Transportation Systems*; USA, 2009, pp. 1–7

[3] Raghavan S. S., Khaligh A. 'Impact of plug-in hybrid electric vehicle charging on a distribution network in a smart grid environment'. *Innovative Smart Grid Technologie*; 2012, pp. 1–7

[4] Shafiee S., Fotuhi-Firuzabad M., Rastegar M. 'Investigating the impacts of plug-in hybrid electric vehicles on power distribution systems'. *IEEE Transactions on Smart Grid*; 2013, vol. 4, pp. 1351–1360

[5] Green R. C., Lingfeng W., Alam M. 'The impact of plug-in hybrid electric vehicles on distribution networks: a review and outlook'. *IEEE Power and Energy Society General Meeting*; 2010, pp. 1–8

[6] Gan L., Xiao-Ping Z. 'Modeling of plug-in hybrid electric vehicle charging demand in probabilistic power flow calculations'. *IEEE Transactions on Smart Grid*; 2012, vol. 3, pp. 492–499

[7] Lassila J., Haakana J., Tikka V., Partanen J. 'Methodology to analyze the economic effects of electric cars as energy storages'. *IEEE Transactions on Smart Grid*; 2012, vol. 3, pp. 506–516

[8] Shireen W., Patel S. 'Plug-in hybrid electric vehicles in the smart grid environment'. *IEEE Transmission and Distribution Conference and Exposition*; 2010, pp. 1–4

[9] Roe C., Meliopoulos A. P., Meisel J., Overbye T. 'Power system level impacts of plug-in hybrid electric vehicles using simulation data'. *IEEE Energy 2030 Conference*; USA, 2008, pp. 1–6

[10] Lund H., Kempton W. 'Integration of renewable energy into the transport and electricity sectors through V2G'. *Energy Policy*; 2008, vol. 36, pp. 3578–3587.

[11] Larsen E., Chandrashekhara D. K., Ostergard J. 'Electric vehicles for improved operation of power systems with high wind power penetration'. *IEEE Energy 2030 Conference*; 2008, pp. 1–6

[12] Locment F., Sechilariu M., Forgez C. 'Electric vehicle charging system with PV Grid-connected configuration'. *IEEE Vehicle Power and Propulsion Conference*; 2010, pp. 1–6

[13] Breucker S. D., Jacqmaer P., Brabandere K. D., Driesen J., Belmans R. 'Grid power quality improvements using grid-coupled hybrid electric vehicles PEMD 2006'. *The 3rd IET International Conference on Power Electronics, Machines and Drives*; 2006, pp. 505–509

[14] Marano V., Rizzoni G. 'Energy and economic evaluation of PHEVs and their interaction with renewable energy sources and the power grid'. *IEEE International Conference on Vehicular Electronics and Safety*; 2008, pp. 84–89

[15] Koyanagi F., Uriu Y. 'A strategy of load leveling by charging and discharging time control of electric vehicles'. *IEEE Transactions on Power Systems*; 1998, vol. 13, pp. 1179–1184

[16] Yilmaz M., Krein P. T. 'Review of the impact of vehicle-to-grid technologies on distribution systems and utility interfaces'. *IEEE Transactions on Power Electronics*; 2013, vol. 28, pp. 5673–5689

[17] Cvetkovic I., Thacker T., Dong D., Francis G., Podosinov V., Boroyevich D. *et al.* 'Future home uninterruptible renewable energy system with vehicle-to-grid technology'. *IEEE Energy Conversion Congress and Exposition*; 2009, pp. 2675–2681

[18] Ashourpouri A., Sheikholeslami A., Shahabi M., Niaki S. A. N. 'Active power control of smart grids using plug-in hybrid electric vehicle'. *2nd Iranian Conference on Smart Grids*; 2012, pp. 1–6

[19] Sekyung H., Hee H. S., Sezaki K. 'Design of an optimal aggregator for vehicle-to-grid regulation service'. *Innovative Smart Grid Technologies*; 2010, pp. 1–8

[20] White C. D., Zhang K. M. 'Using vehicle-to-grid technology for frequency regulation and peak-load reduction'. *Journal of Power Sources*; 2011, vol. 196, pp. 3972–3980

[21] Clement-Nyns K., Haesen E., Driesen J. 'The impact of charging plug-in hybrid electric vehicles on a residential distribution grid'. *IEEE Transactions on Power Systems*; 2010, vol. 25, pp. 371–380

[22] Rios A. D. L., Goentzel J., Nordstrom K. E., Siegert C. W. 'Economic analysis of vehicle-to-grid (V2G)-enabled fleets participating in the regulation service market'. *IEEE Innovative Smart Grid Technologies*; 2012, pp. 1–8

[23] Chenye W., Mohsenian-Rad H., Jianwei H., Jatskevich J. 'PEV-based combined frequency and voltage regulation for smart grid'. *IEEE Innovative Smart Grid Technologies*; 2012, pp. 1–6

[24] Andersson S. L., Elofsson A. K., Galus D., Göransson L., Karlsson S., Johnsson F. *et al.* 'Plug-in hybrid electric vehicles as regulating power providers: case studies of Sweden and Germany'. *Energy Policy*; 2010, vol. 38, pp. 2751–2762

[25] Kempton W., Tomić J. 'Vehicle-to-grid power fundamentals: calculating capacity and net revenue'. *Journal of Power Sources*; 2005, vol. 144, pp. 268–279

[26] Ota Y., Taniguchi H., Nakajima T., Liyanage K. M., Baba J., Yokoyama A. 'Autonomous distributed V2G (vehicle-to-grid) satisfying scheduled charging'. *IEEE Transactions on Smart Grid*; 2012, vol. 3, pp. 559–564

[27] Göransson L., Karlsson S., Johnsson F. 'Integration of plug-in hybrid electric vehicles in a regional wind-thermal power system'. *Energy Policy*; 2010, vol. 38, pp. 5482–5492

[28] Wang J., Liu C., Ton D., Zhou Y., Kim J., Vyas A. 'Impact of plug-in hybrid electric vehicles on power systems with demand response and wind power'. *Energy Policy*; 2011, vol. 39, pp. 4016–4021

[29] Ackermann T. *Wind Power in Power Systems.* England: John Wiley and Sons Ltd; 2005

[30] Abdin E. S., Xu W. 'Control design and dynamic performance analysis of wind turbine-induction generator unit'. *IEEE Transactions on Energy Conversion*; 2000, vol. 15(1), pp. 91–96

[31] Mahmud M. A., Pota H. R., Hossain M. J. 'Dynamic stability of three-phase grid-connected photovoltaic system using zero dynamic design approach'. *IEEE Journal on Photovoltaics*; 2011, vol. 12(4), pp. 564–571

[32] Yazdani A., Fazio A. R. D., Ghoddami H., Russo M., Kazerani M., Jatskevich J., Struz K., Leva S., J Martinez. A. 'Modelling guidelines and a benchmark for power system simulation studies of three phase single stage photovoltaics systems'. *IEEE Transaction on Power Delivery*; April 2011, vol. 26(2), pp. 1247–1264

[33] Tan Y. T., Kirchen D. S., Jenkins N. 'A model of PV generation suitable for stability analysis'. *IEEE Transaction on Energy Conversion.*; 2004, vol. 19, pp. 748–755

[34] Nosrat A., Pearce J. M. 'Dispatch strategy and model for hybrid photovoltaic and tri-generation power systems'. *Applied Energy*; 2011, vol. 88(9), pp. 3270–3276

[35] Yazdani A., Das P. P. 'A control methodology and characterization of dynamics for a photovoltaic (pv) system interfaced with a distribution network'. *IEEE Transaction on Power Delivery*; 2009, vol. 24(3), pp. 1538–1551

[36] Ceraolo M. 'New dynamic models of lead acid batteries'. *IEEE Transaction on Power Delivery*; 2000, vol. 15(4), pp. 1184–1190

[37] Barsali S., Ceraolo M. 'Dynamic models of lead acid batteries: implementation issues'. *IEEE Transaction on Energy Conversion*; 2002, vol. 17(1), pp. 16–23

[38] Islam F. R., Pota H. R., Mahmud M. A., Hossain M. J. 'Impact of PHEV loads on the dynamic performance of power system'. *20th Australasian Universities Power Engineering Conference (AUPEC)*; New Zealand, 2010, pp. 1–5

[39] Guille C., Gross G. 'A conceptual framework for the vehicle-to-grid (V2G) implementation'. *Energy Policy*; 2009, vol. 37(11), pp. 4379–4390

[40] Takagi M., Iwafune Y., Yamaji K., Yamamoto H., Okano K., Hiwatari R. 'Electricity pricing for PHEV bottom charge in daily load curve based on variation method'. *Innovative Smart Grid Technologies (ISGT)*; 2012, pp. 1–6

[41] Mets K., Verschueren T., Haerick W., Develder C., De Turck F. 'Optimizing smart energy control strategies for plug-in hybrid electric vehicle charging'. *Network Operations and Management Symposium Workshops (NOMS Wksps)*; 2010, pp. 293–299

[42] Chakraborty S. V., Shukla S. K., Thorp J. 'A detailed analysis of the effective-load-carrying-capacity behavior of plug-in electric vehicles in the power grid'. *Innovative Smart Grid Technologies (ISGT)*; 2012, pp. 1–8

[43] Koyanagi F., Uriu Y. 'A strategy of load leveling by charging and discharging time control of electric vehicles'. *IEEE Transactions on Power Systems*; 1998, vol. 13, pp. 1179–1184

[44] Kempton W., Kubo T. 'Electric-drive vehicles for peak power in Japan'. *Energy Policy*; 2000, vol. 28, pp. 9–18

[45] De Los Rios A., Goentzel J., Nordstrom K. E., Siegert C. W. 'Economic analysis of vehicle-to-grid (V2G)-enabled fleets participating in the regulation service market'. *Innovative Smart Grid Technologies (ISGT)*; 2012, pp. 1–8

[46] Peterson S. B., Whitacre J. F., Apt J. 'The economics of using plug-in hybrid electric vehicle battery packs for grid storage'. *Journal of Power Sources*; 2010, vol. 195, pp. 2377–2384

[47] Hossain M. J., Pota H. R., Mahmud M. A., Aldeen M. 'Robust control for power sharing in microgrids with low-inertia wind and PV generators'. *IEEE Transaction on Sustainable Energy*; May 2014

[48] Lo E. W. C. 'Review on the configurations of hybrid electric vehicles'. *3rd International Conference Power Electronics Systems and Applications*; 2009, pp. 1–4

[49] Bayindir K. Ç., Gözüküçük M. A., Teke A. 'A comprehensive overview of hybrid electric vehicle: powertrain configurations, powertrain control techniques and electronic control units'. *Energy Conversion and Management*; 2011, vol. 52, pp. 1305–1313

[50] HILTech Developments Limited (application notes) vehicle hybrid powertrain architectures; 2006

[51] Ehsani M., Gao Y., Gay E. S., Emadi A. *Modern Electric, Hybrid Electric, and Fuel Cell Vehicles*; 2009. CRC Press, Boca Raton, ISBN 0-8493-3154-4

[52] Soylu S. *Urban Transport and Hybrid Vehicles*; 2010. Sciyo, Janeza Trdine 9, 51000 Rijeka, Croatia, ISBN 978-953-307-100-8

[53] Livint G., Horga V., Ratoi M. 'Distributed control system for a hybrid electric vehicle implemented with CAN open protocol–Part I'. *Bulletin of the Polytechnic Institute of Iasi*; 2009. Tom LIV (LVIII), FASC. 4, ISSN 1223-8139, pp. 1019–1026

[54] Ali E. *Handbook of Automotive Power Electronics and Motor Drives*. CRC Press, Taylor & Francis Group, LLC, ISBN 0-8247-2361-9; 2005

[55] Livint G., Gaiginschi R., Horga V., Drosescu R., Chiriac G., Albu M., Ratoi M., Damian I., Petrescu M. *Vehicule Electrice Hibride*; 2006, Casa de Editura Venus, Iasi, Romania

[56] Husain I. *Electric and Hybrid Vehicles Design Fundamentals*. CRC Press, ISBN 0-8493-1466-6; 2003

[57] Brandl M., Gall H., Wenger M., Lorentz V., Giegerich M., Baronti F., Fantechi G., Fanucci L., Roncella R., Saletti R., Saponara S., Thaler A., Cifrain M., Prochazka, W. 'Batteries and battery management systems for electric vehicles'. *Design, Automation & Test in Europe Conference & Exhibition (DATE)*; 2012, pp. 971–976

[58] Baronti F. 'Design and verification of hardware building blocks for high-speed and fault-tolerant in-vehicle networks'. *IEEE Transaction of Industrial Electronics*; March 2011, vol. 58(3), pp. 792–801

[59] Wolfs P. 'An economic assessment of "second use" lithium-ion batteries for grid support'. *Proc. 20th Australasian Universities Power Engineering Conference (AUPEC)*; December 2010, pp. 1–6

[60] Manenti A. 'A new BMS architecture based on cell redundancy'. *IEEE Transaction of Industrial Electronics*; 2011, vol. 58(9), pp. 4314–4322

[61] Cao J., Schofield N., Emadi A. 'Battery balancing methods: a comprehensive review'. *Proc. IEEE Conf. Vehicle Power and Propulsion*; 2008, pp. 1–6

[62] Baronti F. 'Hierarchical platform for monitoring, managing and charge balancing of LiPo batteries'. *Proc. Vehicle Power and Propulsion Conference*; September 2011, pp. 1–6

[63] Piller S., Perrin M., Jossen A. 'Methods for state-of- charge determination and their applications'. *Journal of Power Sources*; 2001, vol. 96(1), pp. 113–120

[64] He H. 'State-of-charge estimation of lithium-ion battery using an adaptive extended Kalman filter based on an improved Thevenin model'. *IEEE Transaction on Vehicular Technology*; May 2011, vol. 60(4), pp. 1461–1469

[65] Codeca F., Savaresi S., Rizzoni G. 'On battery state of charge estimation: a new mixed algorithm'. *Proc. IEEE Int'l Conf. on Control Applications*; September 2008, pp. 102–107

[66] Baronti F. 'State-of-charge estimation enhancing of lithium batteries through a temperature-dependent cell model'. *Proc. IEEE Applied Electronics*; September 2011, pp. 1–5

[67] http://www.batterypoweronline.com/images/PDFs_articles_whitepaper_ appros/argus.pdf

[68] Bradley T. H. 'The effect of communication architecture on the availability, reliability, and economics of plug-in hybrid electric vehicle-to-grid ancillary services'. *Journal of Power Source*; 2009, vol. 195(5), pp. 1500–1509

[69] Hossain E., Han Z., Poor H. V. *Smart Grid Communications and Networking*. Cambridge University Press; 2012

[70] Gagnon R., Sybille G., Bernard S., Paré D., Casoria S., Larose C. 'Modeling and real-time simulation of a doubly-fed induction generator driven by a wind turbine'. *International Conference on Power Systems Transients (IPST)*; 2005

[71] Onar O. C., Kobayashi J., Erb, D. C., Khaligh, A. 'A bidirectional high-power quality grid interface with a novel bidirectional noninverted buck-boost converter for PHEVs'. *IEEE Transactions on Vehicular Technology*; June 2012, vol. 61(5), pp. 2018–2032

[72] Liu X., Wang P., Loh P. C. 'A hybrid AC/DC microgrid and its coordination control'. *IEEE Transactions on Smart Grid*; June 2011, vol. 2(2), pp. 278–286

Chapter 6

PEV Charging Technologies and V2G on Distributed Systems and Utility Interfaces

Domagoj Leskarac, Chirag Panchal*, Sascha Stegen* and Junwei Lu**

Abstract

The rapid expansion of plug-in electric vehicles (PEVs) has led to the requirement for advanced charging and power delivery techniques. Smart transformers, AC and DC fast charging techniques have the ability to connect a vehicle to the distribution network at very high power levels. If this connection is not properly utilised, a detrimental impact on the distribution network could occur. By utilising the vehicle-to-grid technique alongside smart scheduling for charging, PEVs can support the distribution network. Improving the compatibility between different types of electric vehicles and the distribution grid can be achieved by diversifying charging methods to a multi-platform charging solution. Advanced wireless charging methods can allow PEVs to travel long distances without stopping or requiring a charging station. This chapter will discuss different types of charging technologies and the utilisation of vehicle-to-grid support on a distributed network.

6.1 Introduction

6.1.1 Overview

As petroleum-based fossil fuels become more expensive to mine and the issues of climate change become more predominant, change in international regulations, such as the Kyoto Protocol (1997) and local government legislation, has forced traditional internal combustion engines (ICE) to subsequently become more expensive to keep running [1, 2]. Most European countries and Australia agreed to binding targets to reduce carbon emissions and adjust energy consumption habits. To create a sustainable transportation system, vehicles must substantially reduce the fuel consumption and decrease emissions. This can be achieved by utilising electrified powertrains which can be combined with ICEs to create hybrid electric

*Griffith University, Queensland, Australia

vehicles (HEV), which meet the KYOTO and national set efficiency and emissions targets [3]. By integrating substantial storage into a HEV, and a connection to the mains power grid, the capability of travelling greater distances without the need to run the ICE increases. This in turn creates the plug-in hybrid vehicles (PHEV) sector. PHEV have the potential to not only reduce fuel consumption and emissions, but also reduce the dependence of oil [4]. By further increasing the energy storage component it is possible to completely remove the ICE to create a purely battery electric vehicle (BEV), or electric vehicle (EV). The differences between the topologies are visualised in Table 6.1.

ICE vehicles and EV-based vehicles are perceived to be two separate achievements in history, but historically, the ICE vehicle and EV were developed alongside one another (Table 6.2). The success of the early EVs, such as the 1888 Flocken Elektrowagen, was hindered by the lack of high-density energy storage solution. By the turn of the century, inventors, such as Ferdinand Porsche, began combining the electric motor developments with the ICEs to produce the first hybrid drive vehicles. Even with the assistance of an ICE, these types of vehicles were extremely heavy, complicated and expensive. This in turn allowed the ICE-based vehicles popularity to dominate the market. By the early 1900s, Ford's Model T was the most common vehicle, second only to the horse-drawn carriage. This hold on the market held its course until the introduction of the first mass-produced hybrid vehicle, the Toyota Prius. The Prius had many advantages over its predecessors: lightweight construction materials, microprocessor control, safety features and most importantly improved battery technology. The Prius was in many sights taking a turn away from purely ICE-powered vehicles. Elon Musk, the CEO of Tesla, took further advances in battery storage technology (Lithium-based batteries) to develop the first high-performance pure EV – the Tesla Roadster. This vehicle was capable of reaching and exceeding equivalent ICE-powered vehicles of its time. Michael Mauer, continuing on with Ferdinand Porsche's original research, developed one of the most advanced hybrid drive vehicles in the world. The Porsche Spyder 918 captures Ferdinand's concept of a hybrid vehicle and transforms it into a performance machine of the 21st century. Vehicles with energy storage and the ability to connect to the electrical grid; such as PHEVs and EVs, will be commonly referred to as plug-in electric vehicles (PEVs) in this chapter.

PEV are projected to reach 2% of the global market by 2020 [5], and potentially reach 25% of the U.S. (California) new car sales market by 2020 [6]. This increase in PEV can only be possible with the reduction of energy storage cost, with some estimations placing the cost for energy storage down to \$300 kWh in 2020 from approximately \$600 kWh in 2014 [5]. This, however, will only increase the potential for more PEV growth. Such PEV penetration will result in a growing requirement to provide sufficient charging stations in diversified locations to keep the vehicles adequately charged (grid-to-vehicle (G2V)) (Figure 6.1(a)). Depending on location, PEV may resemble scattered dynamic loads in cases such as residential premises, or large dynamic loads in areas such as parking lots. Large dynamic loads can cause issues to the distribution network if charging is not managed at appropriate rates even with relatively low levels of penetration, whereas scattered dynamic loads will require a larger penetration to cause the same level of disturbance to a network.

Table 6.1 Topology comparison between different vehicle drivetrains

Vehicle type	Illustration	Description
ICE	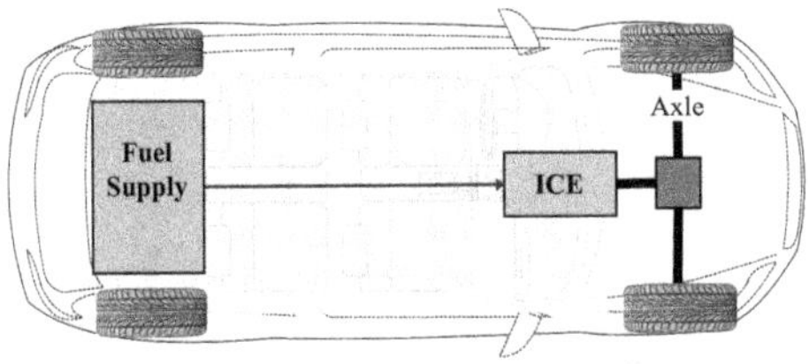	• Single drive through ICE • Liquid hydrocarbon fuel supply
Series-hybrid	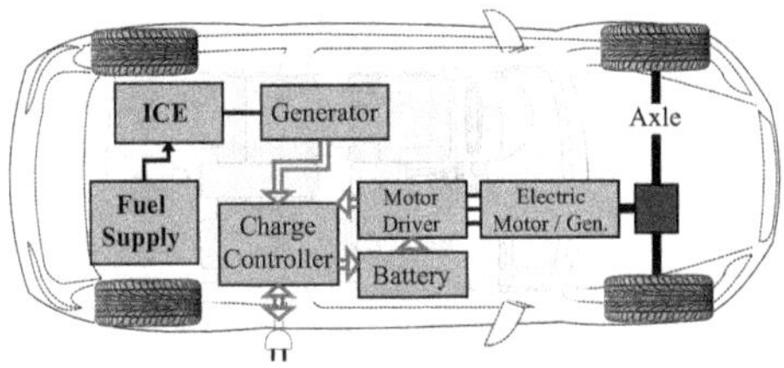	• Electric drive through an electric motor • Battery energy supply with grid tie options. Liquid hydrocarbon fuel and generator to extend range
Parallel-hybrid	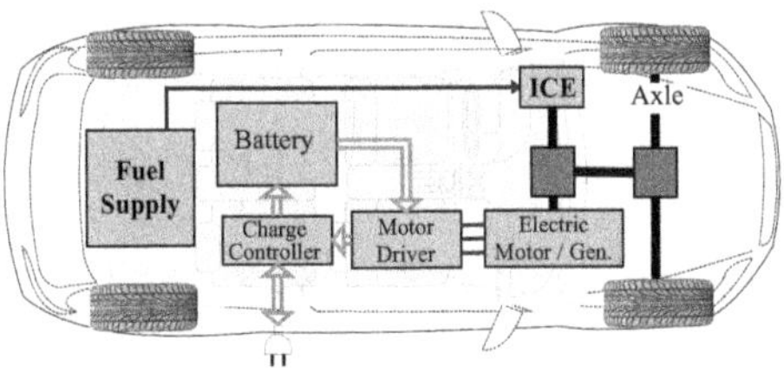	• Dual drive options through electric motors and ICE • Battery energy supply with grid tie options. Liquid hydrocarbon fuel for higher velocity driving
Series-parallel hybrid	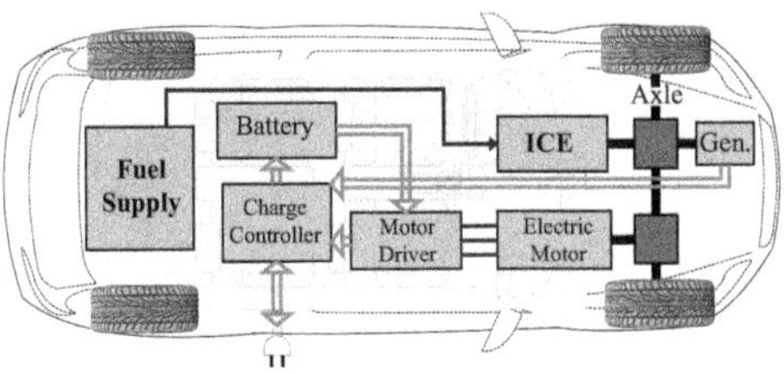	• Dual drive options through electric motors and ICE • Battery energy supply with grid tie options. Liquid hydrocarbon fuel for higher velocity driving
Battery electric vehicle	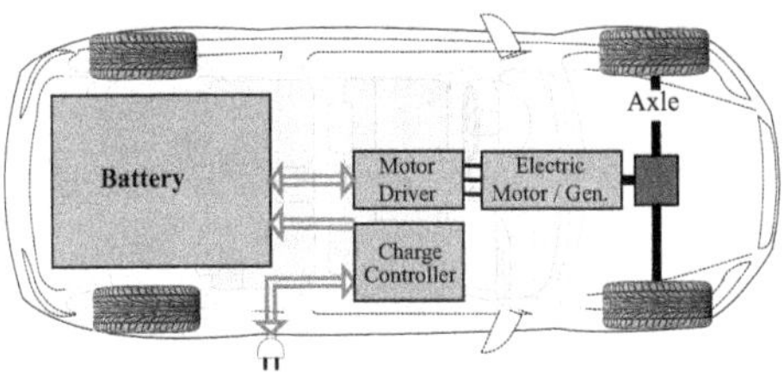	• Electric drive through an electric motor • Large battery energy supply with grid tie options

Table 6.2 Brief history of automobile technologies

Year	Inventor	Vehicle name	Vehicle drive type/fuel type	Vehicle specifications	Advantages	Drawbacks
1801	Richard Trevithick (9)	Steam Carriage (Puffing Devil)	Steam drive/wood, peat or coal	• Mass: 1520 kg • Top speed: 14.5 km/h • Operating pressure: 145 psi	First successful means of mechanical motion not bound by tracks	Slow, limited range, required large amounts of wood to burn, high-pressure tanks, risk of fire
1829	Sir Goldsworthy Gurney (10)	The Goldsworthy Gurney Steam Carriage	Steam drive/ coal oil	• Top speed: 32 km/h	Prolonged means of transportation, increased payload, fuel storage size greatly reduced	Slow, large pressurised tank, heavy
1886	Karl Benz	Benz Patent-Motorwagen	ICE/petroleum or kerosene	• Engine: 854 cc single cylinder – 1.5 kW • Top speed: 16 km/h	First modern vehicle with a four-stroke ICE. Maiden voyage ran just shy of 200 km	Maximum of two passengers, slow moving (16 km/h max)
1888	Andreas Flocken (11)	Flocken Elektrowagen	Electric motor/lead acid (Pb) battery	• Mass: 450 kg • Engine: 0.7 kW • Battery storage: approx. 270 Wh • Range: Approx. 30 km	First successful vehicle driven by electricity	Low mileage, considerable charging times, limited payloads, limited speed
1900	Ferdinand Porsche	Lohner-Porsche Mixte Hybrid	Series-hybrid/Pb battery and petroleum	• Mass: 1500 kg • Battery storage: 21.6 kWh • Motors: four electric motors 1.9–5.2 kW each • Generator: Two 2.6 kW ICE • Cost: 15,000 Austrian Crowns	First vehicle to feature both electric motors and an ICE. Seats two of four persons	Heavy and slow (16 km/h), extremely expensive for the time

1909	Hennery Ford's Motor Company	Model T	ICE/petroleum, kerosene or ethanol	• Mass: 540 kg • Motor: 2.9 L inline – 4–15 kW • Top speed: 70 km/h • Fuel economy: 11 L–18 L/100 km • Cost: $825 USD (approx. $21,710 USD in 2014)	First mass produced low-cost vehicle, with a top speed of 72 km/h	Poor safety features, relatively poor fuel consumption
1997	Toyota	Toyota Prius – NHW10	Series-parallel hybrid/nickel-metal hydride (NiMH) and petroleum	• Mass: 1250 kg • Battery: 1778 Wh • Electric motor: 30 kW, 305 Nm • ICE: 1.5 L inline-4 – 43 kW, 102 Nm • Fuel economy: 5.6 L/100 km • Cost: $17,000–$20,000 USD	First mass produced low-cost HEV. Electric drives for low speeds and take off then ICE engagement for high speeds. Regenerative breaking is achieved with the electric motor	Poor performance, boot space sacrificed for batteries. Sold at a loss for the first years to kick start the market
1999	Honda	Insight	Parallel hybrid/NiMH battery and petroleum	• Mass: 827 kg • Battery: 580 Wh • Electric motor: 10 kW, 49 Nm • ICE: 1 L inline-3 – 50 kW, 92 Nm • Fuel economy: 3.1 L/100 km • Cost: 50,000 AUD	Two motors work together to assist performance and reduce fuel consumption. Regenerative breaking is achieved with the electric motor. Excellent fuel efficiency	Battery packs required replacement frequently, high initial cost
2006	Elon Musk	Tesla Roadster	Electric motor/lithium-ion and distribution grid	• Mass: 1235 kg • Battery: 53 kWh • Electric motor: 215 kW, 400 Nm • Range: 400 km • Cost: 109,000 USD	First pure electric sports car. Could achieve 320 km on a single charge while having an acceleration of 4 seconds to 100 kph	High initial cost

(*Continues*)

Table 6.2 (Continued)

Year	Inventor	Vehicle name	Vehicle drive type/fuel type	Vehicle specifications	Advantages	Drawbacks
2013	Michael Mauer	Porsche 918 Spyder	Plug-in – parallel hybrid, series-parallel hybrid, electric vehicle/ lithium-ion, distribution grid and petroleum	• Mass: 1700 kg • Battery: 6.8 kWh • Engine: 4.6L V8 – 453 kW, 528 Nm • Rear motor: 208 kW • Front motor: 115 kW • Combined torque: 1274 Nm • EV range: 31 km • Fuel economy: 3.5 L/100 km • Cost: Approx. $950,000 USD	Utilises multiple types of hybrid technologies to generate hyper-car performance and hybrid efficiencies. Is capable of running purely on EV mode	Extremely high cost, heavy

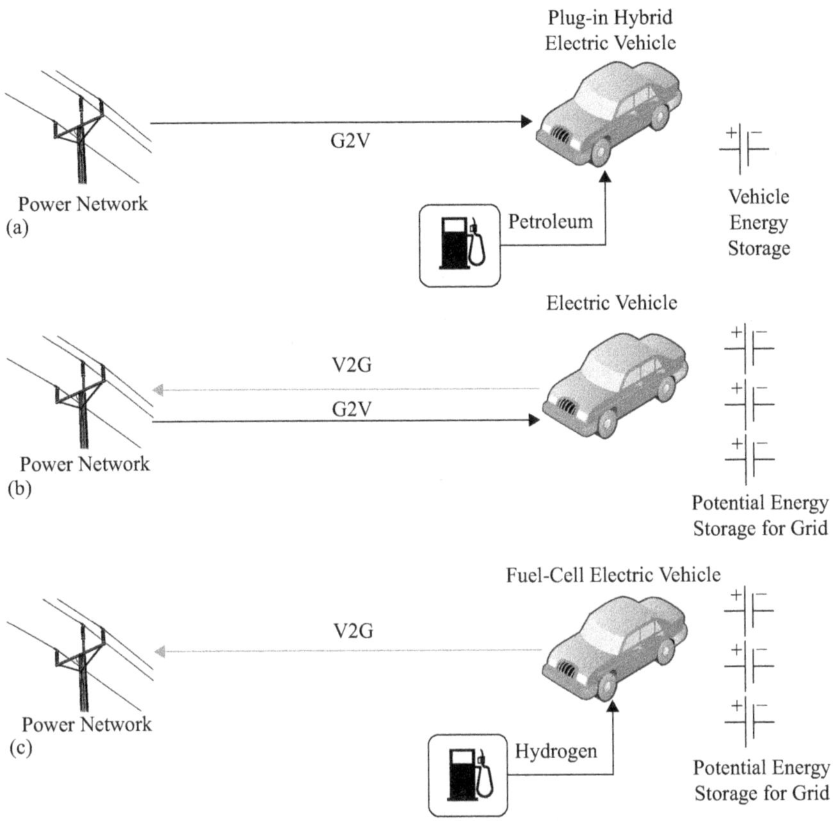

Figure 6.1 (a) G2V with PHEV, (b) G2V and V2G with EV, (c) V2G with FCEV

Methods available to kerb this issue involve implementing smart charging techniques via means of an aggregator [7, 8]. Such an aggregator will control the charging characteristics of all compatible PEVs connected on the same network.

6.1.2 Vehicle-to-Grid Concept and PEV Communication Requirements

Having the ability for PEV to charge and discharge on a distribution network may prove to be extremely beneficial. This concept is referred to as vehicle-to-grid (V2G) connectivity and can be seen in Figure 6.1(b) and (c). The capability of providing energy back to a grid is extremely beneficial for distribution networks especially those with nearby renewable energy production [12–15]. With large populations adopting air-conditioning systems in developed countries, summer peak loads may exceed the rated capacity of the distribution network. In cases such as these, energy distributors generally switch on additional generators, commonly diesel or gas, to supplement the load. These generators are generally quite expensive to operate and are not generally located at the source of energy consumption. With a controlled V2G network, these generators will have a lesser role in supplying power to the grid during peak times. The PEVs connected at the time could be used to supplement the load over the peak period of energy consumption at the load. This in essence will reduce the overall loss of energy transmission, which increases with energy consumption and distance to and from the generator and load. A V2G network is not only beneficial to a grid during the several peak days of the year, but can be beneficial throughout a working day by flattening consumption curves and shifting daily peaks. This could in turn reduce energy prices by reducing uncertainty within a network.

There are a number of different types of EVs which are not only driven by batteries, but rather alternative energy storage means, such as fuel-cell electric vehicle (FCEV) [16]. These vehicle technologies do not utilise batteries as their primary energy storage system, but are capable of releasing electricity at will, resulting in potential for V2G connectivity. The internal topology of an FCEV will resemble that of a series-hybrid, but without the ICE and generator, rather a fuel cell stack. Even though an FCEV is mentioned in this chapter, other types of hydrogen-powered vehicles could have topologies similar to those described in Table 6.1. In the case of the FCEV shown in Figure 6.1(c), the FCEV may not have a large physical internal battery, but if a connection to the grid was possible, energy can be produced by the fuel cell and exported to the network. This will in turn simulate a battery pack comparable to that of a pure EV. These vehicles can be utilised purely as EVs with V2G connectivity for grid support. Benefits can be seen not only to the distribution network in terms of grid support, stabilisation and load shifting, but also to the supplier of the stored energy. Figure 6.1(b) illustrates the case where a PEV is capable of bidirectional power flow. Bidirectional power flow will allow a PEV to charge and discharge onto a network as required. Charging from a network may be as a result of topping up the charge capacity, or in response to the production of renewable energy systems (RES). The combination of both G2V and V2G capabilities will prove to be extremely useful once large-scale RES and PEVs become a common occurrence.

As distributors will no longer require switching on additional high-cost generators, distributed or local energy storage from PEV has the potential to return a profit. The lure of a profit may seem beneficial to some users, but improper management of V2G functionality may present some risks to the internal storage of the vehicle. Charge control management is a necessity for all V2G-enabled PEVs, with set rate and depth of discharge limits for V2G functionality. These precautions must be taken to prolong the life expectancy of the PEV batteries. This brings communications into the spotlight. Communications between the grid and a PEV is crucial in ensuring the V2G concept, and PEV as a whole is successful. A basic level of communication can improve the network substantially enough to reduce the impact of large clusters of PEV in a localised area and improve grid stability simultaneously. To minimise the effects and potentially prevent an overload situation on a distribution network, PEV charging must be monitored and controlled [17]. Such monitoring and control systems should include an intelligent two-way communications system between the on-board energy storage and energy distributor. This type of communications system will enable the network to cope with multiple PEVs by manipulating charge and discharge characteristics of PEV along with only allowing charging during specific off-peak times of the day [18]. Expanding on the intelligent control mechanisms of charging a G2V, a V2G implementation could offset peak demand loads.

6.1.3 Distributed Generation and the Smart Grid

Figure 6.2 represents typical power characteristics of commercial office building combined with a characteristic curve of a PV system. Figure 6.2(a) describes the typical characteristics of a solar installation with intermittent cloud cover, whereas Figure 6.2(b) shows cloudless production characteristics. As can be seen, there is large variability in the production of energy with intermittent cloud cover and a predictable curve during a cloudless day, but with both production characteristic curves exceeding the building supply at peak PV production. The building load is constantly varying around a characteristic path, determined by the time of day, with the most dramatic changes occurring during early morning and late afternoon. In the case of this building, the characteristics can be correlated with the building's automated air-conditioning and lighting systems switching on and off. By connecting PEV onto this network, either a destructive or a constructive characteristic can be obtained. A destructive PEV would simply increase the maximum real power consumption, whereas a constructive PEV has the capability of reducing peak demand. This is achieved by peak load shifting, and in the case of Figure 6.2, excess PV absorption.

A small-scale PEV network with limited depth of discharge for V2G purposes (Figure 6.2(a)) has the potential to reduce and essentially flatten the peak daily load

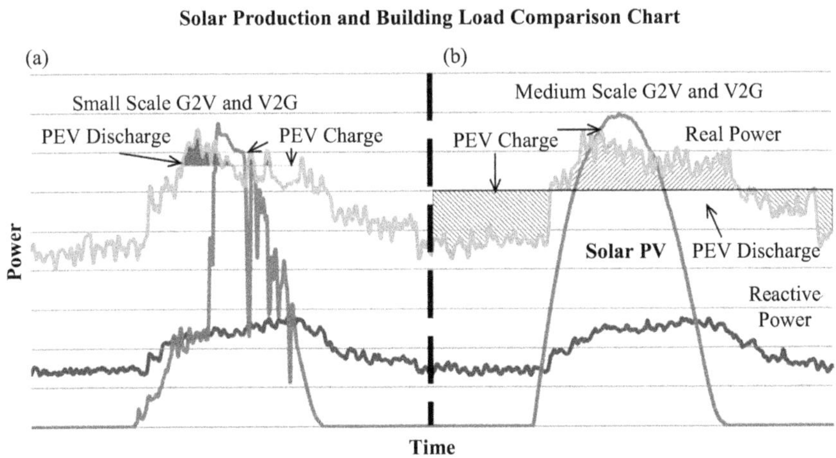

Figure 6.2 Overlay of office building load power characteristics alongside PV production and PEV characteristics. (a) Intermittent clouds on the PV production with a small-scale G2V and V2G performance. (b) Cloudless day PV production with a medium-scale G2V and V2G performance

of the building [19]. Increasing the size of the available energy, either through an increased number of PEV or through an increase in the depth of discharge allowable for V2G purposes (Figure 6.2(b)), greatly increases the peak load shifting effect. With large PEV penetration, the daily load cycle can be represented by a constant load. This would benefit both the energy distributors in providing a static load and the consumers with the reduction in energy prices.

This concept will allow a PEV's on-board energy storage to be utilised not only as a consumer of energy, but also as a supplement to the grid, commercial building or a household during times of peak energy demand. In turn, this would stabilise the energy consumption characteristics of a grid, as noted in Figure 6.2. This supplement will ease the pressure from distributed generators and potentially help rectify some of the issues caused by RES. Optimising such a system and providing a method of communication between the stored energy and grid is a major necessity to construct an efficient V2G system. To maintain a working life of PEV batteries, limits should be in place. These limits would ensure that battery charge is maintained at a level where the vehicle still has a reasonable range. By limiting the discharge on the batteries, their life cycle can also be prolonged.

The formation of a microgrid or a smart grid is a necessity to contain the projected growth of PEV penetration on a distribution grid. A microgrid is defined as a distributed resource island system by IEEE standard 1547.4 [20]. By this definition, a microgrid is capable of self-sufficiently producing enough energy to sustain a localised distribution network as an independent island. A smart grid is as of yet undefined uniformly by major organisations, but the different definitions have several common elements [21–25]. Therefore, a smart grid can be defined by including characteristics described in Table 6.3.

A microgrid has the main, centralised power plants supported by smaller power plants, which are decentralised across the distribution network. In some

Table 6.3 The smart grid characteristics

Sensors	Has the ability to measure load and generator characteristics
Digital communications	Enables a network to be observed and relay information regarding grid integrity
Control	Has the ability to act upon varying load and generator capabilities
Automated	Has the ability to self-regulate autonomously and dynamically
Integrated generation	Accommodates and facilitates RES, PEVs and energy storage systems
Robust and efficient	Provides multiple sources of generation to loads, following the most efficient path
Reliability, security and quality	Provides clean and consistent power free of harmonics at unity power factor (PF)
Marketable network	Generates a sustainable network which can adapt in the long term

cases, these power plants are installed by the energy distributors to maintain energy supply and stability. In other cases, consumers install small power plants to power small sections of load in the case of a power outage [26]. In the case of distributed generation or energy storage, energy can be delivered at the point of use, therefore increasing overall efficiencies on functioning transmission networks [27]. Distributed generation is a major component in a microgrid and in smart grids, where energy can be produced independently to that from the major power plants. Not only can distributed generation provide power to a local network during a disaster, but improvements can be made to the entire distribution network in terms of stability and reliability [26].

With the current trend of implementing renewable energy resources (RES) onto buildings and in large-scale environments, the multiple generation points push a traditional network into a situation where it can no longer maintain stable power delivery [1]. The concept of a smart grid revolves around an advanced and controlled method of energy delivery, production and consumption, where each generator, load and distributor has adjustable characteristics depending on each other's requirements [28]. Depending on location, existing infrastructure and financial incentives, different levels of a smart grid can be created. A smart grid can be achieved by utilising a power quality stabiliser, static synchronous compensator (STATCOM), communications systems, distributed generation and energy storage in conjunction with a smart controller to manage energy flow. Several of these systems can be combined into single units that can be utilised as off-board DC charging systems for PEVs [2, 29]. These systems have the potential to create distributed-STATCOMs (D-STATCOMs) that ensure power factor correction (PFC) for large variable reactive loads, which has been proven to be useful in maintaining unity PF and improving power stability [30]. Applying this across a distribution network has the potential to create the backbone infrastructure of a sophisticated smart grid.

The use of high-frequency (HF) pulse width modulation (PWM) in modern day appliances and power supplies not only has caused a similar effect with regard to PF, but also includes a large issue with HF harmonics or total harmonic distortion (THD) [31]. This can be attributed to the high operating frequencies and non-linear current consumption of PWM converters. THD on a network can cause abrupt pulses and noise, causing sensitive electronic equipment to malfunction or be permanently damaged [32]. By integrating a D-STATCOM capable device onto a network, these issues with PF and THD can be rectified, reducing overall power consumption and increasing the life expectancy of appliances.

6.1.4 Charging Diversity and Utility Interfaces

Connecting PEVs to a utility grid can occur with two electrical energy transfer methods, contact and contactless [33–35], both with optimum use case scenarios. Contact charging can provide a quick and simple solution to ensuring the PEV is adequately charged while stationary. On average, a commuter vehicle will remain parked for approximately 20–22 hours in a day [7]. During these times, the energy storage capacity of an PEV can be charged through means of RES such as PV. Charging an PEV during this time could offset the energy fluctuations from solar modules due to intermittent clouds, if solar production data could be obtained. Three locations have been identified where the predominant charging of PEVs will occur, including, but not limited to, areas such as homes, streets and parking lots. Upon newly released standards, each area will have different levels of charging capabilities, dependant on existing local infrastructure and charging requirements. The electrical infrastructure around these charging locations will be capable of supporting a relatively small PEV penetration. To support the projected PEV penetration by 2020, one estimation predicts that approximately 20 billion dollars will be spent globally to upgrade the electrical infrastructure [36]. The wireless charging method could also be utilised while the vehicle is in motion or stationary [37]. All charging stations must conform to newly released standards governing plugs, communication, voltages and level of charging capacity. Managing PEV charging characteristics as individual entities and as a part of large systems will be a crucial task to ensure grid stability. This topic will be discussed in detail in Section 6.6.

6.1.4.1 Communication between PEV and Utility

Wireless communication between an PEV and a utility network is a must. Communication between an PEV and a consumer is also required to keep the consumer up to date with his/her vehicle's current state of charge. This will become more important when the consumer enables the V2G functionality, to ensure there is enough energy storage within the vehicle for the upcoming trip from the user. With the current trend of wireless payment methods becoming increasingly popular, payment of energy used by the PEV can be achieved in the same or similar manner [16, 38].

6.1.5 Local, Central and Distributed Generation

There are two main energy storage configurations, central storage and local storage, each with their positives and negatives. Figure 6.3 illustrates two scenarios on the same medium voltage (MV) distribution network, depicting central storage on the left and local storage on the right. In this network, an assumption of eight residential dwellings with a total of three units of PEV energy storage is taken. Assuming that each low voltage (LV) distribution network has a normalised energy distribution, centralised storage has several benefits that make it a viable option as a supportive supply system for the LV network. Its benefits range from a lesser cost for infrastructure development required to support the energy storage to an ease of maintenance, as all storage devices are in the same location. PEVs may help increase the available energy storage, creating temporary centralised (a large PEV car park) and localised (home charging station) environment.

Energy storage on utility grids, specifically centralised storage shown in Figure 6.3, has been in use since the 1990s to help manage power demand [14]. The energy storage used is dependent on the location (suburban or semi-rural) of installation and load requirements. If a large area is available, alternative energy storage techniques can be utilised. Installations of this nature occur predominantly to deal with summer peak loads on networks where station transformers are not capable of supplying the required peak power for prolonged periods of time. Depending on the scale of storage implemented and location, power outage durations could be reduced or not be rectified entirely. As the energy storage will

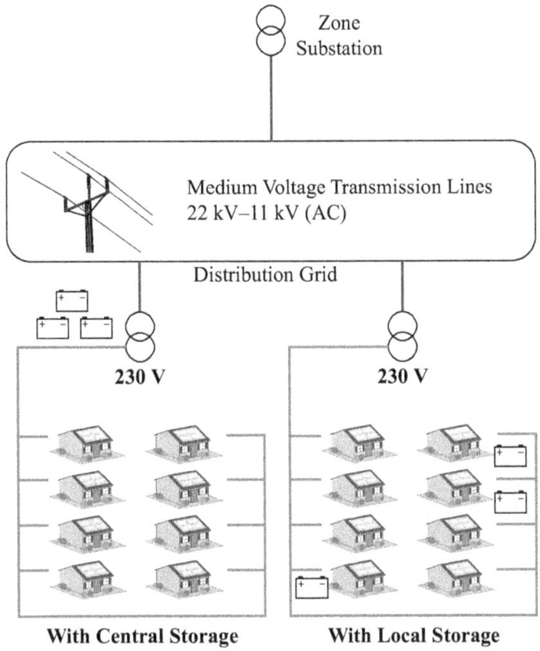

Figure 6.3 Energy storage configurations – left: centralised storage; right: localised storage

be located away from the end users, the transmission losses need to be considered when designing the system. The transmission lines which feed the loads also need to be considered. If these lines cannot carry extra power to supplement peak loads, infrastructure upgrades may be required.

Local storage can be defined as a single entity in a network that contains storage, which is shown in Figure 6.3. Like a central storage solution, many benefits to a grid can be achieved by local storage. As the storage is concentrated at the point of consumption, efficiency gains in transmission can clearly be achieved.

With RES installations such as solar PV predominantly being installed on residential premises, energy storage of the excess production can be directly fed into the local storage solution [39]. The stored energy can then be sold to neighbours during peak load times, increasing the cost benefit to households with energy storage solutions. This power sharing technique will almost always be more efficient than increasing power production from the non-renewable power plants down the transmission network. The use of PEVs as energy storage devices can become problematic when state of charge (SOC) requirements comes into play. The SOC of an PEV should not dip below the average commuter travelling distance or to a point where the battery pack degrades. This may result in a useful battery size of only 10–20% from maximum charge. By implementing energy storage with power flow management systems in localised or centralised configurations, peak loads, off-peak RES excess power production, transmission reliability and power outage issues can be mitigated [12, 14]. This can cause an intestinal islanding effect, where some loads are not affected by a local or feeder fault [28]. Smart control of PEV energy storage combined with RES can create the beginnings of an integrated smart grid network.

6.2 Current PEV Charging Standards

Standards refer to influential groups or bodies which govern PEV charging technologies and ratings. Table 6.4 describes current legislation which governs PEV charging capacities, plug types, communication protocols, voltages ratings and safety requirements [40–44].

Table 6.4 Organisations and relevant standards associating to PEV

Organisation	Relevant standard(s)	Standard definition
Society for Automobile Engineers (SAE)	J1772	Conductive connector
	J1773	Inductively coupled charging
	J2847, J2836, J2931	Communications
	J2864	Power quality
	J2293	Energy transfer systems
	J2344, J1766, J2578	Safety considerations
Institute of Electrical and Electronic Engineers (IEEE)	1547	Grid tied connections
	P1809	Electric transport infrastructure
	P2100	Wireless power and charging systems
	P2030	The smart grid technologies
National Electric Code (NEC)	625	EV charging systems and safety
	626	EV parking space characteristics
National Fire Protection Association (NFPA)	70	National electrical code
	70B	Electrical equipment maintenance
	70E	Electrical hazards
Underwriters Laboratories Inc. (UL)	2231	EV supply circuits safety
	2251	Plugs, receptacles and couplers for EV
	2202	EV charging system equipment
	2594	EV supply equipment
Deutsches Institut fuer Normung (DIN)	43538	Battery system specifications
	EN 50620	EV cable requirements
	VDE 0510–11	Safety requirements for batteries
International Electromechanical Commission (IEC)	TC 21	Cells and batteries
	TC 64	Electrical installations and protection
	TC 69	EV and electric industrial trucks
	TC 22/SC 3	Electrical and electronic equipment
	61851–2–3	EVs conductive charging systems
Japan Electric Vehicle Association (JEVS)	C601	Plugs and receptacles for EV charging
	D001–002	Battery characteristics
	D701–709	Test procedure of batteries
	G101–105	Quick charging systems
	G106–109	Inductive charging system

6.2.1 Socket Types

There are still several sockets in use [42] to charge on-board energy storage. These sockets in some instances differ due to the different types of power being delivered, but more often than not are referenced to the geographical location. For example, the Australian Standard for PEV sockets and plugs relates the socket to the amount of power draw. The maximum parameters are in reference for voltages not exceeding 500 V AC and a rated current not exceeding 63 A for three-phase connections. For single-phase connections a maximum of 70 A can be utilised [42]. As the diversity of standard single-phase voltages varies between 100 V and 240 V between 50 Hz and 60 Hz (Table 6.5), differences between connectors must be realised. The most common and current standards around the world regarding PEV plugs can be visualised in Table 6.6. Only three-phase and DC plugs are shown in Table 6.6, as the single-phase connectors are simply that of the country of origin. DC charging can benefit over AC power delivery as voltages and currents are not dependent on geographic locations. This in turn creates a stable voltage platform to where all PEV can normalise. The common methodology between all the plug types is power delivery and communications between the vehicle and the charger or grid connection. The communication interface can be noted by two smaller contacts on either side of the main power contacts. The communication interface allows for enabling charging currents, state-of-charge identification and fault detection [43].

The limiting factor for power delivery will either be the grid connection or the plug interface. Expanding to include two types of energy transfer, AC and DC currents, will allow for much greater power and energy transfer. A combined AC and DC plug (Table 6.6) shows a combined DC/AC charging option, which can utilise two energy delivery methods to increase power delivery to the PEV.

Table 6.5 Current global voltage ratings and frequencies

Continent	Voltage range	Frequency	Exceptions
North America	100–130 V	60 Hz	
	220–240 V	50 Hz	Greenland
South America	220–240 V	50 Hz	
	220–240 V	60 Hz	
	100–130 V	60 Hz	
Africa	220–240 V	50 Hz	
	100–130 V	60 Hz	Liberia
Europe	220–240 V	50 Hz	
Asia	220–240 V	50 Hz	
	220–240 V	60 Hz	
	100–130 V	60 Hz	
	100–130 V	50 Hz	East Japan
Australia/Oceania	220–240 V	50 Hz	

Table 6.6 International socket types for PEV charging systems

Power type	Type 1	Type 2	GB/T
AC (3 phase)	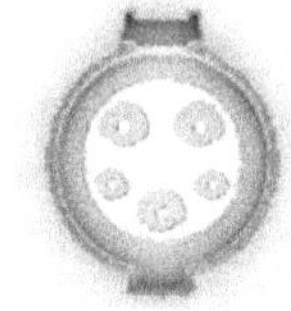 SAE J1772/IEC 62196-2	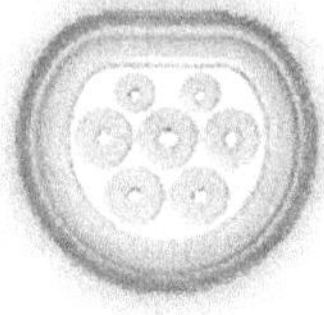 IEC 6219-2	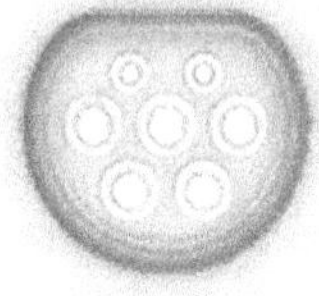 GB/T Part 2
DC	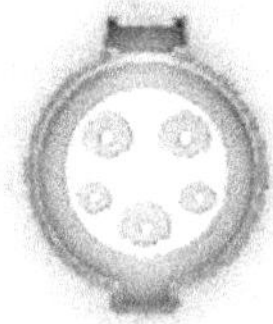 IEC 62196-3	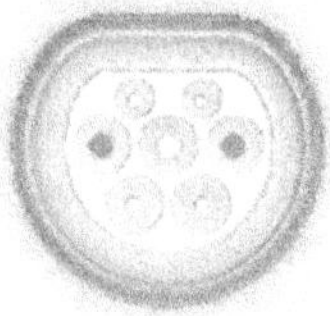 IEC 6219-3	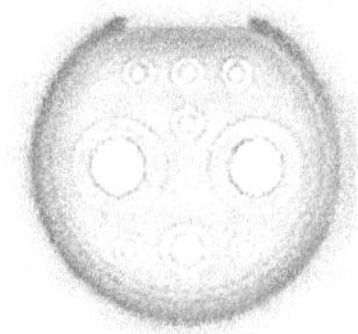 GB/T Part 2/IEC 6219-3
Combined AC and DC	 IEC 6219-3	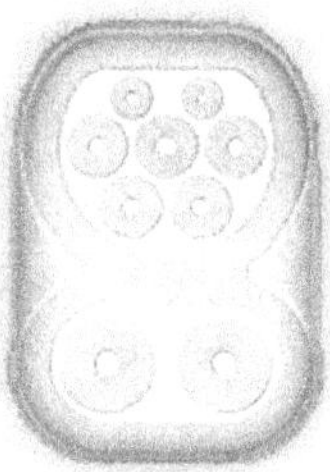 IEC 6219-3	

6.3 Contact-Based PEV Charging

Two distinct methods are used to charge PEV: the direct contact method and the wireless method. The direct contact charging method includes two subset power types of AC and DC. AC charging varies between single-phase and three-phase networks, while the DC charging type relies on a single charging voltage. Both AC and DC charging methods utilise a voltage control method for charging PEV, where the voltage source is maintained and the current supply varied [45]. Focusing on contract charging brings upon several different charging topologies. These include AC methods, DC methods, and a combination of the two along with a wireless interface. Each method has varying charging characteristics which depend on location of use and power requirements [45]. As PEV's energy storage is typically in the form of a battery, a DC charging supply is required. The most common outlet is in the form of AC, which results in the need for energy conversion equipment, internal or external to that of the PEV.

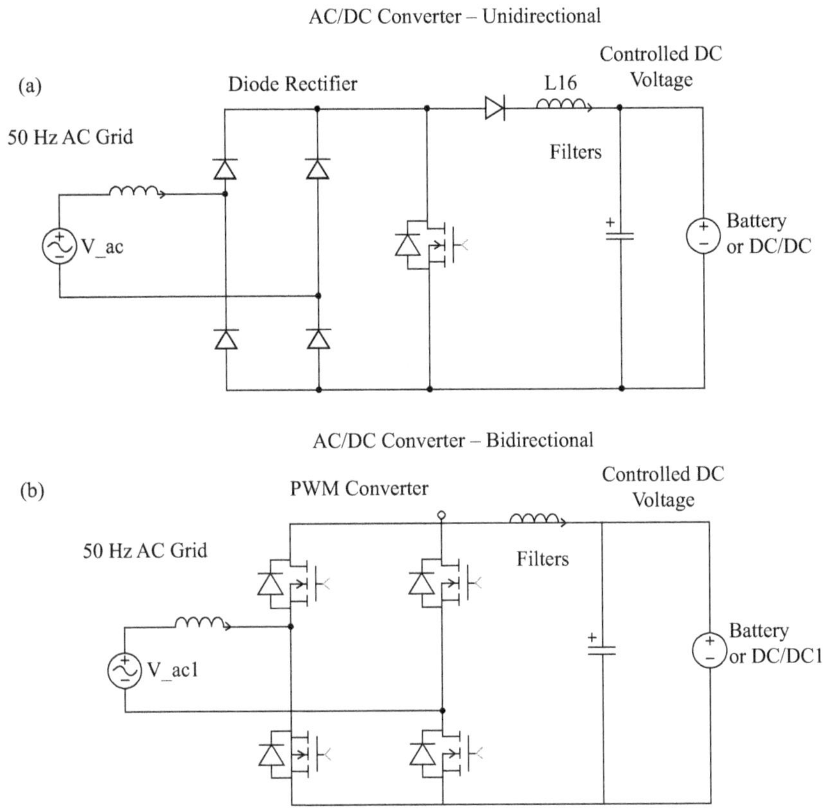

Figure 6.4 Simple rectifiers utilised for G2V: (a) unidirectional; (b) bidirectional

6.3.1 Rectifier Topologies for G2V

The simplest form of a charging mechanism for PEV's incorporates the unidirectional AC/DC converter; Figure 6.4(a). In this converter topology, four diodes and filters are utilised to rectify the AC waveform with a single MOSFET switch to control frequency. To be able to create a bidirectional converter, multiple MOSFET switches must be used such as in Figure 6.4(b). In this configuration, the removal of the diodes allows current to flow to and from the battery supply. This configuration allows for a single stage conversion process assuming the battery voltage and grid supply have an appropriate or similar voltage rating.

6.3.2 Inverter Topologies for V2G

Typical inverters utilise the half-bridge converter, which is a combination of a buck and a boost converter. One key feature of a half bridge converter is the ability to allow power to flow in both directions – from source to load and from load to source. This circuit has the capability of changing current flow from positive through zero to negative, while always maintaining a positive input voltage.

By connecting this circuit to a DC motor, single-quadrant controls can be obtained. By varying the duty cycle of the switches, speed control of a motor can be obtained [46]. In the half-bridge, the AC voltage which would be produced will be in the order of a peak-to-peak equalling (ideal case) of the input voltage.

Two major methods are used to convert DC to grid stable AC – the transformer-based inverter and the transformerless inverter. Both systems require the use of switching to convert the DC input to AC. Depending on the switch and switching mechanism, different types of wave outputs could be achieved. The simplest output wave is the square wave with sharp edges, which can be achieved by utilising the topology outlined in Figure 6.5(a). A square wave output is not desirable when used in transformers or applied back onto the grid. The sharp edges introduce harmonics onto the underlying sine wave, thus distorting it relative to the amount of power supplied. Harmonics in a system can cause an efficiency reduction and distort the carrier waveform. By introducing multiple MOSFET type switches, as in Figure 6.5(b), a *modified sine-wave* or *quasi sine-wave* can be created and therefore

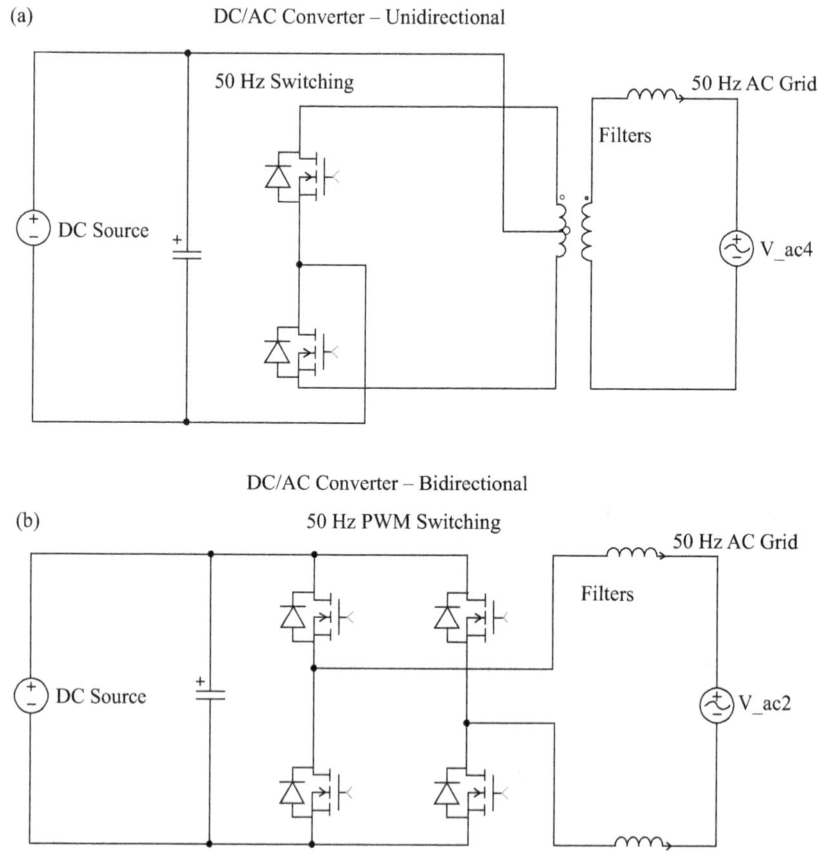

Figure 6.5 Simple inverters utilised for V2G: (a) unidirectional; (b) bidirectional

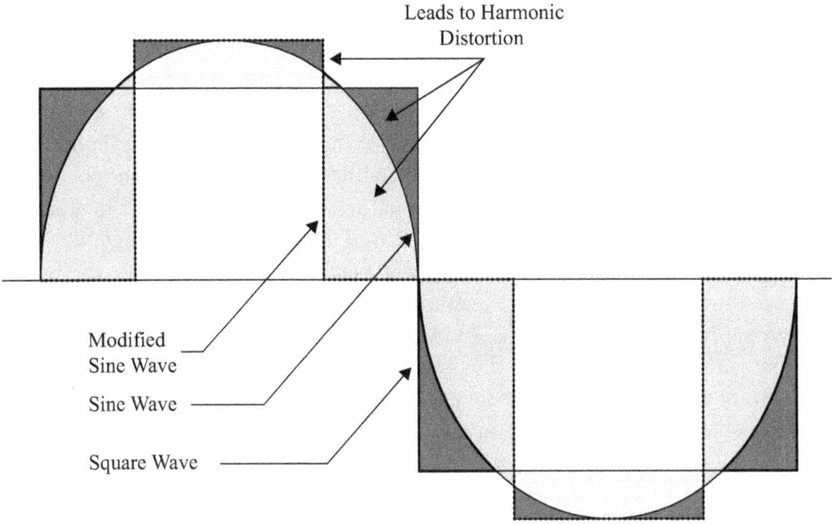

Figure 6.6 Comparison between square wave, modified square wave and sine wave

a reduction in the harmonics can be achieved. Implementing multiple switches to control *high and low* power flow is a necessity to achieve a controlled bidirectional inverter. To improve the efficiency of the inverter, adjusting the duty cycle of the switches to generate a momentary zone where no power is being delivered creates the zero-voltage switching phenomenon; Figure 6.6. To achieve a *pure-sine* wave, adequately sized inductors and capacitors must be used to filter the sharp switching edges.

6.3.2.1 Traditional Transformer-Based Inverter

The full bridge is connected to either a low frequency or an HF transformer which will either step up or step down the voltage to be converted to mains AC. Typically, modern transformer-based inverters utilise a PWM control to create the grid frequency, though in some cases transformers are situated on the mains AC conversion side. The operating frequency of the mains grid is either 50 Hz or 60 Hz; depending on location. This results in the physical size of the transformers being quite large, as the necessary magnetising inductance is dependent on frequency. The use of such transformers is not appropriate in PEVs due to the weight and size of the device.

6.3.2.2 Transformerless Inverter

Transformerless inverters operate similarly to the transformer-based inverters; however they utilise switching converters to buck or boost the panel DC voltage to an acceptable mains voltage. By incorporating a full-bridge converter and enabling zero voltage switching, isolation can be provided without resorting to a transformer [47]. Traditionally transformerless inverters had issues with high cost, as the necessary high voltage solid-state switch-gear – commonly insulated-gate bipolar transistor

(IGBT) or metal-oxide semiconductor field-effect transistors (MOSFETs) were not cost effective at the time. The reduction of cost between these switch types has lead this technique to become a more viable alternative to traditional transformer-based inverters. By removing the transformer, several kilograms in weight savings and inverter size reduction can be achieved. The advanced switching techniques which are implemented on modern systems have the ability to operate at higher voltage and current ranges compared to a more traditional transformer systems. Efficiencies of transformerless inverters can easily exceed that of the traditional topology, even with the addition of numerous additional switching stages.

Since transformerless inverter technologies can utilise a full-bridge topology to generate the low-frequency mains AC waveform, some PF correction techniques could be implemented. With the introduction of standard AS4777.2 [48], PV inverters must operate between a PF range not limited to unity. This gives distributors the ability to approve inverter technologies such as the transformerless to PF correct and support voltages on the networks [48].

6.3.3 DC/DC Converters

Common DC/DC converters utilise HF switching components to control output voltage and current. The voltage can either be of a higher order or lower depending on the application. The process of reducing or increasing DC voltage is done through a similar circuit, which are known as *buck* and *boost* converters respectively. A representation of a DC/DC buck converter is illustrated in Figure 6.7(a), where a single MOSFET switch and diode are utilised to reduce the voltage through the inductor. The ability to increase and decrease the voltage of a DC supply can be done by integrating both the buck and boost converter topologies together to form a *half-bridge* converter; Figure 6.7(b). The half-bridge can operate in two quadrants, however if four-quadrant control is required, two half-bridge converters can be utilised to create a *full-bridge* conversation stage. The full-bridge converter has a capability of operating in all four quadrants. Four-quadrant control, in the case of an electric motor, implies that the converter has the ability to accelerate and brake in the forward direction and accelerate and brake in the reverse direction. All of the DC/DC converters can be controlled by changing the duty cycle of the switching and adjusting the operating frequency.

6.3.4 Full-Bridge Converters with STATCOM Capabilities

To increase the capabilities of DC/DC converters, the full-bridge topology can be used. Figure 6.8(a) describes a unidirectional converter which utilises a diode-based rectifier to return to a DC source. Figure 6.8(b) illustrates the required topology to achieve bidirectional power flow between the load and the source. This structure can be applied to an electrical network where accuracy and tuning are required in multiple power flow directions [49]. Depending on the application, it may be beneficial to combine a DC transformation method with an AC transformation method. By doing so, benefits between both systems can be achieved, such as HF operation (improved efficiency and size reduction) and isolation. Utilising a full-bridge structure, a peak-to-peak voltage waveform of twice the input voltage can be obtained

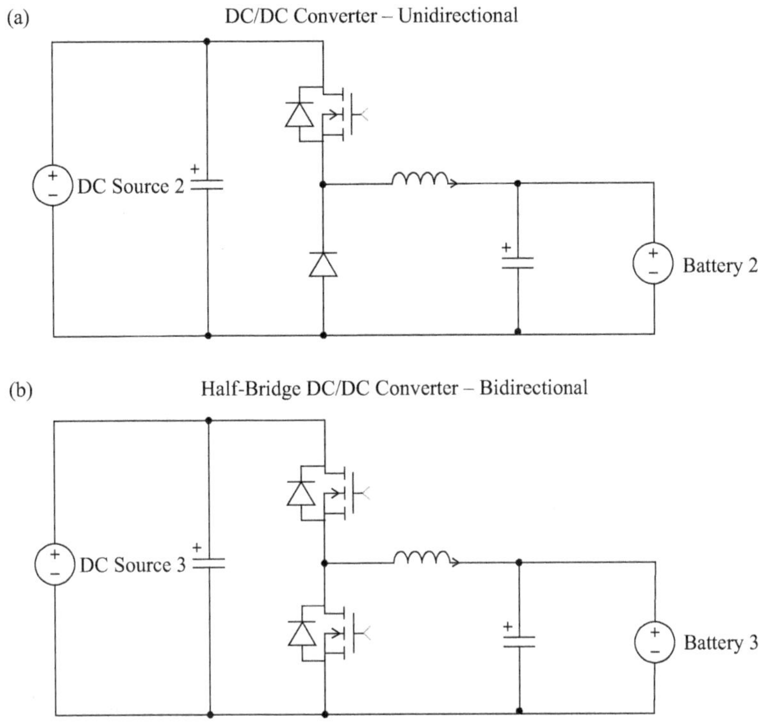

Figure 6.7 Simple DC/DC converter topologies

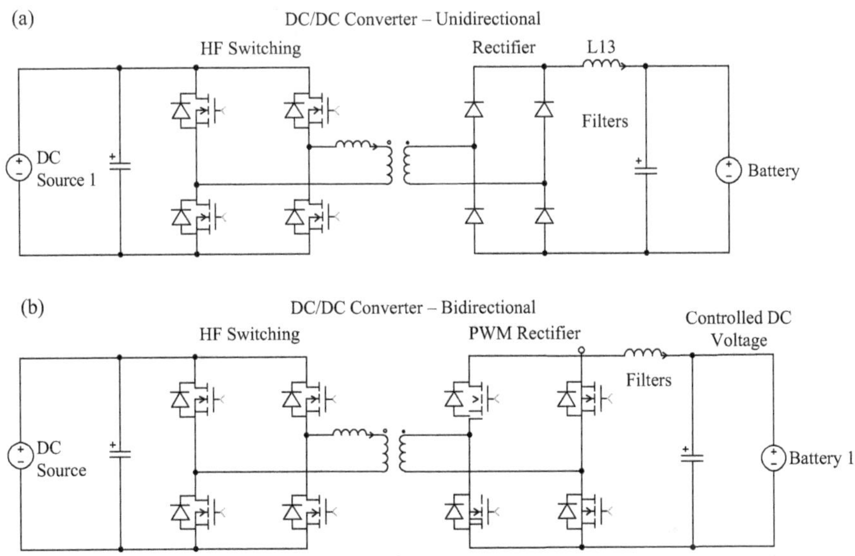

Figure 6.8 Full-bridge DC/DC converters

under ideal circumstances. PF correction is a necessary implementation of the transmission network. Depending on the location, size and level of PF correction required, different techniques can be implemented. These techniques include switching inductors or capacitors onto the network to correct the PF by means of mechanical or electronic switches. Relatively simple control can be achieved by creating a SVC device. To achieve four-quadrant control, a STATCOM device can be used. A STATCOM inherently will be more complicated, but will allow for greater control of the reactive power. In a three-phase system, load balancing can be achieved simply through VAR control [50]. If a common DC bus is available, load balancing can be achieved through means of an integrated circuit control system.

6.3.5 Internal STATCOM

Power IGBT, MOSFETs and now SiC MOSFETs are proving to become contenders to a now traditional static volt-amp-reactive compensation (SVC) technique [51]. STATCOMs benefit over SVC devices in their response time for reacting to a change in PF or balancing between loads. This is because STATCOMs can operate internally at much higher frequencies over SVC devices, thus resulting in sub-cycle analysis and mitigation of upcoming issues [52]. Commonly STATCOMs are based on the voltage source inverter (VSI) topology, with some based on the current source inverter (CSI) topology [53, 54]. VSI topology converters convert DC into a single- or three-phase output, where the voltage, amplitude, frequency and phase can be controlled [54, 55]. CSI topologies have advantages in harmonic filtering and require smaller DC energy storage components compared to a VSI topology [55]. VSI topologies have a greater advantage though, as the topology can allow the coupling transformer to form part of, or completely form, the inductance required for the AC filter. When STATCOM functionality is available as part of a charging system in a PEV, a number of benefits, including grid stability management, can be achieved [56]. As there is a large potential to have multiple PEVs charging simultaneously, the grid stabilisation effect can be amplified. This is because multiple PEV STATCOM controllers and topologies can be implemented together to share support, or partially the reactive power requirements in relation to the network requirements [57]. STATCOMs require symmetrical capacitive and inductive ranges to compensate reactive power and a real energy storage component to cover all four-quadrants; described in Figure 6.9. PEV's have an inerrant advantage as they already have a large real power storage system implemented in their battery banks. This makes V2G systems favorable for dynamic four-quadrant grid support systems. A STATCOM has the added ability to control the output current independently to the AC system voltage [53]. The DC component of the PEV can be utilised as the voltage source, and can control the real power output of the device and accept DC inputs from RES.

6.3.6 PEV On-board Charging Designs

There are many different positions for a charging system to be implemented on a PEV. The most common location is as a separate unit rectifier which can accept single- or three-phase power to charge the on-board energy storage. Depending on the type of motor used in the PEV, another location for AC charging can be by the motor

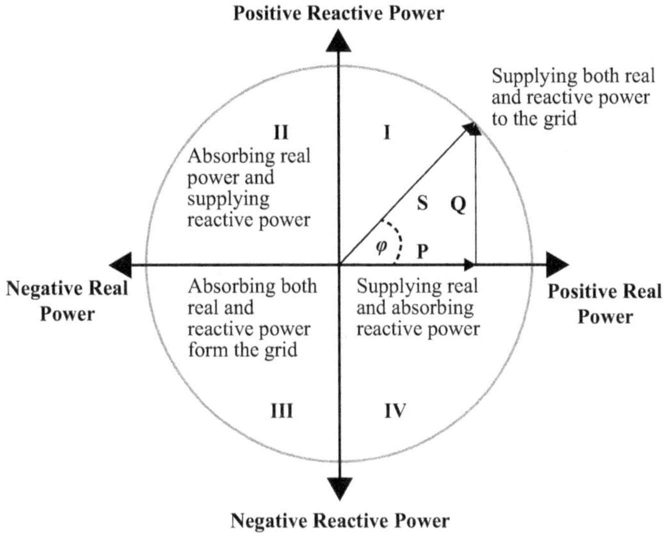

Figure 6.9 Four-quadrant power control

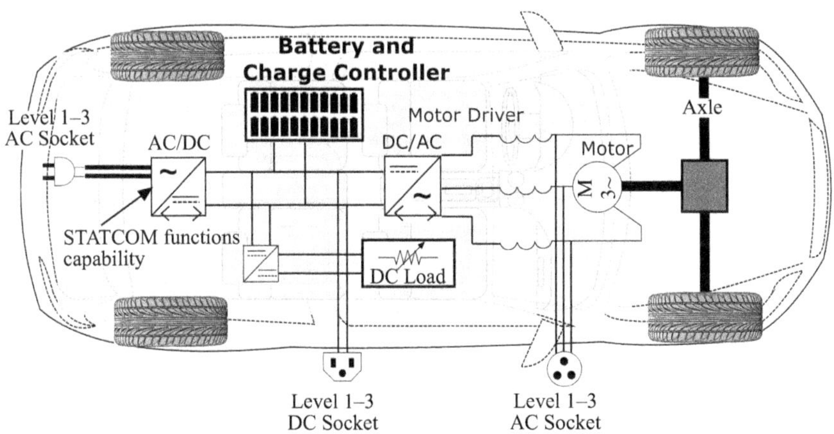

Figure 6.10 Basic componentry of a BEV with STATCOM and motor driver charging port

driver unit. If the motor is a three-phase induction machine which operates close to three-phase voltages, the motor driver can act as a rectifier as well as the driver for the motor. To achieve this, the motor driver has to be capable of bidirectional power flow. DC charging typically occurs through the use of an external AC/DC converter, which supplies power to the PEV as a potential which can be directly placed across the internal PEV DC bus. Figure 6.10 illustrates these charging mechanisms. Charging protection circuitry is placed along every point where charging can occur. This protection includes overcurrent, voltage, frequency and short circuit protection. These protection protocols are omitted from this diagram, but are included on a typical PEV.

6.4 Smart Transformer for Charging Station in a Microgrid

A traditional electrical grid consists of several static and active components which transport energy to a number of loads from a single point of generation. In this type of network, the generator is controlled by the demand from the load [1]. In a typical power network, there are several large generators producing electricity at a pre-set voltage and frequency, only varying power output on energy requirements. From these large generators, energy is transported through the extra high voltage, high voltage and MV distribution infrastructure. As the majority of consumer's appliances operate at LV distribution network, the distribution transformer must be able to supply sufficient real and reactive-to-diverse loads [58]. A typical distribution transformer will have an average capacity between 50 kVA and 500 kVA with single-phase voltages between 110 V and 240 V and a frequency of 60 Hz and 50 Hz, depending on geographic location (Table 6.5). This relatively low voltage and high power capacity can result in large resistive losses on the transmission network. This coupled with unbalanced, reactive and high harmonic distortion loads reduce the efficiency of distribution by approximately 6–7%. Figure 6.11 illustrates the traditional distribution network with both AC and DC loads. To supply DC loads, additional AC/DC converters are required. These converters add another step to the electric conversion process, ultimately reducing the overall efficiency of the network.

With the pending increase of PEV on distribution networks, the requirement for high-power and higher efficiency transformers is a necessity. Supplying both AC and

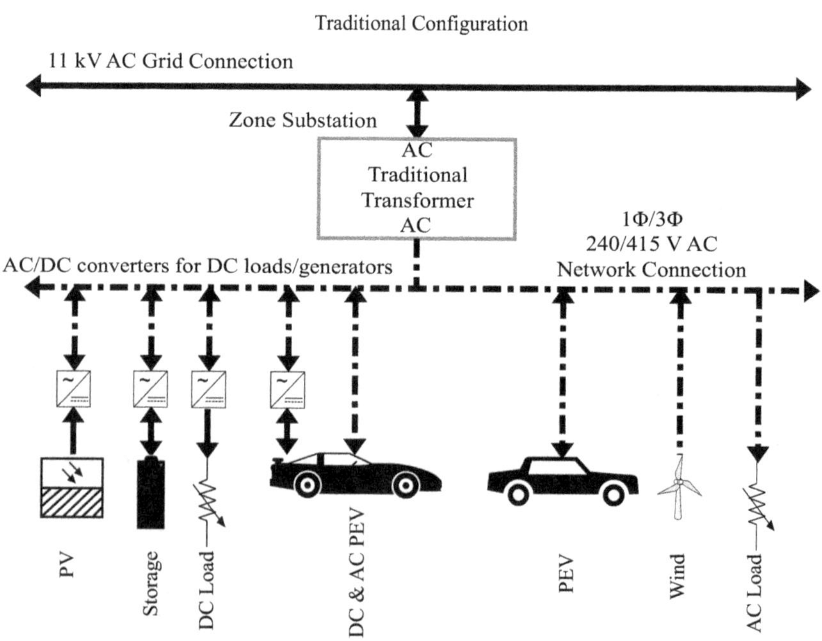

Figure 6.11 Traditional (current) transformer-based distribution and charging infrastructure

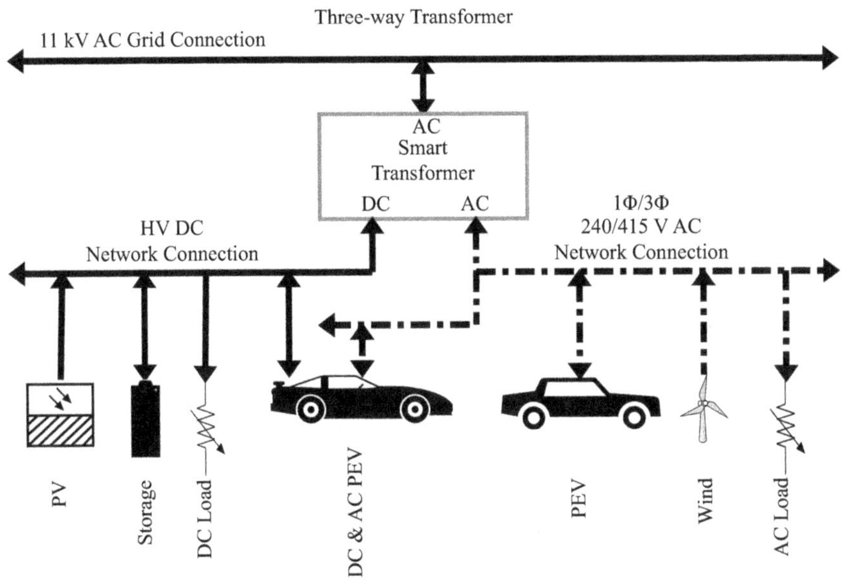

Figure 6.12 Utilisation of smart transformer on distribution and charging network

DC (PEVs) loads will become a normal function for these transformers. Not only will the transformers have to cope with an increased load, but with the onset of more RES on the network, the potential to feed back into the MV distribution network. A smart transformer is thought to have the capability of providing such benefits as a replacement to the traditional transformer distribution network. The smart transformer will be capable of supplying both DC and AC loads, accepting reverse power flow from those loads and energising MV transmission network – shown as *11 kV AC grid connection* in Figure 6.12. The smart transformer concept utilises solid-state converters, described in sections 6.3.1 to 6.3.4, to remove the necessity of a low-frequency (50–60 Hz) transformer. The removal of such transformers results in reduction in size and weight and can increase the efficiency of the overall conversion process [51]. By having internal points where there is a DC bus, DC power can be exited from the transformer as its own supply. Incorporating four-quadrant power controlling full-bridge converters with STATCOM functionality (sections 6.3.4 and 6.3.5), in the smart transformer, reactive power, harmonics and load balancing compensation can occur [56]. The topology of a smart transformer is shown in Figure 6.13.

6.5 Contact Charging Safety Considerations

6.5.1 AC Charging

AC charging techniques are relatively easy to accomplish, due to the historical AC transmission network. With AC transmission, a zero-crossing point of the power wave is present, at which point an arc will simply self-extinguish.

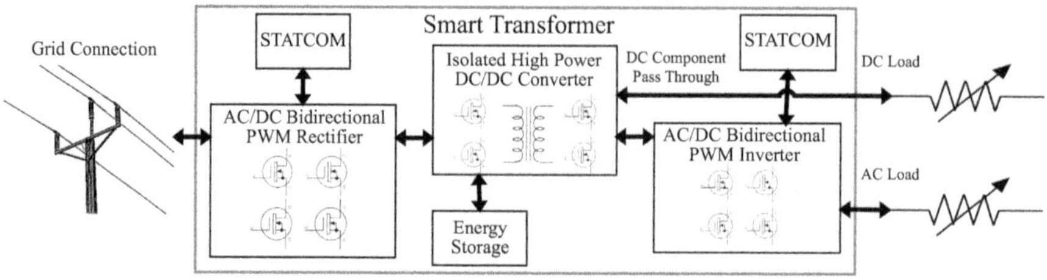

Figure 6.13 Basic structure of a smart transformer

6.5.2 DC Charging

Direct contact DC charging has been problematic in the past, majorly due to inadequate fault-sensing equipment and switchgear. Recent developments in silicon microelectronics and integrated circuit design have allowed for DC to return as a major player in energy transmission. Thyristors, IGBTs, MOSFETs and modern-aged silicon carbide (SiC)–based MOSFETs have allowed DC switchgear to enter the high-voltage power networks. Modern SiC MOSFETs have the capability of efficiently switching upwards of 2 kV in a relatively small form factor.

6.5.2.1 DC ARC Faults

A DC arc can be described as a sudden current flow over an air gap between two conductors, or a single conductor to ground. Just like an AC arc, if not extinguished, the arc can cause fire or even explosions depending on the environment [59]. A DC arc differs in many characteristics over an AC arc. An AC arc has a point in time where no power flows, also known as the zero-crossing point. This in combination with an appropriate open contact resistance is generally enough to cause the AC arc to extinguish. A DC arc does not have a zero-crossing point, rather a constant potential, if constant power is available [59, 60]. There are many causes for DC arcing, but in reference to an EV, two types of arcing are most common: a series arcing and a parallel arcing. Both types of DC arcs are difficult to detect but can be dangerous if not mitigated. Series arc faults generally occur inside a loose conductors or broken wires, due to corrosion or poor installation [60]. When an arc occurs, a semi-conductive layer of copper oxide is formed. This layer forms a location for an arc to occur but not adversely affect series battery voltage or current.

Standard inverter technologies of single phase can mitigate a series arc from becoming too extensive. They do this by comparing maximum open circuit voltage to that of the maximum power point voltage. If the voltage drop over the arc gap becomes greater than that of the open circuit voltage, then the inverter switches the string off [59–61]. This is only possible for single-phase inverters with a relatively low string input voltage. A parallel arc occurs between two parallel conductors of different potentials. Parallel arc faults to ground occur due to a short where the negative cable is grounded and insulation of the positive cable is damaged. Cross-shorting parallel arc faults occur when conductors of different potentials cross and short. Numerous methods including RF sensing and detailed voltage measuring are used to detect such a fault in a system, but these methods commonly lead to false alarms. Faults can reach in excess of 1,000 °C, which in any case can cause fires. Methods to mitigate DC arcing faults should be implemented on every system, even if false alarms are given to reduce this risk.

6.6 Wireless Charging Systems for PEV

6.6.1 Wireless Power Characteristics

Contactless charging methods have the potential to integrate into infrastructure easier when compared to contact charging solutions. This is because there are only a small amount of developed wireless charging systems (WCS) and therefore little variance between countries. The term wireless in wireless power transfer (WPT) and WCS refers to the absence of a physical (copper) connection from the energy supplier – typically an electrical grid – to the load. Instead of free flowing electrons charging a battery through a cable, electromagnetic (EM) energy in the form of magnetic or electric fields are used to bridge the charging gap.

WCS have become one of the famous and innovative technologies in the last decade for various applications, both low power and high power. Low-power WCS include wireless biomedical devices [62] and compact electronic products such as smartphones [63–65], while high-power WCS include concepts for EVs [66] or BEVs [67] and PHEVs [68]. Currently, three important factors – high cost, limited range and limited power transfer – are preventing EVs to become more user-friendly [67]. In order to make WPT technologies more accepted, the disadvantages must be rectified and the inherent advantages promoted. A major advantage of WCS is the ability to increase the range of EVs without substantial impact on the weight or cost.

To transmit energy, a WCS utilises a standard power supply (mains/battery energy storage) and converts it to the necessary voltage/current/frequency to allow for wireless power transfer to occur. Figure 6.14 illustrates this process in the form of an induction charging method for a PEV. In this illustration, the mains AC power is converted into DC using AC/DC rectifiers, filters and PFC circuitry. The filtered DC is fed into HF AC converters to convert the DC into an appropriate HF AC source for the primary winding. To reduce losses and improve performance,

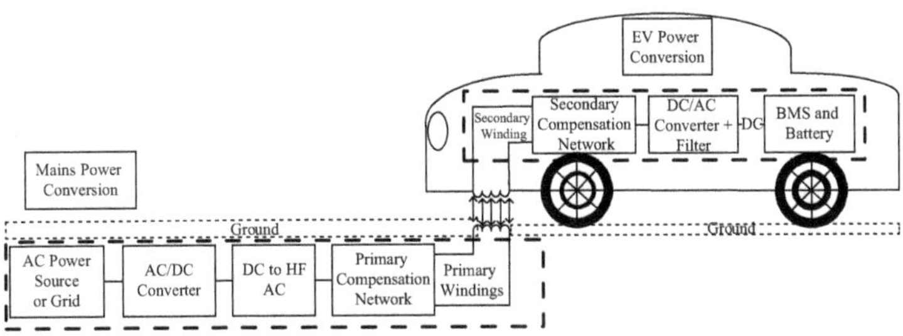

Figure 6.14 Basic diagram of a wireless charging system for PEVs

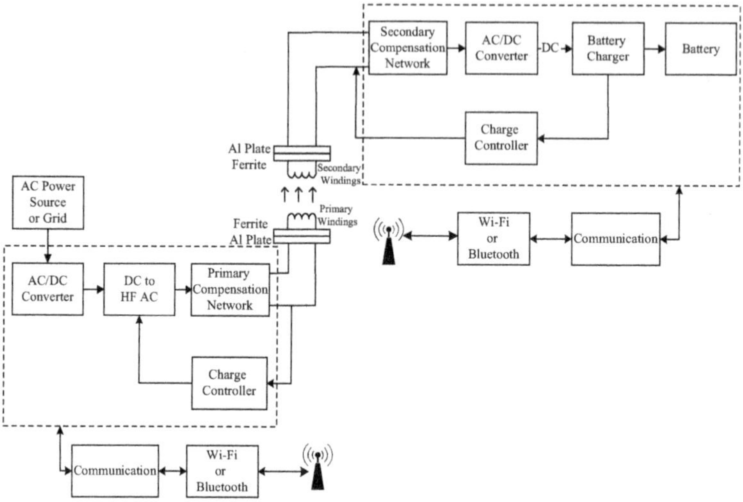

Figure 6.15 Advanced closed loop WCS for PEVs

primary and secondary compensation circuits should be employed. As the primary and secondary windings are inherently separated by an air gap defined by minimum vehicle ride height legislations, typically 100 mm, the energy transfer characteristics are typical to that of an air-core transformer. A varying current in the primary winding creates magnetic field which generate varying electromotive force (EMF) or voltage in the receiver coil. On the receiving side, the PEV power conversion starts at the secondary winding. The HF AC voltage is rectified and filtered into a DC source and then transferred to the battery management system (BMS). This device includes the battery charger, protection circuits and SOC information. From this point onwards, DC power can be directed into the battery banks. In general, the receiver coil is installed underneath the car at a distance from the ground, while other electronics are inbuilt into the car.

Efficiency is the Achilles-heel of a traditional WCS. Advanced WCS tend to utilise methods of controlled feedback loops to more accurately provide energy transfer details to increase efficiency. Figure 6.15 illustrates an advanced closed loop for WCS. In comparison with traditional WCS, the feedback has two important features: dual internal charge controllers and wireless communication systems. By providing two charge controllers, one on the transmitter and another on the receiver, the charging characteristics can be managed on both sides of the wireless transfer medium. In-built wireless communication systems allow communication between the source and receiver with a number of different parameters. These parameters include power transfer, efficiency, charging level, load level and control parameters. The information exchange can be employed utilising universal wireless communication protocols such as Wi-Fi, Bluetooth and near-field communication (NFC). The placement and isolation of communication devices from the internal WCS high voltage are one of the challenging tasks for a designer, but it is solved with the help of technology advances.

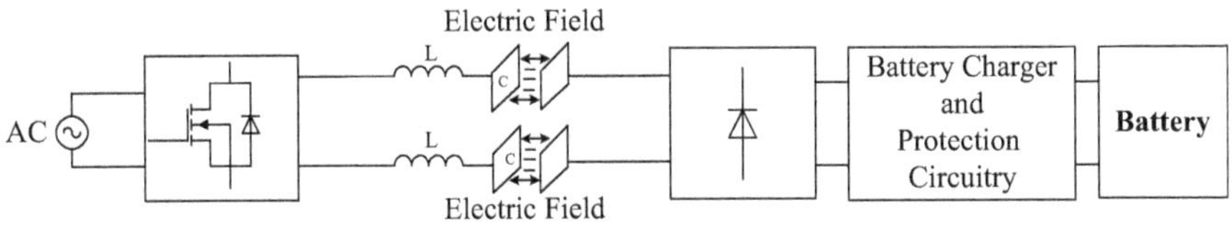

Figure 6.16 Basic diagram of capacitive wireless power transfer (CWPT)

6.6.2 Methods of Wireless Power Transfer

Three major WCS are in use today: capacitive, magnetic gear and EM field transfer [69, 70]. Each WCS has both desirable and not so desirable aspects of their energy transfer techniques, particularly in reference to PEVs.

6.6.2.1 Capacitive Wireless Power Transfer

CWPT technology has been utilised to transfer power wirelessly through the use of capacitive coupling. A basic block diagram of CWPT is presented in Figure 6.16. This diagram illustrates where the electric field is utilised to transfer power. CWPT can transfer power through metal barriers [71–73] even if there is a strong magnetic field interference [74]. In addition, this method can transfer power and data over a same network with minimal electromagnetic radiation [75]. This technology has advantages such as simple and light-weight design, easy structure of couplings, low cost and position flexibility [73, 75–77]. Due to the nature of the CWPT, the capacitive coupling between the two transfer plates diminishes exponentially with respect to distance. This results in a limited power transfer density [77]. As a result, this technology would be largely ineffective in transferring large amounts of power to PEVs due to the typical air gap of 150–200 mm [69]. To combat this issue, three basic techniques can be utilised: operation at HF (kHz to MHz), utilisation of a high-voltage power source and reducing the coupling impedance. First two techniques can be easily achieved, but reducing the AC coupling impedance is difficult and problematic. Utilising an inductor-capacitor (LC) series resonant circuit will improve the coupling efficiency, but the LC resonant circuit is extremely susceptible to minor changes to coupling parameters. This is because the LC resonant circuit must be tuned to a specific value to reduce the AC impedance, which is directly relative to the distance and position between the primary and secondary coils, along with power draw [77]. Due to these factors, research for CWPT-based products for PEVs has not yielded promising results. CWPT technology though has mostly been suitable for low-power applications such as chargers for wireless toothbrush and cellular phone charging pads [70, 78, 79].

6.6.2.2 Magnetic Gear and Permanent Magnet–Based Wireless Power Transfer

Magnetic gear and permanent magnet (PM)–based wireless power transfer is a technology which utilises mechanically assisted magnetic forces to transfer power from the source. The basic block diagram of magnetic gear and PM-based wireless charging system is shown in Figure 6.17. In the transmitter side, a rotating cylinder-shaped PM rotor is driven by an external rotor which induces current into

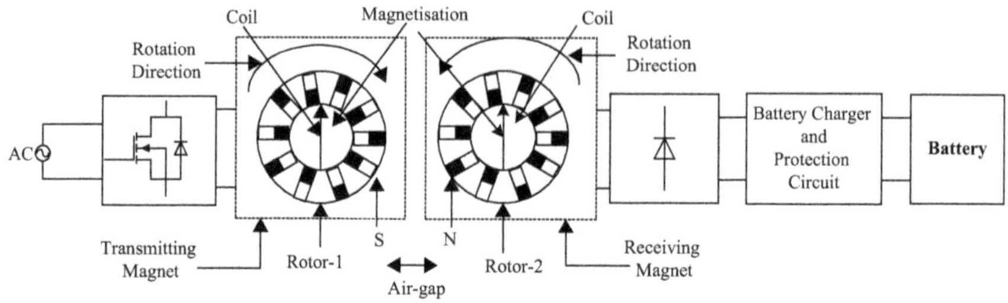

Figure 6.17 Basic diagram of magnetic gear and permanent magnet–based wireless power transfer

transmitting winding. This in turn generates an electromechanical force in the PM. The transmitting winding can be installed either at certain air gap or inside a hollow rotor. At the receiver side, a similar rotor is positioned within a specific air-gap distance. The generated magnetic field through the *Transmitting Magnetic* is also received by the *Receiving Magnet*. As a result, the receiver magnet rotates synchronously with the transmitting magnet. The receiver operates as a generator and delivers power to the rectifier and battery through other charger and protection circuits [69, 70]. Li [80] proposes the first magnetic gear and PM-based WPT prototypes, which were developed for two different applications. First prototype was demonstrated to transfer 1.6 kW of power with 150 mm air gap for PEVs and the second was developed to transfer 60 W power with 100 mm air gap for medical implants. Both prototypes were able to transfer power with the efficiency of 81% at 150 Hz. Another magnetic gear and PM-based WPT was proposed to transfer 6.6 W output power and 1 W supplying with 30 mm distance from the skin for the biomedical implants [81].

6.6.2.3 Electromagnetic Field–Based Wireless Power Transfer

WPT can also be possible in time-varying electromagnetic fields: near-field (6.1) or far-field (6.2). Transferring large amounts of power through the far-field is a challenging task, as the far-field requires line of sight to be able to transfer any meaningful power [82]. In addition, transferring large amounts of energy over large distances may generate health and safety issues. This is due to the concern regarding high-power and HF wireless transmissions producing strong electromagnetic fields. Unlike the far-field, rear-field-based technologies can transfer sufficient power with higher efficiency over shorter distances in non-radiative regions for PEVs [69, 70]. There are two main types of near-field-based WPTs: inductive power transfer (IPT) and resonant inductive power transfer (RIPT).

$$N_{field} = \frac{\lambda}{2\pi} \tag{6.1}$$

$$F_{field} = \frac{2D^2}{\lambda} \tag{6.2}$$

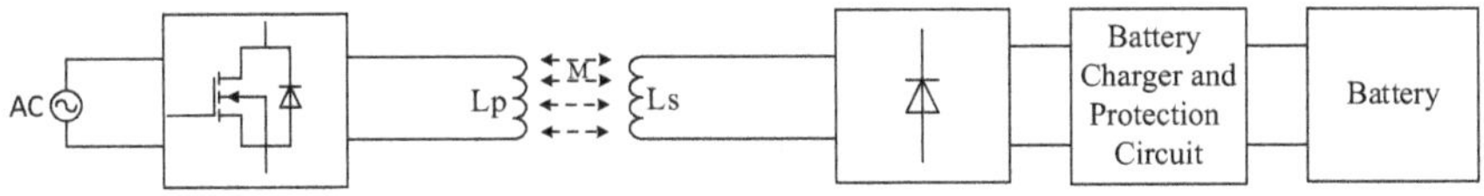

Figure 6.18 Basic diagram of inductive power transfer

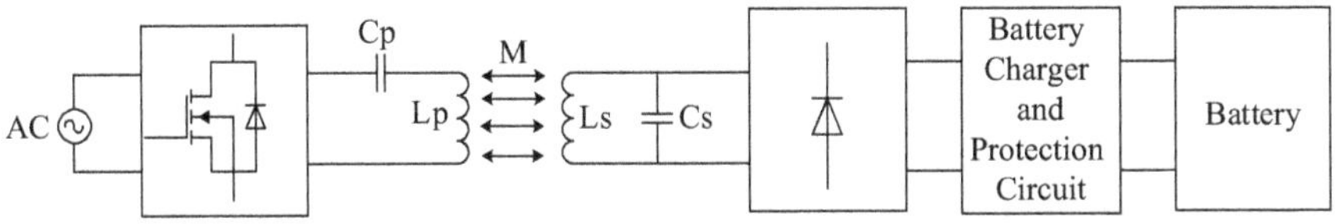

Figure 6.19 Basic diagram of series-parallel resonant inductive power transfer (RIPT)

6.6.2.4 Inductive Power Transfer

Transferring energy through inductive coupling methods is not a new idea, rather an ongoing idea originally conceived by Nikola Tesla in 1914 [83, 84]. Since then many experiments have been performed and many designs have been proposed for industrial, public and private uses [85, 86]. The basic block diagram of IPT is shown in Figure 6.18. IPT consists of two transmission coils: one transmitting coil and one receiving coil, also known as primary winding and secondary winding. When an alternating current is applied to the primary coil, a continually varying magnetic flux field is generated on the transmission side. In the receiver side, this continually varying magnetic field induces electrons to flow in the direction of the varying field, resulting in a current potential. The coils operate at a self-resonant frequency. This AC potential is converted to DC using rectifier and utilised to provide a power to a battery or electronics products. IPT has been an immerging method to transfer power wirelessly in various applications such as electric toothbrush, mobile phones and early PEVs in the power ranges between milliwatts and some kilowatts. For example, GM EV1 [1996] and Chevrolet S-10 [1997] [87] are IPT-based PEVs where the charging primary coil (paddle) of the *Magne Charger* is placed at the centre of the receiver coil. This design is able to charge contactless with a power rating of 6.6 kW or 50 kW [70].

Disadvantages of traditional IPT include their limited transfer distances, poor efficiency and lower power transmission capacity. The efficiency and power transfer level can significantly drop if there is a misalignment or an increase in distance between the transmitter and receiver [88]. In practical applications, the distance and power requirements may vary depending on applications. For example, PEVs require to charge their battery bank wirelessly with the power of several kilowatts and from the distance of 150–200 mm.

6.6.2.5 Resonant Inductive Power Transfer

Nowadays, the resonant inductive power transfer (RIPT) is the most famed WPT technology for PEVs and BEVs. As shown in Figure 6.19, it is also known as coupled magnetic resonant IPT. This technology was developed from the original designed by Nikola Tesla. In comparison with IPT, this technology uses a resonance function to transfer power efficiently from the source to receiver. The resonant transfer occurs when the transmitter and receiver coils have same frequency and characteristic impedance. At this stage, the coupling becomes most efficient; therefore energy transfer becomes most efficient. The coils can be either loosely or strongly coupled. In comparison with strongly coupled, loosely coupled magnetic resonance can operate at larger distances but transfer energy at lower efficiencies and cause high electromagnetic emissions [89, 90]. The efficiency problems can be solved by utilising compensation networks and employing advanced tools called high-quality resonators. Electromagnetic emission problems can be solved by using well-combined magnetic materials and core structures [66]. The power transfer mostly depends on a number of factors including air gap, WPT efficiency, coupling factor and quality factor [91–93]. Other factors also affect the power transfer ability, including coil alignment, coil design, ferrite material and structures.

- ***Air gap:*** the distance between the transmitter and receiver. If the distance increases between the transmitter and receiver, the air gap becomes too large. The air gap is proportional to the power loss.
- ***WPT efficiency:*** The ratio of received power at the secondary winding and the transmitting power from the primary winding is called efficiency (η). The efficiency can be measured from primary coil to secondary coil or for the entire system. It can be calculated using (6.3).

$$\eta = \left(\frac{P_{out}}{P_{in}}\right) \times 100\% \tag{6.3}$$

  where P_{out} is the output power at the receiver and P_{in} is the input power at the transmitter.
- ***Coupling factor:*** The magnetic field generated by the transmitting coil is received by the secondary coil. The coupling factor (k) depends on the amount of magnetic field received by the receiver coil. The value is normally lies between 0 and 1. The value of 0 states none of the magnetic field receiver, while 1 states that the entire magnetic field is received at the receiver side. The advantages of higher coupling factor include higher transfer efficiencies and a reduction in heat and other losses. The coupling factor can be calculated using (6.4).

$$k = \frac{L_m}{\sqrt{L_p \cdot L_s}} \tag{6.4}$$

  where L_m is the mutual inductance, L_p and L_s are the primary (transmitter coil) and secondary (receiver coil) inductances.
- ***Quality factor:*** The ratio of reactance components and resistance of the inductor is called Quality factor (Q). The resistance of the inductor depends on many other parameters such as core loss, skin effect and other resistances. It is also known as Q-factor. Quality factor can be defined using (6.5).

$$Q = \frac{X_L}{R} = \frac{\omega \cdot L}{R} = \frac{2\pi \cdot f \cdot L}{R} \tag{6.5}$$

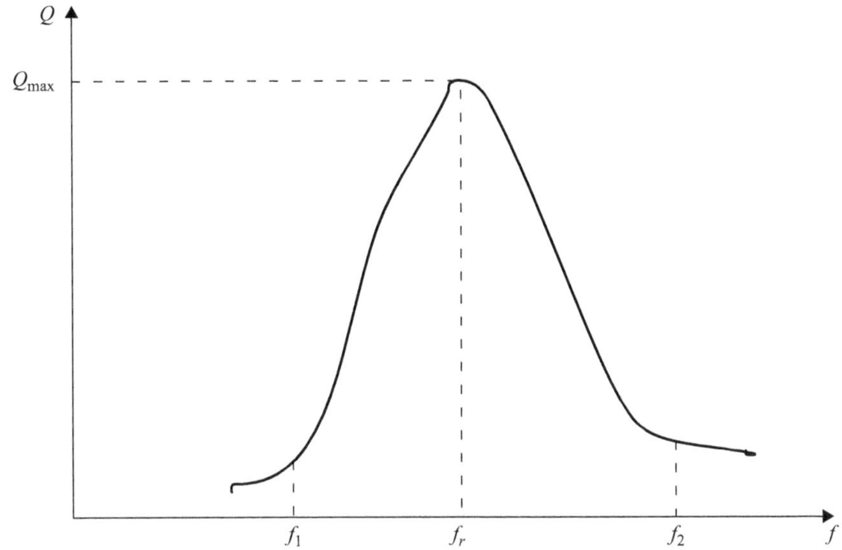

Figure 6.20 Relationship between resonant frequency and Q-factor

where X_L is the reactance and R is the resistance of inductor, and f and L are the frequency and inductance of the inductor.

As shown in Figure 6.20, higher Q-values can be optimised at the resonant frequency. When the switching frequency becomes equal to resonant frequency, the maximum power transfer occurs. The resonant frequency can be calculated using (6.6).

$$f_r = f_s = \frac{1}{2\pi\sqrt{L_s \cdot C_s}} \tag{6.6}$$

The resonant frequency depends on self-Inductance (L_s) of the secondary winding and capacitance (C_s) of the secondary compensation circuit of the coil. But it may be difficult to find the same resonant frequency for different shapes and sizes of the transmitter and receiver coil.

The PF in traditional IPTs is highly dependent on the load and inductances of the primary and secondary windings. The impedance of the source becomes more inductive with increasing frequency, resulting in smaller inductors, but this has an adverse effect on the PF. The PF will approach zero as the frequency approaches infinity, which has a significant effect on power transfer efficiency. This is the major drawback of the traditional IPT. To overcome this problem, the traditional IPT requires capacitive compensation networks in primary as well secondary windings. As shown in Figure 6.21 [66, 90, 94], there are four basic topologies of compensation networks: series-series (SS), series-parallel (SP), parallel-series (PS) and parallel-parallel (PP).

The parameters and characteristics of different types of resonant compensation topologies defined by Figure 6.21 are presented in Table 6.7 [94–96]. An important

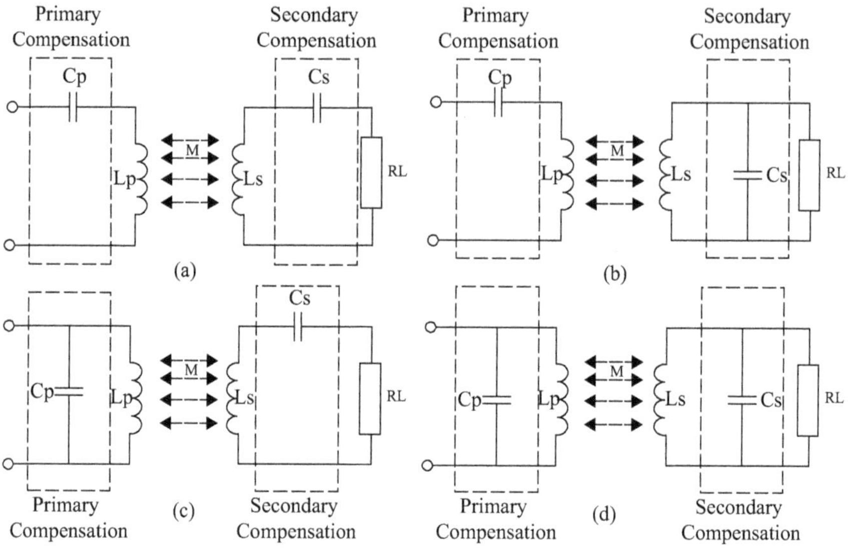

Figure 6.21 Four basic compensation networks: (a) series-series; (b) series-parallel; (c) parallel-series; (d) parallel-parallel

Table 6.7 Parameters and characteristics of four different topologies

Topologies	Primary compensation capacitance (C_p)	Secondary compensation capacitance (C_s)	AC source	Efficiency	Frequency change effect on efficiency	PF change effects on flexible frequency
SS	$C_p = \frac{1}{\omega_0^2 \cdot L_p}$	$C_s = \frac{1}{\omega_0^2 \cdot L_s}$	Voltage	High	Low	High
SP	$C_p = \frac{1}{\omega_0^2 \left(L_p - \frac{M^2}{L_s}\right)}$	$C_s = \frac{1}{\omega_0^2 \cdot L_s}$	Voltage	Low	High	Low
PS	$C_p = \frac{1}{\omega_0^2 \cdot L_p}$	$C_s = \frac{1}{\omega_0^2 \left(L_s - \frac{M^2}{L_p}\right)}$	HV or current	High	Low	High
PP	$C_p = \frac{1}{\omega_0^2 \left(L_p - \frac{M^2}{L_s}\right)}$	$C_s = \frac{1}{\omega_0^2 \left(L_s - \frac{M^2}{L_p}\right)}$	HV or current	Low	High	low

benefit of using primary compensation network is that it reduces the source apparent power (VA) ratings by operating at unity PF. The secondary coil compensation increases power transfer efficiency. The capacitors in both compensation networks; primary and secondary, store and supply reactive power to and from the IPT windings. This reduces the reactive power draw from the source supply, also resulting in a reduction of HF harmonics. In addition, the resonant frequency could be controlled and changed by the compensation networks. RIPT operates in kilohertz frequency ranges which can bring additional benefits such as lower electromagnetic interference (EMI), reduced size of the components and higher overall efficiency. By implementing PFC tuning circuitry, the resonant frequency of the IPT can be at unity PF whilst providing the maximum power transfer point at the height of an average vehicle [70]. This method can provide the maximum transfer efficiency for both the distribution network and vehicle.

6.6.2.6 IPT Coil Structures

Winding structure is very important in transformer and inductor design. In WPT, the air-core transformer concept is utilised to transfer energy from the transmitter to the receiver. In order to improve the performance, different types of winding structures have been proposed such as spiral, square, hexagonal and fractal, as shown in Figure 6.22. These structures have been used to design WPT devices for low- and high-power applications such as mobile phones, laptops, medical implants and PEVs. Coils that have the capability of being created on a printed circuit board (PCB) (Figure 6.22(a–c) and (g–h)) can be manufactured in a bifilar (multi-layer) design. This design would increase the inductance of the coils without increasing size or adding much weight [97, 98].

The spiral circular structure (Figure 6.22(a)) [99, 100] has proved to be a valuable coil design first described by Nicola Tesla in the 1800s. As there are no sharp edges, induced eddy currents between the windings are left to a minimum. To change the magnetic field lobes, the central diameter size can be adjusted; decreased or increased. Decreasing the central diameter size will cause a sharp spike in magnetic field distribution contained by the central diameter. This can provide excellent coupling between the primary and secondary coils of a WCS, but only if the alignment is correct. By increasing the central diameter size, a larger central flux distribution field will be created. This field will be at a lower magnitude compared to that of a smaller central diameter, but alignment will be less of an issue. This though allows for greater misalignment protection. Figure 6.22(b) [99, 100]

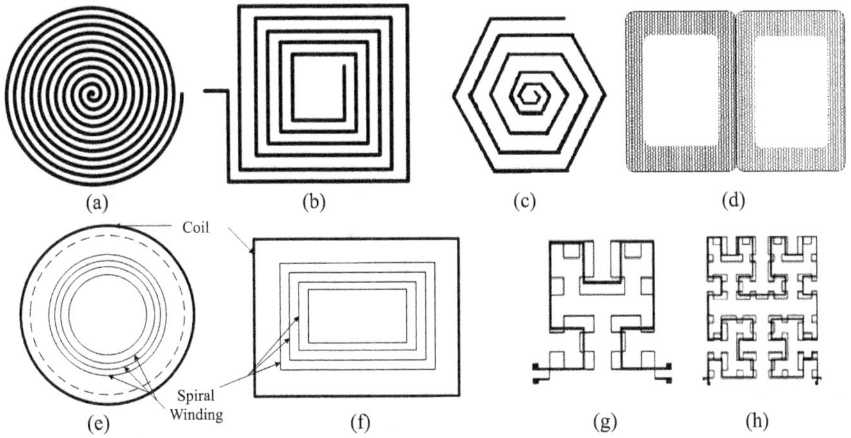

Figure 6.22 Different coil structures: (a) circular; (b) square; (c) hexagonal; (d) double D; (e) hybrid circular; (f) hybrid square; (g) fractal 1–4 ratio; (h) fractal 4–16 turn ratio

illustrates a square inductor topology. This topology can benefit when placed in an array due to the flush fitting sides. The parallel lines between the turns mutually benefit one another, increasing the inductance of the topology. The sharp turns around the corners generate eddy currents, which in turn heat up the corners. This effect increases the impedance of the windings and therefore is not extremely suitable for high-power applications. For low-power applications, the hot-spot effect is not so severe, and the benefit of combining this structure in an array is desirable to reduce primary and secondary coupling misalignments.

Figure 6.22(c) illustrates a hexagonal spiral winding structure. This structure has benefits over other designs when it is placed in an array. A circular coiled array will have zones between the coils where no energy is being transmitted. A square or rectangular structure inherently has sharp turns around the edges, which can cause eddy currents to be induced onto the edges, reducing efficiency and increasing temperatures. The hexagonal structure removes the dead zones and reduces the angles between turns, increasing the WPT and improving efficiency. This type of structure has been extremely popular for portable devices [63]. In order to improve performance and power transfer, in comparison with traditional circular coils, double D–shaped coils were developed (Figure 6.22(d)). This topology combined with quadrature structure (coils within coils) has been proposed to decrease misalignment loss, support large air gaps and improve overall efficiency. Additional advantages of this topology includes high power delivery with low EMI, higher coupling, superior *X*/*Y* tolerance and easy interoperability with other manufacturers' charging pads [101, 102]. Figure 6.22(e) and (f) illustrates advanced hybrid structures in circular and rectangular spiral windings. Xun and Hui propose these topologies for planar contactless battery charging [103]. In this design, the primary coil's magnetic field is supported by that of smaller internal spiral windings. This combination of coils generates a magnetic flux distribution which is uniform in magnitude. Fractal-based HF planar WPT transformer designs (Figure 6.22(g) and (h)) have the capability of providing a charging platform that is not dependent on the position on the secondary coil [104]. This can be done by increasing the ratio of the number of fractals to the point where number of magnetic field lobes created by the structure is sufficient to ensure WPT in any charging orientation.

6.6.2.7 Ferro-magnetic Core Material and Structures

In WCS for PEVs, the power transfer between transmitters to receivers is from several watts to hundreds of kilowatts to charge the battery bank. The magnetic flux which is generated in this power ranges would be significantly high. If there is no shielding, the magnetic flux can spread to multiple directions. It not only reduces the coupling efficiency between two windings, but also may bring health and safety related issues. In order to solve this problem, magnetic cores can play a significant role in wireless charging systems for PEVs. There are two main functions of the ferromagnetic material. First, the ferromagnetic structure can provide a controlled path and guide for magnetic flux to flow from the primary windings (transmitter) to secondary windings (receiver). It also creates a well-defined magnetic path length for the magnetic flux. In addition, it increases not only mutual inductance but also self-inductance of two or more magnetically coupled coils [105, 106]. As a result, ferromagnetic core materials

(a)

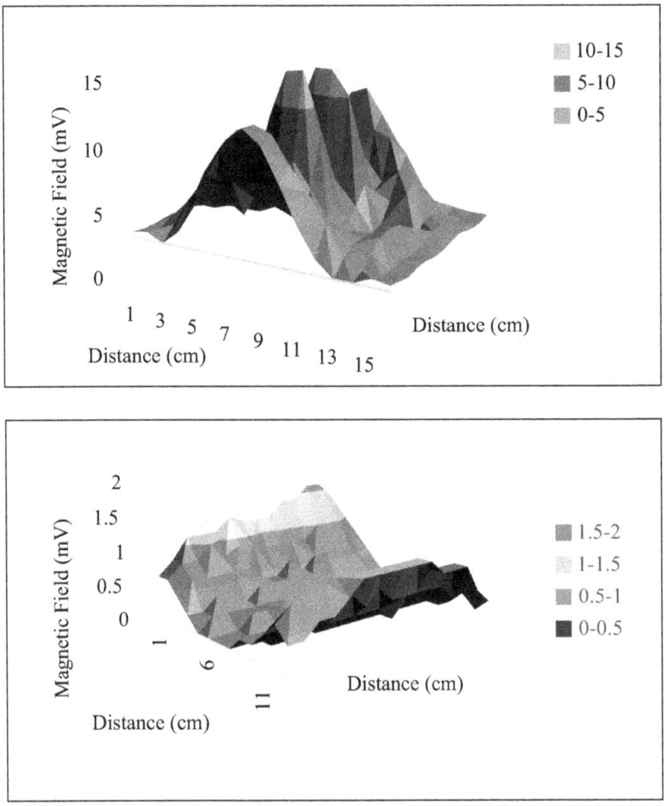

(b)

Figure 6.23 Near-field measurements of HFHPD-WPT prototype (a) without and (b) with ferrite core at secondary side

provide excellent shielding by reducing magnetic flux leakage around the primary and secondary coils. As an example, a 60 W HF high-power density (HFHPD) WPT was developed to demonstrate the magnetic flux distribution and leakage, with and without a ferromagnetic core [97, 98]. A condition of with and without magnetic material at the secondary (receiver) side is presented in Figure 6.23. Figure 6.23(a) shows near-field measurement of the HFHPD-WPT prototype without and with ferrite core at secondary side. Figure 6.23(b) displays the magnetic flux distribution when the magnetic material is utilised at the secondary side. This illustrates the flux leakage at the sides of the core material. To validate the results, the finite element method (FEM) was employed in no-load and load conditions, as shown in Figure 6.24 [97, 98]. In order to minimise the magnetic flux leakage to avoid EMC problems, U-shaped magnetic ferrite material was proposed. As shown in Figure 6.25, the U-shaped edges capture the magnetic flux and divert it to the secondary winding to increase the magnetising impedance of the system. It helps to improve magnetic flux distribution and reduces stray flux [97, 98].

The selection of a core material depends on size, weight, temperature rise, core loss, flux density, permeability, operating frequency and cost. Different ferrite

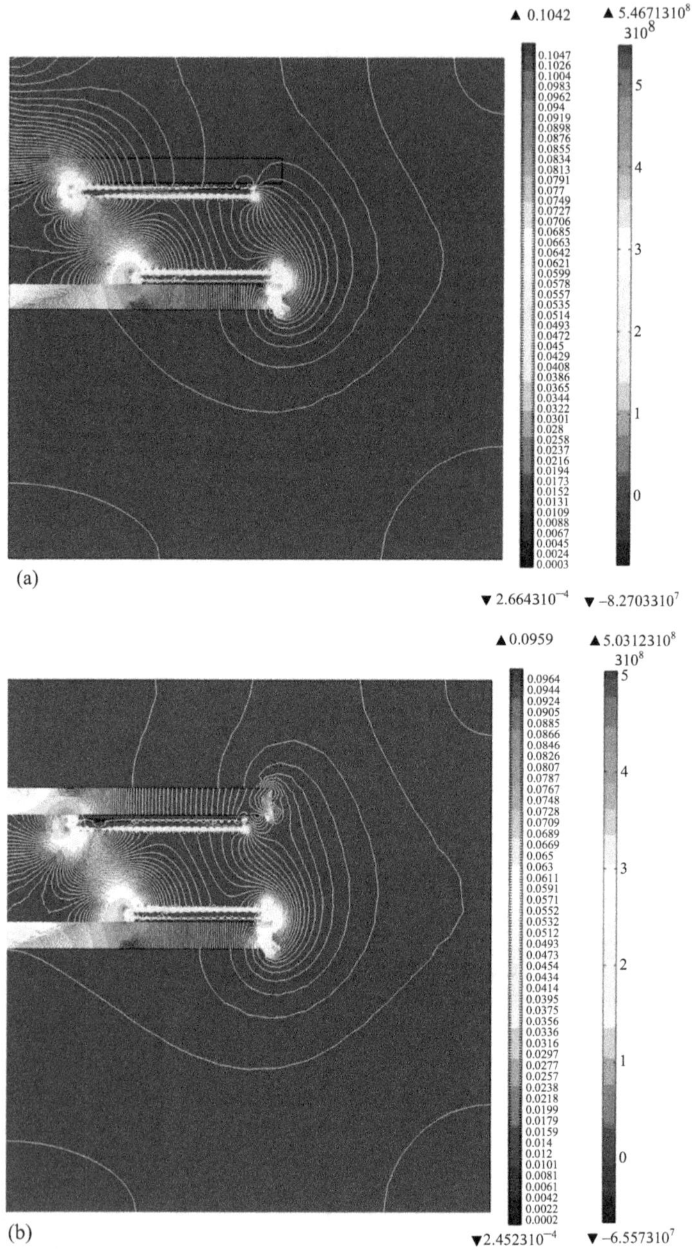

Figure 6.24 Simulation results of HFHPD-WPT prototype (a) without and (b) with ferrite core at secondary side

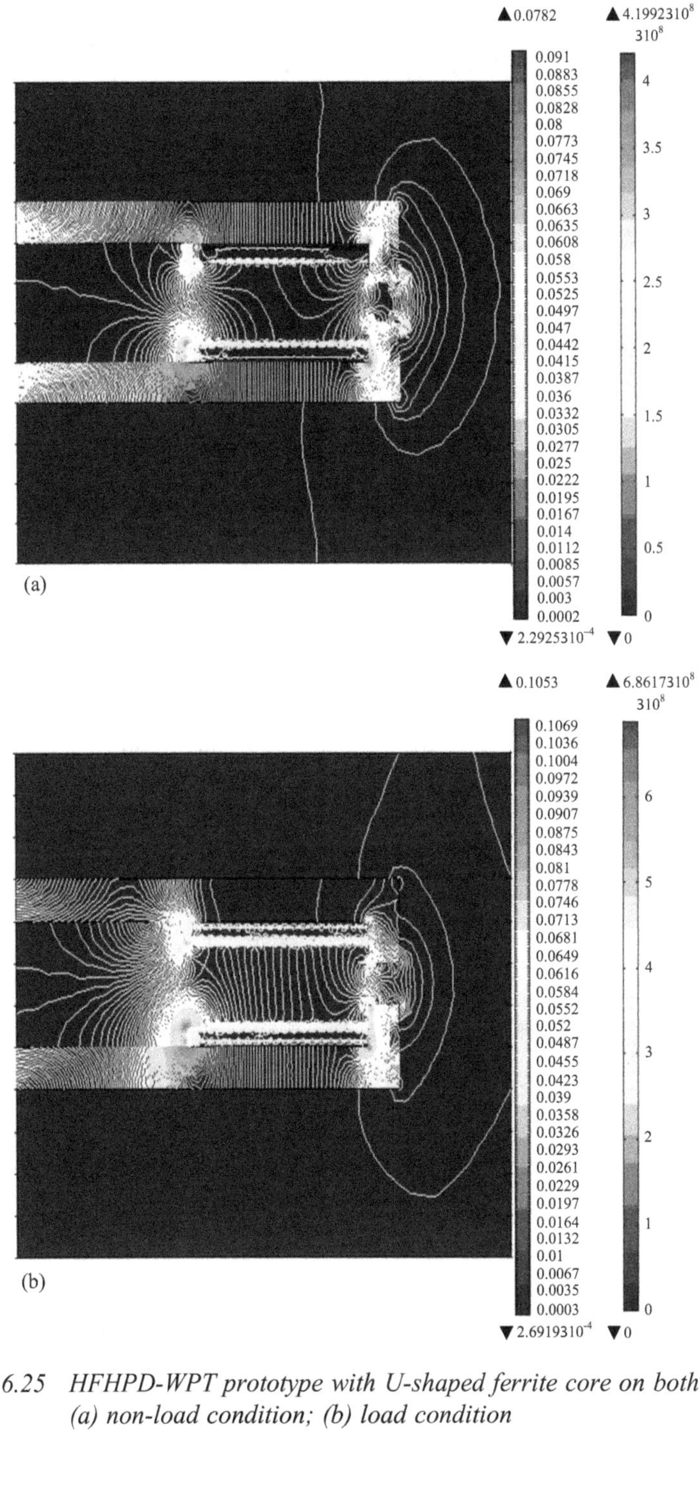

Figure 6.25 HFHPD-WPT prototype with U-shaped ferrite core on both sides: (a) non-load condition; (b) load condition

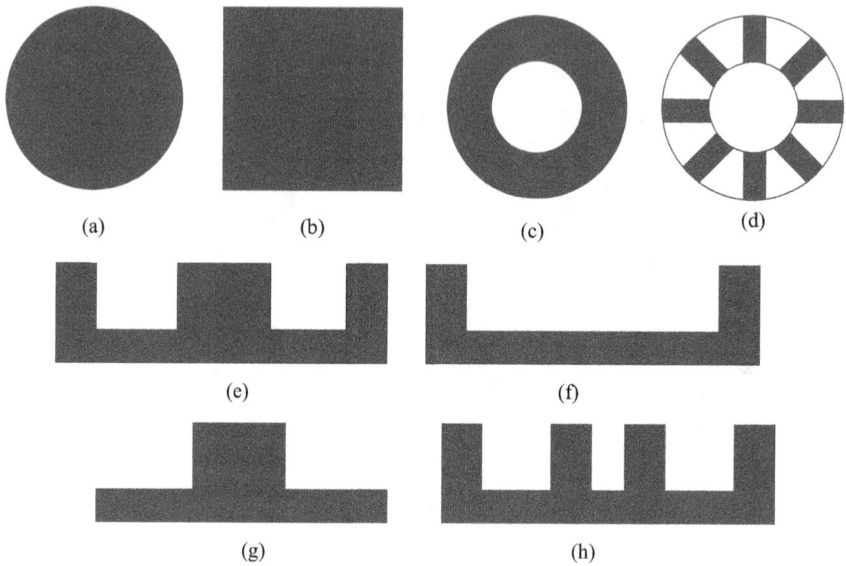

Figure 6.26 Different ferrite structures: (a) circular; (b) square; (c) hollow circular; (d) circular arranged ferrite bars; (e) E or ETD core; (f) U-shaped; (g) T-shaped; (h) modified E-core

shapes and sizes have been used in many static and dynamic applications in transmitter side as well receiver side as a pick-up module. Some of the examples of ferrite structures such as circular, square, hollow circular, circular arranged ferrite bars [34] [107], E or ETD core and U shaped [97, 98, 106] which are utilised in WCS for PEVs are presented in Figure 6.26.

As shown in Figure 6.26(g) and (h), T and modified E-core shaped ferrite cores have been proposed to develop the Shaped Magnetic Field in Resonance (SMFIR) technologies. These core structures have been utilised in static and dynamic WCS for PEV busses and trains [33, 106, 108–110].

There are numerous ferromagnetic materials which can be used as core materials for WCS. These typically have higher permeability figures around 1,000–2,500. Softer magnetic materials such as manganese zinc (Mn-Zn) and nickel zinc (Ni-Zn) are most suitable because they have higher magnetic relative permeability. In addition, they meet operating frequency ranges from kilohertz to megahertz, where the WCS for PEVs typically operates [91].

Skin Depth

When designing an HF transformer, HF losses must be investigated and considered. These losses primarily include the skin effect losses and the proximity effect. The skin effect determines the relative surface area of a conducting wire when exposed to a high switching concentric magnetic field [30]. As the frequency increases, the induced electromagnetic fields increase, causing non-uniformity within the current path.

This in turn increases the eddy currents within the conductor, effectively reducing the current flow through the conductor to a thin skin at the surface. This is represented in the form of (6.7) [91, 92, 111].

$$\delta = \sqrt{\frac{\rho}{\pi \cdot \mu \cdot f}} \tag{6.7}$$

where δ is the skin depth, ρ is the resistivity of conductor (1.68×10^{-8} Ωm), μ is the permeability and f is the oscillating frequency.

The proximity effect is caused when an HF current carrying wire is in the proximity of another wire – the current carrying wire inducing an electromagnetic current into it. As in a planar transformer, both wires would have current flowing through them, causing eddy currents to circulate through both conductors. This effect can have the potential to dramatically reduce the overall efficiency of a transformer. To curb this effect, adequate spacing between the conductors must be considered in the design [91, 92, 111].

Winding Resistance

Winding resistance is one of the important factors in the coil design because it determines the efficiency and power transfer capability of the wireless transformer. It can be calculated using (6.8), where the conductor resistivity and total length of the conductor are divided by the cross-sectional area of the conductor [87].

$$R_{DC} = \frac{\rho \cdot l}{A} \tag{6.8}$$

where R_{DC} is the DC winding resistance, ρ is the resistivity of conductor (1.68×10^{-8} Ωm), l is the total length of the conductor and A is the cross-section of the conductor.

Overall, the designing process of WCS for PEVs depends on many factors, such as air gap, efficiency, WPT methods, size, cost and several others. Table 6.8 shows the summary of different methods for WCS for PEVs [66, 69, 86, 112, 113].

6.6.3 Standards of Wireless Charging Systems for PEVs

In WCS for PEVs, it is important to consider issues related to design and safety to create user-friendly technology. This technology must require complying with international standards in order to prevent any instant long-term health and safety issues on human being, flora and fauna. As shown in Table 6.9 [114–119], many international and national organisations such as International Electrotechnical Commission (IEC), Society for Automobile Engineers (SAE), Institute of Electrical and Electronic Engineers (IEEE), International Commission on Non-Ionizing Radiation Protection (ICNIRP) and Japan Electric Vehicle Association (JEVS) are working to develop standards of WCS for PEVs.

The standards define three main areas: WCS design, communication and safety. These standards could help design and rectify the dilemma related to this

Table 6.8 Summary of wireless charging methods for PEVs

	PEVs			
Feature	**Capacitive WPT**	**Magnetic gear and permanent magnet–based WPT**	**Electromagnetic-based WPT**	
Medium of power transfer	Electric field	Electro-mechanical force	Electromagnetic field (near field)	
			Inductive power transfer	Magnetic resonance
Power transfer level	Low	Medium/High	Medium/High	Medium/High
Complexity of design	Medium	High	Medium	High
Operating frequency ranges (kHz)	100–500	100–500	10–50	10–150
Electromagnetic interference (EMI)	Medium	High	Medium	Medium/High
Development cost	Low	High	Medium	High
Power transfer distance (air gap)	Low	Medium	Low	Medium/High
Size/Volume	Low	High	Medium	Medium/High
Efficiency	Low	Low	Medium	Medium/High
Capabilities of charging EVs	Maybe	Yes	Maybe	Yes

Table 6.9 International standards of WPT for PEVs

Organisation		Relevant Standard(s)	Standard definition
Society for Automobile Engineers (SAE)	SAE	J2847/6	Wireless charging communication between PHEV and the utility grid
		J2836/6	Specific requirements of communication between the on-board charging and wireless EV supply equipment (W-EVSE) for detection, charging process and control
		J2931/6	Digital communication for wireless charging between EVSE, its utility, advance metering infrastructure (AMI) and home area network (HMI)
		UL2750	Wireless charging safety
		J2954	Wireless charging frequencies Transmitter and receiver positioning Power transfer efficiency EV's speed by power levels in dynamic WCS
Institute of Electrical and Electronic Engineers (IEEE)	IEEE	P2100.1	Wireless power and charging systems
		C95.1-2005	Safety level with respect to human exposure to radio frequency (RF) EMFs from 3 kHz to 300 kHz
		C95.7-2005	Radio-frequency safety programmes from 3 kHz to 300 GHz
		C95.3.1-2010	Measurements and computations of electric, magnetic and EMFs with respect to human exposure to such fields from 0 Hz to 100 kHz
International Electromechanical Commission (IEC)	IEC	TC69/JPT 61980-1	Part-1 General requirements of WPT system for EV
		TC69/JPT 61980-2	Part-2 Specific requirements for communication system between EV and infrastructure in WPT system for EVs
		TC69/JPT 61980-3	Part-3 Specific requirements for the magnetic field power transfer
Japan Electric Vehicle Association	JEVA	G106–109	General requirements, manual connection, software interface for inductive charging system for EVs
International Commission on Non-Ionizing Radiation Protection (ICNIRP)	ICNIRP	ICNIRP 1998	Guidelines on electric and magnetic field exposure limit or restriction in occupational and public areas (0 Hz–300 GHz)
		ICNIRP 2010	Updated guidelines on electric and magnetic field exposure limit or restriction in occupational and public areas (1 Hz–100 kHz)

technology. In the WCS design standards, the organisations will announce the power transfer levels between the transmitter and receiver, charging pad installation positions; parking and road, efficiency, operating frequency, minimum ground clearance and compatibility with other manufactures.

The Society of Automotive Engineers (SAE) has recently announced initial operating power level international standards (SAE TIR J2954) for WPT, which is presented in Table 6.10 [116]. Power classes have been divided into three power levels which depend on applications with greater than 90% efficiency (grid to battery input) and 85 kHz central frequency band ranging from 81.38 kHz to 90 kHz. The power classes 3.6 kW and 19.2 kW will be utilised in low-duty vehicle charging at home or fast charge points respectively. The 150 kW power class is defined for use in commercial vehicles such as busses.

Standards for electromagnetic compatibility (EMC) and electromagnetic interference (EMI) have been announced by IEEE C95.1, C95.7-2005, IEEE C95.3.1-2010 (3 kHz–300 GHz) and ICNIRP [1998, 2010]. Utilising these standards, several organisations such as the FCC, ANSI and NCRP have created electric and magnetic field exposure guidelines for occupational (controlled) and general public (uncontrolled) environments. These standards provide a specific absorption ratio (SAR) measurement and assessment methods for body parts and limbs with respect to radio frequency electromagnetic fields. For example, ICNIRP suggests non-ionising radiation protection limits for occupational (100 μT) and general public (27 μT) for the frequency range 3–100 kHz [120]. Table 6.11 [114, 121, 122] describes several

Table 6.10 SAE TIR J2954 international standards of WPT for light-duty electric vehicles

	WPT power class		
	WPT1	**WPT2**	**WPT3**
Maximum input WPT power ratings	3.6 kW	19.2 kW	150 kW
Application	Low-duty vehicle	Low-duty vehicle fast charge	Bus

Table 6.11 Examples of safety certificates or compliance for industry-based prototypes of WPT

Industries/Organisations	**Safety certificates/compliance/evaluation**
OLEV, KAIST	ICNIRP (62.5 mG at 20 kHz)
WiTricity	FCC regulation, IEEE and ICNIRP
EVATRAN Plugless Power	Idaho National Laboratory (INL), FCC, ICNIRP
WAVE	FCC regulation, IEEE and ICNIRP

commercial or university-based organisations which have performed safety analysis tests for their prototypes to measure exposure limit of electromagnetic fields in human environments, complied with international standards for EMC/EMI.

6.6.4 Application of Wireless Charging Systems for PEVs

Depending on their applications, WCS for PEVs can be segregated into the two fundamental scenarios to charge the battery bank into the car: static WCS and dynamic WCS.

6.6.4.1 Static Wireless Charging System

Static WCS for PEVs can bring many benefits to users. Unlike a wired charger, the static wireless charger does not require any physical connection with the car. It means that it completely performs autonomous function by utilising minimum driver's participation. It can easily remove the hassle of the PEV users who forget to plug-in their cable to charge the vehicle. In addition, it can significantly improve the trip hazard and electric shock by removing additional cables and cords. The static wireless charging system for PEVs is illustrated in Figure 6.27. Like other WCS, it consists of transmitter and receiver parking charging pad, charging module and level indicators. Most suitable areas to install the static wireless charging system is the parking areas in home and at shopping centres and by *park'n'ride* initiatives. This technology is no longer in its infancy and is being adopted by companies on a global scale. These companies include Evatran, Momentum, WiTricity and many more that have proposed or developed commercial products for PEVs, as shown in Table 6.12.

WCS charging time may vary depending on level of charging, SOC and energy storage system capacity. To improve the performance in terms of range of PEVs, a large battery bank or numerous replacement batteries are required.

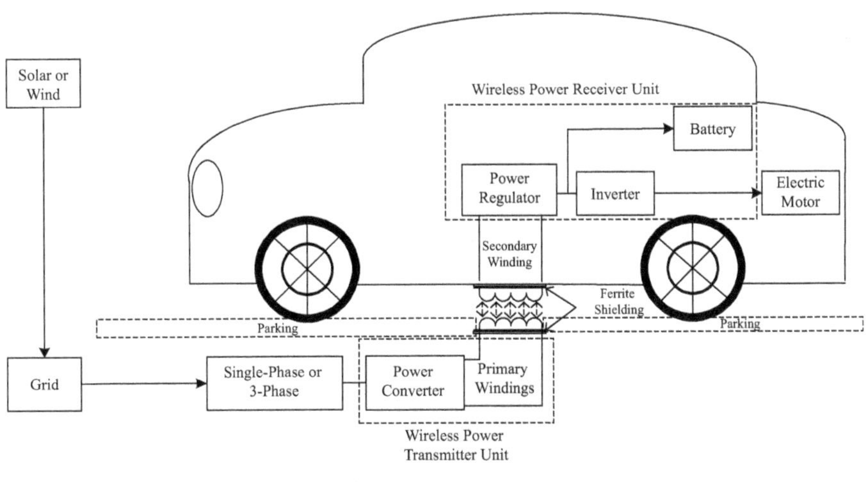

Figure 6.27 Static wireless charging system for PEVs incorporated with RES

Table 6.12 Summary of static WCS for PEVs

Inventor(s)	PEVs					
	'Known as' and prototypes	Power transfer (kw)	Operating frequency (kHz)	Air gap (mm)	Weight (kg)	Efficiency (%)
Evatran and Oak Ridge National Lab (ORNL) (commercial) [122]	Plug-less power and car	3.3 (level 2)	19.5	100	16	86–89
Hunter H. Wu, Aaron Gilchrist, Ky Sealy, Daniel Bronson, Energy Dynamics Laboratory (EDL) and Utah State University Research Foundation (USURF) [112, 125, 126]	Dual control	5 (level 2)	20	152–267	24	>90
Wireless advanced vehicle electrification [WAVE] and Utah Power Electronics Laboratory (UPEL)	Bus	5(2011) 25(2013) 50(2014–15)	20	250	–	>90
Oak Ridge National Laboratory (ORNL) [127]	Cars	6.6 and 10 200 to 260	22	160	–	85
California PATH Program, Institute of Transportation Studies, University of California, Berkeley, CA [1989–96] [128, 129]	Roadway powered vehicle project	–	–	76	–	60
University of Auckland, New Zealand [85, 86, 129, 130]	Inductively coupled power transfer (ICPT) • Cars • Train	 2–5 30	 10–40	 100–300 45	 Variable	 85

Waseda advanced Electric Micro Bus (WEB), Waseda University, Japan and Showa Aircraft Industry Co. Ltd [2010–11] [129, 131]	• Non-contact rapid charging inductive power • Electric-micro bus	30–150	22	100	35	92
The University of Georgia, Athens, GA and Southern Company Services Inc., Birmingham, AL [132, 133]	Lab prototype	3	20	160		>80
Germany Bombardier Co.	Three phase for train	–	–	65		92
WiTricity Corporation [WiT 300] commercial [134]	'Park-and-charge' system Car	3.3	85	100–200	9.6	90
Momentum Dynamic wireless power (under development) [135, 136]	Under development • Car • Light commercial trucks • Medium duty commercial trucks • Heavy vehicles (buses)	310 (Development) 30 (Development) 60 (Planned) 100 (Planned)	–	300	–	91
Qualcomm Halo^{TM} technology (under development) [137]	WEVC (under development) Cars	3.3 6.6 7 20	85	150	–	>90
Delphi Wireless charging System with start-up WiTricity & MIT Licence (under development) [138]	Cars	3.3	–	200	–	90
Inverto, Flanders drive, Mombardier [139]	Single-phase car* Three-phase car**	3.6* 22**	125–145* 140*	100	25	>90
Conductix-Wampfler [140]	Bus	60–180		40		>90

*Single phase operation; **three phase operation.

One method which can be utilised to overcome the range limitation without installing a large battery bank is the utilisation of the quasi-dynamic WCS. This type of WCS can be installed at traffic lights or high traffic start-stop lanes in urban areas to allow wireless enabled motorists to charge while they are in motion. This method is more convenient for commercial battery operated vehicles such as busses and taxies in terms of cost and time savings. This method may require an advanced monitoring system [123]. In addition, this technology can easily be incorporated with ERS such as solar and wind and quite easy to install on the existing infrastructure framework [124].

6.6.4.2 Dynamic Wireless Charging System

Dynamic or in-motion wireless charging system is one of the potential future technologies to extend the range of PEVs without increasing mass and cost of the vehicle due to large battery bank requirements, as shown in Figure 6.28. They have been also known as 'Roadway Powered EVs' (RPEVs) or 'On-Line EVs (OLEV)' [141]. In the design of dynamic or in-motion WCS, the primary transmitter is installed on the road way with high-density traffic with a high-voltage three-phase AC or DC power supply. The secondary coil/receiver is fitted underneath of the vehicle. By only switching on the charging coils as the PEV travels across them, the PEV can be charged while in motion [142]. This type of installation has the capability of reducing battery bank requirements in the dynamic wireless PEV by approximately 20% (1/3–1/5) in comparison with traditional PEVs [70]. This design could be utilised in a specific locations and pre-defined high traffic routes [143]. The operational frequency of the dynamic charging is normally very low in kilohertz range [109, 144, 145].

Even though this type of concept has its benefits, there are still some drawbacks and challenges with design. First, the main coil requires special embedded power tracks underneath the roadway in order to charge the vehicle. To meet this requirement, initial infrastructure cost would be very high. Second, it may be difficult for some wider or narrower vehicles to drive on the transmitter tracks, as this size discrepancy could create misalignment with the receiver and transmitter coils, reducing efficiency. An issue may arise as multiple WCS enabled vehicles commence heavy charging simultaneously. If the infrastructure network is not sufficient enough, power capacity may result in some vehicles running out of power mid-trip. Even with some drawbacks, the benefits of the dynamic wireless charging system can easily be noted. The utilisation of such technology is applicable in many applications, such as personal rapid transport (PRT), light rail and medium rail systems, PEV buses and cars. Table 6.13 provides some of the prototypes which have been developed by industries and research universities for dynamic WCS. A recent and advanced system of the 100 kW OLEV prototype which operates at 20 kHz frequency, developed with 80% efficiency and 200 mm air-gap, has been demonstrated. This technology has great potential, but some of the health and safety related issues need to be considered before a wider adoption of the technology is conducted. These factors will be discussed and analysed in section 6.6.5.

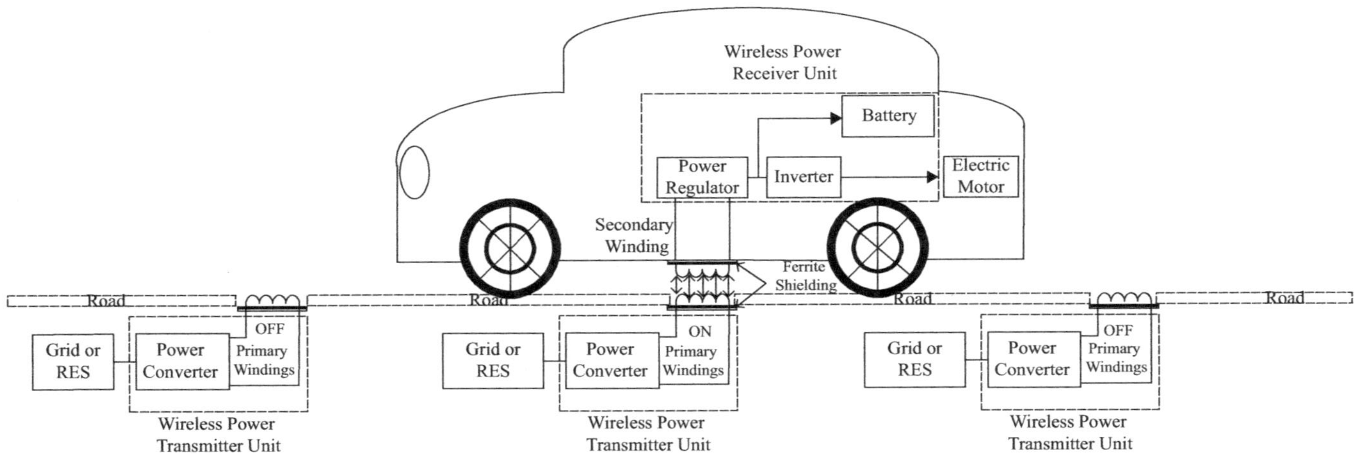

Figure 6.28 Dynamic wireless charging system for PEVs

Table 6.13 Summary of dynamic WCS for PEVs

Inventor(s)	**PEVs**					
	'Known as' and prototypes	**Pick-up power (kw)**	**Operating frequency (kHz)**	**Air gap (mm)**	**Weight (kg)**	**Efficiency (%)**
University of Auckland, New Zealand [85, 86, 146]	ICPT • Cars • Train	20–30	12.9	500	Variable	85
KAIST University, Korea	OLEV					
• 1st Generation	• Small car	3^{*}		10^{*}	20^{*}	80^{*}
• 2nd Generation	• Bus	6^{**}		170^{**}	80^{**}	72^{**}
• 3rd Generation	• SUV	15^{***}	20	170 for SUV, 200 for Bus, 120 for train***	110^{***}	71 for SUV, 83 for bus, 74 for train***
• 4th Generation [33, 106, 108, 109, 121, 147–153]	• Train	25^{****}		200^{****}	80^{****}	80^{****}
KAIST University, Korea [2012–14] [153, 154]	OLEV • Bus and train****	100^{****} (20 kW × 5)	20	200–260****	80^{****}	85^{****}
Oak Ridge National Laboratory (ORNL) [68, 155, 156]	In-motion wireless charging system • Car	–	22–23	125–175	–	90

Flanders Drive with industries and universities [139]	• Bus	80	20	100	–	88–90
EV System Lab. and Research Testing Section No. 1, Nissan Research Centre, Nissan Motor Co., Ltd, Japan [157]	Wireless dynamic charging system • 2 seater car	1	90	100	–	>90
North Carolina State University, USA [34, 120, 158]	Dynamic wireless charging • Small Lab prototype	0.3	100	170	–	7–90

*First generation; **second generation; ***third generation; ****fourth generation.

6.6.5 Wireless Safety Considerations

WCS for PEVs has the ability to remove the requirement of cables for conventional chargers but it may bring additional exposure to non-ionising electromagnetic fields (EMFs) [159], so it is necessary to define safe operating region in both normal and abnormal conditions. The abnormal conditions mean that a human or animal is underneath the vehicle while the PEV is charging, self-inflicted or forced [66]. It may be difficult to identify the 'radiation zone' for the specific WCS not only underneath the car but in areas around it. In addition, an investigation needs to be conducted on the effect of such charging devices onto patients with implantable electronic devices in their body, and their health and age status [160, 161]. The safety considerations of WCS for PEVs have concerns in three main areas: EMF exposure, electric shock and the risk of fire on objects in the charging path [88, 162, 163].

In the high-power WCS (200 kW or up to 200 A), there would be chances to exceed voltage potential between the primary and secondary windings. In order to prevent exposure to the human and fauna, it would require covering or shielding to the exposure areas in environmental conditions such as desert, heat, rain and dry weather, and absorption and seismic activities. Another hazard which is associated with high power transfer is the fire because of leakage, insulation fault or any other electrical failure. Furthermore, fire safety is one of the potential hazard for high power transfer WCS because high current flow or excessive heating may create potential threat as the WCS is mostly installed underneath the ground in public areas [88, 162, 163].

The electric shock and risk of fire can be eliminated with proper design and precaution. However, EMF issues change for a WCS depending on the power level and resonant frequency of the system. The requirements of the power transfer in WCS for PEVs are from some kilowatts to hundreds of kilowatts. In order to provide sufficient power at the secondary, receiving, side the WCS needs to generate strong magnetic field which can only be produced by applying high current at the primary, transmission, side. For this operation, the inductive charging unit should have a sufficient size to tolerate the high currents and provide strong magnetic fields in comparison with acceptable misalignment [164]. The nominal air gap between the primary and secondary windings of WCS for an PEV is between 150 mm and 280 mm due to the varying heights of a vehicle [165, 166]. Due to the larger air gap between the two windings and in case of misalignment, the leakage fluxes would be high without any proper shielding or protection. The leakage flux increases as the power requirement of the WCS also increases.

Another important issue regarding EMF is the operating frequency. Many experiments have been performed to analyse EMF exposure of the WCS for PEVs, which operates between kilohertz to megahertz ranges. It is known that every frequency has a significant impact on two important factors: EMC and EMI [167]. The experiment results indicated that the WCS could generate large amounts of EMI at HF [37, 164, 168]. It is necessary that this technology must be investigated to solve the dilemma of the effect of electromagnetic radiation on human beings, flora and fauna.

The advances in the investigation can bring more freedom to design and support the in international standardisation for this technology [169, 170]. The human exposure guidelines have been announced by IEEE C.95.1 2005 [171], ICNIRP 1998 (0 Hz–300 GHz) [172], and ICNIRP 2010 (0 Hz–100 kHz) [173]. There are many other international organisations which provide standards for different areas of the human body with regards to EMF exposure, which are elaborated in section 6.6.3. Many experiments have been performed to investigate the best operating frequency ranges and design parameters to reduce these adverse effects. For example, the 3.3 kW static WCS for PEVs at 145 kHz frequency, which was developed by WiTricity, examined a human leg using FEM simulation at around 600 mm distance. Their measured electric field and SAR values were generalised to the basic public EMF exposure restriction levels [166]. Another Level 2, 6.6 kW WCS for PEVs at greater than 90% efficiency was demonstrated and their EMF exposure limits meet the international standards for average body exposure in public zone. The human body template was used to record average four measurement points in validation of the measured results [68]. A prototype of 100 kW dynamic WCS for OLEV, which had 62.5 mG electromagnetic field (EMF) exposure at 200 mm air gap and 20 kHz frequency, was demonstrated [109, 144]. In order to bring wide acceptance, WCS for PEVs require easy to use international standards to provide environmental safety.

6.7 Conclusion

For plug-in electric vehicles to become successful, infrastructure, methods and types of charging must be diverse enough to appeal to the consumer. The development of advanced off-board charging stations and wireless charging stations can help the issue of charging diversification, and also help support the electrical grid. This can be achieved by using charging techniques such as in smart transformers with bidirectional power flow and four-quadrant power factor correction. Communications between vehicles and the distribution grid via such a smart transformer can allow for V2G support to occur. Such support not only is useful for the distribution grid, but also can provide monetary incentives to the owner of the PEVs. This is possible due to the on-demand energy generation inherent to battery storage systems. The converter systems that are used in electric vehicles must be high power, efficient and extremely light to increase performance and driving range. This creates incentives for the utilisation of existing internal parts, such as the inductors on the AC motors, to assist in the power conversion process.

References

[1] Ipakchi A., Albuyeh F. Grid of the future. *Power and Energy Magazine*. 2009 March–April:52–62.

[2] Mwasilu F., Justo J.J., Kim E.-K., Do T.D., Jung J.-W. Electric vehicles and smart grid interaction: A review on vehicle to grid and renewable energy sources integration. *Renewable and Sustainable Energy Reviews*. 2014;34(0): 501–16.

[3] Bilgin B., Emadi A. Electric motors in electrified transportation: A step toward achieving a sustainable and highly efficient transportation system. *Power Electronics Magazine, IEEE*. 2014;1(2):10–17.

[4] Duvall M., Knipping E., Alexander M., Tonachel L., Clark C. *Environmental Assessment of Plug-In Hybrid Electric Vehicles*. Palo Alto, CA; 2007.

[5] Trigg T., Telleen P. *Global EV Outlook—Understanding the Electric Vehicle Landscape to 2020*. Paris, France: International Energy Agency, April 2013.

[6] Williams B., DeShazo J., Ben-Yehuda A. *Early Plug-in Electric Vehicle Sales: Trends, Forecasts, and Determinants*. Southern California: UCLA Luskin Center for Innovation, 2012 July. Report No. 1.

[7] Gulle C. *A Conceptual Framework For The Vehicle-to-Grid (V2G) Implementation*. Urbana, IL: University of Illinois, *Urbana-Champaign*; 2007.

[8] Yilmaz M., Krein P.T. Review of the impact of vehicle-to-grid technologies on distribution systems and utility interfaces. *Power Electronics, IEEE Transactions on*. 2013;28(12):5673–89.

[9] Buchanan C.D. *Mixed Blessing: The Motor in Britian*. London: Leonard Hill; 1857.

[10] Porter D.H. *The Life and Times of Sir Goldsworthy Gurney: Gentleman Scientist and Inventor, 1793–1875*. Illustrated ed. Leigh, NE: Lehigh University Press; 1998.

[11] Schrader H. *Deutsche Autos: 1885–1920*. Stuttgart: Motorbuch Verlag; 2002.

[12] Denholm P., Ela E., Kirby B., Milligan M. *The Role of Energy Storage with Renewable Electricity Generation*. Golden, CO: National Renewable Energy Laboratory, Energy USDo; 2010 January. Report No. NREL/TP-6A2-47187.

[13] Fu Q., Hamidi A., Nasiri A., Bhavaraju V., Krstic S., Theisen P. The role of energy storage in a microgrid concept. *IEEE Electrification Magazine*. 2013 December:21–9.

[14] Roberts B.P., Sandberg C. The role of energy storage in development of smart grids. *Proceedings of the IEEE*. 2011;99(6):1139–44.

[15] Alonso O., Galbete S., Sotés M. *Sizing and Management of Energy Storage for a 100% Renewable Supply in Large Electric Systems*. In: Carbone PR, editor. Energy Storage in the Emerging Era of Smart Grids: InTech; 2011.

[16] IEEE Electrification Magazine. *Electrification Magazine, IEEE*. 2014; 2(2):73.

[17] Moses P.S., Masoum M.A.S., Hajforoosh S., editors. Overloading of distribution transformers in smart grid due to uncoordinated charging of plug-in electric vehicles. *Innovative Smart Grid Technologies (ISGT), 2012 IEEE PES*; 2012 16–20 January 2012.

[18] Sovacool B.K., Hirsh R.F. Beyond batteries: An examination of the benefits and barriers to plug-in hybrid electric vehicles (PHEVs) and avehicle-to-grid (V2G) transition. *Energy Policy*. 2009;37:1035–44.

[19] Inange S.-I., Elzinga D. *Modelling Load Shifing Using Electric Vehicles in a Smart Grid Environment*. Working Paper. 2010.

[20] IEEE. IEEE Standards Coordinating Commitee 21. *15474 IEEE Guide for Design, Operation and Integration of Distributed Resource Island Systems with Electric Power Systems*. New York: IEEE; 2011. p. 54.
[21] Kranz J. *Studies on Technology Adoption and Regulation of Smart Grids*. Berlin: Druck und Verlag: epubli GmbH; 2011.
[22] Ekanayake J., Jenkins N., Liyanage K., Wu J., Yokoyama A. *Smart Grid: Technology and Applications*. Wiley; 2012.
[23] European Comission. *Vision and Strategy for Europe's Electricity Networks of the Future*. European SmartGrids Technology Platrform. 2006.
[24] Department of Energy and Climate Change. *Smarter Grids: The Opportunity*. London, UK: 2009 December.
[25] SmartGridGB. *Smart Grid: A Race Worth Winning?* London, UK: 2012 April.
[26] Suryanarayanan S., Kryiakides E. Microgrids: An emerging technology to enhance power system reliability. *IEEE Smart Grid*. 2012 March.
[27] Chiradeja P., editor. *Benefit of Distributed Generation: A Line Loss Reduction Analysis*. Transmission and Distribution Conference and Exhibition: Asia and Pacific, IEEE/PES; 2005.
[28] Ye Y., Yi Q., Sharif H., Tipper D. A survey on smart grid communication infrastructures: motivations, requirements and challenges. *Communications Surveys & Tutorials, IEEE*. 2013;15(1):5–20.
[29] Kesler M., Kisacikoglu M.C., Tolbert L.M. Vehicle-to-grid reactive power operation using plug-in electric vehicle bidirectional offboard charger. *Industrial Electronics, IEEE Transactions on*. 2014;61(12):6778–84.
[30] Hansen L.H., Helle L., Blaabjerg F., Ritchie E., Munk-Nielsen S., Bindner H., *et al. Conceptual survey of Generators and Power Electronics for Wind Turbines*. Roskilde, Denmark: Risø National Laboratory, 2001 December. Report No. Risø-R-1205(EN).
[31] Luszcz J. High frequency harmonics emission in smart grids. 2013. In: *Power Quality Issues* [Internet]. InTech; [277–92]. Available from: http://www.intechopen.com/books/power-quality-issues/high-frequency-harmonics-emission-in-smart-grids.
[32] Lundquist J. *On Harmonic Distortion in Power Systems*. Göteborg, Sweden: Chalmers University of Technology; 2001.
[33] Boyune S., Jaegue S., Seokhwan L., Seungyong S., Yangsu K., Sungjeub J., *et al.*, editors. Design of a high power transfer pickup for on-line electric vehicle (OLEV). *Electric Vehicle Conference (IEVC), 2012 IEEE International*; 2012 4–8 March 2012.
[34] Lukic S., Pantic Z. Cutting the cord: Static and dynamic inductive wireless charging of electric vehicles. *Electrification Magazine, IEEE*. 2013;1(1):57–64.
[35] Yilmaz M., Krein P.T. Review of battery charger topologies, charging power levels, and infrastructure for plug-in electric and hybrid vehicles. *Power Electronics, IEEE Transactions on*. 2013;28(5):2151–69.
[36] (BCG) BCG. *Batteries for Electric Cars—Challenges, Oppertunities and the Outlook to 2020*. BCG; 2010.

[37] Seungmin J., Hansang L., Chong Suk S., Jong-Hoon H., Woon-Ki H., Gilsoo J. Optimal operation plan of the online electric vehicle system through establishment of a dc distribution system. *Power Electronics, IEEE Transactions on*. 2013;28(12):5878–89.

[38] Ovideo R.M., Fan Z., Gormus S., Kulkarni P. The Reign of EVs? An economic analysis from the consumer's perspective. *Electrification Magazine, IEEE*. 2014 24 June 2014:61–71.

[39] Kumar P.T.V.B.N, Suryateja S., Naveen G., Singh M., Kumar P., editors. Smart home energy management with integration of PV and storage facilities providing grid support. *Power and Energy Society General Meeting (PES), 2013 IEEE*; 2013 21–25 July 2013.

[40] United Kingdom: World Nuclear Association; 2014 [updated February 2014; cited 14 March 2014]. Available from: http://www.world-nuclear.org/info/Non-Power-Nuclear-Applications/Transport/Electricity-and-Cars/.

[41] Comission IOfSaIE. Road vehicles—*Vehicle to grid communication interface*. Part 1: General information and use-case definition; 2013.

[42] Limited S.A. *Plugs, socket-outlests, vehicle connectors and vehicle inlets—Conductive charging of electric vehicles*. Part 2: Dimensional compatability and interchangeability requirements for ac pin and contact-tube accessories: SAI Global Limited; 2014.

[43] Limited S.A. *Electric vehicle conductive charging system*. Part 24: Digital Communication between a dc EV charging station and an electric vehicle for control of dc charging: SAI Global Limited; 2014.

[44] Limited S.A. *Electric vehicle conductive charging system*. Part 23: DC electric vehicle charging station: SAI Global Limited; 2014.

[45] Monteiro V., Gonçalves H., Ferreira J.C., Afonso J.L. Batteries charging systems for electric and plug-in hybrid electric vehicles. In: Joao Carmo, editor. *New Advances in Vehicular Technology and Automotive Engineering*. InTech, Croatia; 2012.

[46] Mohan N. *First Course On—Power Electronics and Drives*. Minneapolis, MN: MNPERE Minneapolis; 2003.

[47] Sambuis E., Chon S. *TMS320C2000^{TM} DSP Controllers: A Perfect Fit for Solar Power Inverters*. Texas Instruments; 2006.

[48] AS4777.2—Inverter Requirements. AS4777—Grid Connection of Energy Systems via Inverters: Standards Australia; 2005.

[49] Rashid M.H. *Power Electronics Handbook—Devices, Circuits and Applications*. Massachusetts: Butterworth-Heinemann; 2011.

[50] Otto R.A., Putman T.H., Gyugyi L. Principles and applications of static, thyristor-controlled shunt compensators. *Power Apparatus and Systems, IEEE Transactions on*. 1978;PAS-97(5):1935–45.

[51] Melkonyan A. *High Efficiency Power Supply using New SiC Devices*. Unidruckerei, Germany: University of Kassel; 2007.

[52] Mithulananthan N., Canizares C.A., Reeve J., Rogers G.J. Comparison of PSS, SVC, and STATCOM controllers for damping power system oscillations. *Power Systems, IEEE Transactions on*. 2003;18(2):786–93.

[53] Deng Y. *Reactive Power Compensation of Transmission Lines*. Concordia: Concordia University; 2007.

[54] Noroozian M., Petersson N.A., Thorvaldson B., Nilsson B.A., Taylor C.W., editors. Benefits of SVC and STATCOM for electric utility application. *Transmission and Distribution Conference and Exposition*, 2003 IEEE PES; 2003 7–12 September 2003.

[55] Yunus H.I., Bass R.M., editors. Comparison of VSI and CSI topologies for single-phase active power filters. *Power Electronics Specialists Conference*, 1996 PESC '96 Record, 27th Annual IEEE; 1996 23–27 June 1996.

[56] Kisacikoglu M.C., Ozpineci B., Tolbert L.M. Examination of a PHEV bidirectional charger system for v2g reactive power compensation. *Applied Power Electronics Conference and Exposition (APEC)*; 21–25 February; Palm Springs, CA: IEEE; 2010:458–65.

[57] Moreno-Muñoz A. *Power Quality: Mitigation Technologies in a Distributed Environment*. Springer; 2007.

[58] Thomson B., Adams F., Dixon M., Peters S., Hucker K., Bennet M., *et al. Guide to Connecting a Distributed Generator in Victoria*. In: Victoria S., editor. Victoria; 2013.

[59] Xiu Y., Herrera L., Yi H., Jin W., editors. The detection of DC arc fault: Experimental study and fault recognition. *Applied Power Electronics Conference and Exposition (APEC), 2012 Twenty-Seventh Annual IEEE*; 2012 5–9 February 2012.

[60] Johnson J., Montoya M., McCalmont S., Katzir G., Fuks F., Earle J., *et al.*, editors. Differentiating series and parallel photovoltaic arc-faults. *Photovoltaic Specialists Conference (PVSC), 2012 38th IEEE;* 2012 3–8 June 2012.

[61] Johnson J., Pahl B., Luebke C., Pier T., Miller T., Strauch J., *et al.*, editors. Photovoltaic dc arc fault detector testing at sandia national laboratories. *Photovoltaic Specialists Conference (PVSC),* 2011 37th IEEE; 2011 19–24 June 2011.

[62] Bashirullah R. Wireless implants. *IEEE Microwave Magazine*. 2010;11(7): S14–S23.

[63] Hui S.Y.R., Ho WWC. A new generation of universal contactless battery charging platform for portable consumer electronic equipment. *Power Electronics, IEEE Transactions on*. 2005;20(3):620–7.

[64] Olvitz L., Vinko D., Svedek T., editors. Wireless power transfer for mobile phone charging device. MIPRO, *2012 Proceedings of the 35th International Convention*; 2012 21–25 May 2012.

[65] Ho S.L., Junhua W., Fu W.N., Mingui S. A comparative study between novel witricity and traditional inductive magnetic coupling in wireless charging. *Magnetics, IEEE Transactions on*. 2011;47(5):1522–5.

[66] Li S., Mi C. Wireless power transfer for electric vehicle applications. *Emerging and Selected Topics in Power Electronics, IEEE Journal of.* 2014;PP(99):1.

[67] Lorico A., Taiber J., Yanni T., editors. Effect of inductive power technology systems on battery-electric vehicle design. IECON 2011—*37th Annual Conference on IEEE Industrial Electronics Society*; 2011 7–10 November 2011.

[68] Milller J.M. *Wireless Plug-in Electric Vehicle (PEV) Charging*; 2012:21pp.
[69] Ching T.W., Wong Y.S., editors. Review of wireless charging technologies for electric vehicles. *Power Electronics Systems and Applications (PESA), 2013 5th International Conference on*; 2013 11–13 December 2013.
[70] Musavi F., Edington M., Eberle W., editors. Wireless power transfer: A survey of EV battery charging technologies. *Energy Conversion Congress and Exposition (ECCE), 2012 IEEE;* 2012 15–20 September 2012.
[71] Rose J.A., Cates J.A. Capacitive charge coupling with dual connector assemblies and charging system. U.S. Patent 5 714 864 A February 3, 1998.
[72] Culurciello E., Andreou A.G. Capacitive inter-chip data and power transfer for 3D VLSI. *Circuits and Systems II: Express Briefs, IEEE Transactions on*. 2006;53(12):1348–52.
[73] Chao L., Hu A.P., Covic G.A., Nair N.C. Comparative study of ccpt systems with two different inductor tuning positions. *Power Electronics, IEEE Transactions on*. 2012;27(1):294–306.
[74] Xia C.-Y., Li C.-W., Zhang J., editors. Analysis of power transfer characteristic of capacitive power transfer system and inductively coupled power transfer system. *Mechatronic Science, Electric Engineering and Computer (MEC), 2011 International Conference on*; 2011 19–22 August 2011.
[75] Kline M., Izyumin, I., Boser BaS, S., editors. Capacitive power transfer for contactless charging. *Applied Power Electronics Conference and Exposition (APEC), 2011 Twenty-Sixth Annual IEEE*; 2011, Fort Worth, TX.
[76] Chao L., Hu A.P., Nair N.C., editors. Coupling study of a rotary capacitive power transfer system. *Industrial Technology, 2009 ICIT 2009 IEEE International Conference on*; 2009 10–13 February 2009.
[77] Funato H., Kobayashi H., Kitabayashi T., editors. Analysis of transfer power of capacitive power transfer system. *Power Electronics and Drive Systems (PEDS), 2013 IEEE 10th International Conference on*; 2013 22–25 April 2013.
[78] Liu C., Hu A.P., Dai X., editors. A contactless power transfer system with capacitively coupled matrix pad. *Energy Conversion Congress and Exposition (ECCE)*, 2011 IEEE; 2011, Phoenix, AZ.
[79] Ludois D.C., Erickson M.J., Reed J.K. Aerodynamic fluid bearings for translational and rotating capacitors in noncontact capacitive power transfer systems. *Industry Applications, IEEE Transactions on*. 2014;50(2):1025–33.
[80] Li W. *High Frequency Wireless Power Transmission at Low Frequency using Permanent Magnet Coupling*. Master Applied Science Thesis. Vancouver: 2007.
[81] Suzuki S., Ishihara M., Kobayashi Y. The improvement of the noninvasive power-supply system using magnetic coupling for medical implants. *Magnetics, IEEE Transactions on*. 2011;47(10):2811–4.
[82] Glaser P.E. Method and apparatus for converting solar radiation to electrical power. United States of America patent 3781647. 1973 December 25.
[83] Tesla N. *System of Transmission of Electrical Energy*. New York patent 645576. 1900 March 20.

[84] Tesla N. *Appartus for Transmitting Electrical Energy*. New York patent 1119732. 1914 December 1.

[85] Covic G.A., Boys J.T. Modern trends in inductive power transfer for transportation applications. *Emerging and Selected Topics in Power Electronics, IEEE Journal of.* 2013;1(1):28–41.

[86] Covic G.A., Boys J.T. Inductive power transfer. *Proceedings of the IEEE.* 2013;101(6):1276–89.

[87] Chopra S. *Contactless Power Transfer For Electric vehicle Charging Application* [Master]. Delf University of Technology: Delf University of Technology; 2011.

[88] Jiang H., Brazis P., Tabaddor M., Bablo J., editors. Safety considerations of wireless charger for electric vehicles—A review paper. *2012 IEEE Symposium on Product Compliance Engineering (ISPCE)*; 2012, Portland, OR.

[89] Kurs A.B., Karalis A., Moffatt R., Joannopoulos J.D., Fisher P.H., Solijac M. Wireless power transfer via strongley coupled magnetic resonances. *International Science Journal*. 2007;317:83–6.

[90] Zhen Ning L., Chinga R.A., Ryan T., Jenshan L. Design and test of a high-power high-efficiency loosely coupled planar wireless power transfer system. *Industrial Electronics, IEEE Transactions on*. 2009;56(5):1801–12.

[91] Bossche A., Valchev V.C. *Inductors and Transformers for Power Electronics*. 6000 Broken Sound Parkway NW, Suite 300: 2005 by Taylor & Francis Group, LLC; 2005.

[92] MCLYMAN CWT. *Transformer and Inductor design Handbook*. Third ed. United States of America: Kg Magnetics, Inc.; 2004.

[93] Waffenschmidt E., Research P. Qi and the Wireless Power Consortium: Wireless Power Consortium; 2014 [cited 30 August 2014]. Available from: http://www.wirelesspowerconsortium.com/.

[94] Chwei-Sen W., Covic G.A., Stielau O.H. Power transfer capability and bifurcation phenomena of loosely coupled inductive power transfer systems. *Industrial Electronics, IEEE Transactions on*. 2004;51(1):148–57.

[95] Cederlöf M. *Inductive Charging of Electrical Vehicles—System Study*. Stockholm, Sweden; 2012.

[96] Prasanth V. *Wireless Power Transfer for E-Mobility*. Master of Science Thesis. Delft University of Technology; 2012.

[97] Panchal C., Leskarac D., Lu J., Stegen S., editors. Investigation of flux leakages and EMC problems in wireless charging systems for EV and other mobile applications. *Environmental Electromagnetics (CEEM), 2012 6th Asia-Pacific Conference on*; 2012 6–9 November 2012.

[98] Panchal C., Leskarac D., Lu J., Stegen S., editors. Investigation of magnetic flux distribution of EV wireless charging systems. *Asia-Pacific International Symposium and Exhibition on Electromagnetic Compatibility: APEMC 2013*; 2013; Melbourne, Australia: Engineer Australia.

[99] Bhattacharya S., Tan Y.K., editors. Design of static wireless charging coils for integration into electric vehicle. *Sustainable Energy Technologies (ICSET), 2012 IEEE Third International Conference on*; 2012 24–27 September 2012.

[100] McDonough M., Fahimi B., editors. Comparison between circular and square coils for use in wireless power transmission. *Computation in Electromagnetics (CEM 2014), 9th IET International Conference on*; 2014 31 March–1 April 2014.

[101] Nindl T. *Qualcomm Halo^{TM} WEVC interoperability multi-coil resonant magnetic induction*. [Presentation]. In press 2014.

[102] Ombach G. *Wireless EV charging, optimum operating frequency selection for power range 3.3 and 6.6kW*. [Power Point Presentation], 2013. Available from: http://www.ecartec.com/wp-content/uploads/2015/03/Juli-2014-Qualcomm.pdf

[103] Xun L., Hui S.Y.R. Optimal design of a hybrid winding structure for planar contactless battery charging platform. *Power Electronics, IEEE Transactions on*. 2008;23(1):455–63.

[104] Panchal C., Jun-Wei L. High frequency planar transformer (HFPT) for universal contact-less battery charging platform. *Magnetics, IEEE Transactions on*. 2011;47(10):2764–7.

[105] Hongseok K., Jonghyun C., Seungyoung A., Jonghoon K., Joungho K., editors. Suppression of leakage magnetic field from a wireless power transfer system using ferrimagnetic material and metallic shielding. *Electromagnetic Compatibility (EMC), 2012 IEEE International Symposium on*; 2012 6–10 August 2012.

[106] Jiseong K., Jonghoon K., Sunkyu K., Hongseok K., In-Soo S., Nam Pyo S., *et al.*, Coil design and shielding methods for a magnetic resonant wireless power transfer system. *Proceedings of the IEEE.* 2013;101(6):1332–42.

[107] Budhia M., Covic G.A., Boys J.T. Design and optimization of circular magnetic structures for lumped inductive power transfer systems. *Power Electronics, IEEE Transactions on*. 2011;26(11):3096–108.

[108] Jaegue S., Boyune S., Seokhwan L., Seungyong S., Yangsu K., Guho J., *et al.*, editors. Contactless power transfer systems for on-line electric vehicle (OLEV). *Electric Vehicle Conference (IEVC), 2012 IEEE International*; 2012 4–8 March 2012.

[109] Seungyoung A., Junso P., Taigon S., Heejae L., Jung-Gun B., Deogsoo K., *et al.*, editors. Low frequency electromagnetic field reduction techniques for the on-line electric vehicle (OLEV). *Electromagnetic Compatibility (EMC), 2010 IEEE International Symposium on*; 2010 25–30 July 2010.

[110] Yangbae C., Seongwook P., Jiseong K., Hongseok K., Kiwon H., Joungho K., *et al.*, editors. System and electromagnetic compatibility of resonance coupling wireless power transfer in on-line electric vehicle. *Antennas and Propagation (ISAP), 2012 International Symposium on*; 2012 29 October 2012–2 November 2012.

[111] Wong F.K. *High Frequency Transformer for Switching Mode Power Supplies*. Brisbane: School of Microelectronic Engineering, Faculty of Engineering and Information Technology, Griffith University; 2004.

[112] Wu H.H., Gilchrist A., Sealy K., Israelsen P., Muhs J., editors. A review on inductive charging for electric vehicles. *Electric Machines & Drives Conference (IEMDC), 2011 IEEE International*; 2011 15–18 May 2011.

[113] Musavi F., Edinton M., Eberle W., editors. Wireless Power transfer: A survey of Ev battery charging technologies. *Energy Conversion Congress and Exposition (ECCE), 2012 IEEE*; 2012, Raleigh, NC.

[114] Brecher A., David Arthur P.E. *Review and Evaluation of Wireless Power Transfer (WPT) for Electric Transit Applications*. [Report]. In press 2014.

[115] IEEE. *IEEE Standard for Safety Levels with Respect to Human Exposure Radio Frequency Electromagnetics Fields, 3 kHz to 300 GHz*. IEEE C 951:2005. USA, 2006.

[116] Ponticel P. Wireless charging advances with selection of 85-kHz charging frequency. SAE International; 2013 [cited 1 January 2014]. Available from: http://articles.sae.org/12647/.

[117] ICNIRP Guidelines Electric and Magnetic Fields and Health: EMFs.info; 2014 [cited 20 September 2014]. Available from: http://www.emfs.info/limits/5060-2/.

[118] Scholer R. *SAE J2836^{TM}, J2847, J2931, J2953 PEV Communications Update*. [Presentation]. In press 2012.

[119] Yu S. P2100.1—Standard Specifications for Wireless Power and Charging Systems: IEEE Standards Association; 2013 [cited 20 September 2014]. Available from: http://standards.ieee.org/news/2013/ieee_p2100_1_pma.html.

[120] Kibok L., Pantic Z., Lukic S., editors. Field containment in dynamic wireless charging systems through source-reciever interaction. *Energy Conversion Congress and Exposition (ECCE), 2013 IEEE*; 2013 15–19 September 2013.

[121] Rim CT. *The Development and Deployment of On-Line Electric Vehicles (OLEV)*. [Presentation]. In press 2013.

[122] *Evatran^{TM} Awarded Department of Energy Sub-Contract To Integrate High Power Wireless Charging Technology Into Production Electric Vehicles* [press release]. PR Newswire Association LLC: PR Newswire Association LLC, 18 March 2013.

[123] Tamer N., Mohrehkesh S., editors. Toward a wireless charging for battery electric vehicles at traffic intersections. *Intelligent Transportation Systems (ITSC), 2011 14th International IEEE Conference on*; 2011; Washington, DC.

[124] Turki F., Guetif A., Sourkounis C., editors. Contactless charging electric vehicles with renewable energy. *Renewable Energy Congress (IREC)*, 2014 5th International; 2014 25–27 March 2014.

[125] Wu H.H., Gilchrist A., Sealy K., Bronson D., editors. A 90 percent efficient 5 kW inductive charger for EVs. *Energy Conversion Congress and Exposition (ECCE), 2012 IEEE*; 2012 15–20 September 2012.

[126] Wu H.H., Gilchrist A., Sealy K.D., Bronson D. A high efficiency 5 kW inductive charger for evs using dual side control. *Industrial Informatics, IEEE Transactions on*. 2012;8(3):585–95.

[127] Wireless Charging Research Activities Around the World [Society News]. *Power Electronics Magazine, IEEE*; 2014;1(2):30–8.

[128] Team P. Roadway Powered Electric Vehicle Project Parametric Studies: Phase 3D Final Report: California PATH Program, Institute of Transportation Studies,

University of California, Berkeley, CA, 1996 October. Report No. UCB-ITS-PRR-96-28. Available from http://www.path.berkeley.edu/.

[129] Jin H., Sungwoo L., Changbyung P., Gyu-Hyeoung C., Chun-Taek R., editors. High performance inductive power transfer system with narrow rail width for on-line electric vehicles. *Energy Conversion Congress and Exposition (ECCE), 2010 IEEE*; 2010 12–16 September 2010.

[130] Covic G.A., Boys J.T., Lu H.G., editors. A three-phase inductively coupled power transfer system. *Industrial Electronics and Applications, 2006 1ST IEEE Conference on*; 2006 24–26 May 2006.

[131] Kobayashi K., Pontefract T., Kamiya Y., Daisho Y., editors. Development and performance evaluation of a non-contact rapid charging inductive power supply system for electric micro-bus. *Vehicle Power and Propulsion Conference (VPPC), 2011 IEEE*; 2011 6–9 September 2011.

[132] Gao Y., Farley K.B., Tse Z.T.H., editors. Investigating safety issues related to electric vehicle wireless charging technology. *Transportation Electrification Conference and Expo (ITEC), 2014 IEEE*; 2014 15–18 June 2014.

[133] Gao Y., Farley K.B., Tse ZTH. Investigating Safety Issues of Electric Vehicle Wireless Charging Medical Robotics Lab: University of Georgia; 2014 [cited 1 October 2014]. Available from: http://wireless.uga.edu/EVWC-ITECconference.pdf.

[134] WiT-3300™ Resonator Pair R2.3 WiTricity Corporation: WiTricity Corporation; 2014 [cited 27 September 2014]. Available from: http://www.witricity.com/assets/WiT-3300_R2.3_DS.pdf.

[135] Herron D. Wirelessly Charging Stationary Electric Cars Green-Transportation.info: GreenTransportation.info; 2014 [cited 27 September 2014]. Available from: http://greentransportation.info/ev-charging/wireless/wirelessly-charging-stationary.html.

[136] Cobb J. Momentum Dynamics' Wireless Charging Could Relegate Plugs to History Hybridcars.com: Hybridcars; 2014 [cited 27 September 2014]. Available from: http://www.hybridcars.com/momentum-dynamics-wireless-charging-could-relegate-plugs-to-history/.

[137] Qualcomm Halo Wireless Electric Vehicle Charging Media Centre: Qualcomm Incorporated; 2014 [cited 27 September 2014]. Available from: http://www.qualcommhalo.com/index.php/media-centre.html?id=41#media-centre.

[138] Delphi Wireless Charging System Delphi: Delphi; 2011 [cited 27 September 2014]. Available from: http://delphi.com/shared/pdf/ppd/pwrelec/wireless-charging-system.pdf.

[139] Perik H. *Practical EV Integration Cases for Static and Dynamic Wireless Power Transfer*. [Presentation]. In press 2013.

[140] Charging Electric Buses Quickly and Efficiently: Bus Stops Fitted with Modular Components make "Charge & Go" Simple to Implement Conductix-Wampfler: Conductix-Wampfler; 2013 [cited September 2014]. Available from: http://www.conductix.com.au/en/news/2013-05-29/charging-electric-buses-quickly-and-efficiently-bus-stops-fitted-modular-components-make-charge-go.

[141] Sungwoo L., Jin H., Changbyung P., Nam-Sup C., Gyu-Hyeoung C., Chun-Taek R., editors. On-Line Electric Vehicle using inductive power transfer system. *Energy Conversion Congress and Exposition (ECCE), 2010 IEEE*; 2010 12–16 September 2010.

[142] Jang Y.J., Ko Y.D., Jeong S., editors. Optimal design of the wireless charging electric vehicle. *2012 IEEE International Electric Vehicle Conference (IEVC)*; 2012, South Carolina.

[143] Young Jae J., Young Dae K., Seungmin J., editors. Optimal design of the wireless charging electric vehicle. *Electric Vehicle Conference (IEVC), 2012 IEEE International*; 2012 4–8 March 2012.

[144] Seungyoung A., Joungho K., editors. Magnetic field design for high efficient and low EMF wireless power transfer in on-line electric vehicle. *Antennas and Propagation (EUCAP), Proceedings of the 5th European Conference on*; 2011 11–15 April 2011.

[145] In-Soo S., Jedok K., editors. Electric vehicle on-road dynamic charging system with wireless power transfer technology. *Electric Machines & Drives Conference (IEMDC), 2013 IEEE International*; 2013 12–15 May 2013.

[146] Covic G.A., Elliott G., Stielau O.H., Green R.M., Boys J.T., editors. The design of a contact-less energy transfer system for a people mover system. *Power System Technology, 2000 Proceedings PowerCon 2000 International Conference on*; 2000.

[147] Sungwoo L., Wooyoung L., Jin H., Hyun-Jae K., Changbyung P., Gyu-Hyeong C., *et al.*, editors. Active EMF cancellation method for I-type pickup of on-line electric vehicles. *Applied Power Electronics Conference and Exposition (APEC), 2011 Twenty-Sixth Annual IEEE*; 2011 6–11 March 2011.

[148] Jin H., Wooyoung L., Gyu-Hyeong C., Byunghun L., Chun-Taek R., editors. Characterization of novel inductive power transfer systems for on-line electric vehicles. *Applied Power Electronics Conference and Exposition (APEC), 2011 Twenty-Sixth Annual IEEE*; 2011 6–11 March 2011.

[149] Young Jae J., Young Dae K., Seungmin J., editors. Creating innovation with systems integration—Road and vehicle integrated electric transportation system. *Systems Conference (SysCon), 2012 IEEE International*; 2012 19–22 March 2012.

[150] Jaegue S., Seungyong S., Yangsu K., Seungyoung A., Seokhwan L., Guho J., *et al.* Design and implementation of shaped magnetic-resonance-based wireless power transfer system for roadway-powered moving electric vehicles. *Industrial Electronics, IEEE Transactions on*. 2014; 61(3):1179–92.

[151] Boyune S., Jaegue S., Sanghoon C., Seungyong S., Seokhwan L., Yangsu K., *et al.*, editors. Design of a pickup with compensation winding for on-line electric vehicle (OLEV). *Wireless Power Transfer (WPT), 2013 IEEE*; 2013 15–16 May 2013.

[152] Jaegue S., Boyune S., Seungyong S., Sanghoon C., Yangsu K., Guho J., *et al.*, editors. Design of buried power line for roadway-powered electric vehicle system. *Wireless Power Transfer (WPT), 2013 IEEE*; 2013 15–16 May 2013.

[153] Hwang K., Kim S., Kim S., Chun Y., Ahn S. Design of wireless power transfer system for railway application. *International Journal of Railway*. 2012;5(4):167–74.

[154] Hlyoon. KAIST Develops Wireless Power Transfer Technology for High Capacity Transit: Korea Advanced Institute of Science and Technology; 2013 [updated 19 February 2013; cited 27 September 2014]. Available from: http://www.kaist.ac.kr/_prog/_board/?mode=V&no=10390&code=ed_news&site_dvs_cd=en&menu_dvs_cd=0601&list_typ=B&skey=&sval=&smonth=&site_dvs=&GotoPage=22.

[155] Onar O.C., Miller J.M., Campbell S.L., Coomer C., White C.P., Seiber L.E., editors. A novel wireless power transfer for in-motion EV/PHEV charging. *Applied Power Electronics Conference and Exposition (APEC), 2013 Twenty-Eighth Annual IEEE*; 2013 17–21 March 2013.

[156] Miller J.M., Onar O.C., White C., Campbell S., Coomer C., Seiber L., *et al.* Demonstrating dynamic wireless charging of an electric vehicle: The benefit of electrochemical capacitor smoothing. *Power Electronics Magazine, IEEE*. 2014;1(1):12–24.

[157] Throngnumchai K., Hanamura A., Naruse Y., Takeda K., editors. Design and evaluation of a wireless power transfer system with road embedded transmitter coils for dynamic charging of electric vehicles. *Electric Vehicle Symposium and Exhibition (EVS27), 2013 World;* 2013 17–20 November 2013.

[158] Kibok L., Pantic Z., Lukic S.M. Reflexive field containment in dynamic inductive power transfer systems. *Power Electronics, IEEE Transactions on*. 2014;29(9):4592–602.

[159] Fnato H., Chiku Y., Harakawa K., editors. Wireless power distribution with capacitive coupling excited by switched mode active negative capacitor. *Electrical Machines and Systems (ICEMS), 2010 International Conference on*; 2010 10–13 October 2010.

[160] Chinnachamy P., Daubert J.P., Lakshmanadoss U. Electro-magnetic interference of pacemakers. 2011. In: *Modern Pacemakers-Present and Future* [Internet]. Europe and China: INTECH: 229–52.

[161] Baste V., Riise T., Moen B.E. Radiofrequency electromagnetic fields; male infertility and sex ratio of offspring. *European Journal of Epidemiology*. 2008;23(5):369–77.

[162] Hui S.Y.R., Wenxing Z., Lee C.K. A critical review of recent progress in mid-range wireless power transfer. *Power Electronics, IEEE Transactions on*. 2014;29(9):4500–11.

[163] Christ A., Douglas M.G., Roman J.M., Cooper E.B., Sample A.P, Waters B.H, *et al.*, Evaluation of wireless resonant power transfer systems with human electromagnetic exposure limits. *Electromagnetic Compatibility, IEEE Transactions on*. 2013;55(2):265–74.

[164] Barth H., Jung M., Braun M., Schmülling B., Reker U. Concept evaluation of an inductive charging system for electric vehicles. *3rd European*

Conference SmartGrids and E-Mobility; 17–18 October 2011; Munchen, Germany 2011.

[165] Wu H. *Energy Dynamics Laboratory, Wireless Power Transfer for Electric Vehicles* [PDF], 2012. 10 December 2013. Available from: http://arpa-e.energy.gov/sites/default/files/documents/files/Wu%20Wireless%20Power.pdf.

[166] Kesler M. *Highly Resonant Wireless Power Transfer: Safe, Efficient, and Over Distance*. ©WiTricity Corporation: 2013.

[167] Carobolante F. Qualcomm 10 April 2013. Available from: http://www.qualcomm.com/media/documents/files/psma-roadmap-wireless-power-transfer.pdf.

[168] Ding P.-P., Pichon L., Bernard L., Razek A. Electromagnetic field computation in human body exposed to wireless inductive charging system. *Conference on Computation of Electromagnetic Fields*, COMPUMAG 2013; Budapest, Hungary 2013.

[169] Sang Wook P., Wake K., Watanabe S. Calculation errors of the electric field induced in a human body under quasi-static approximation conditions. *Microwave Theory and Techniques, IEEE Transactions on*. 2013;61(5): 2153–60.

[170] Harris L.-R., Zhadobov M., Chahat N., Sauleau R. Electromagnetic dosimetry for adult and child models within a car: multi-exposure scenarios. *International Journal of Microwave and Wireless Technologies*. 2011;3(6): 707–15.

[171] Gao J. Traveling magnetic field for homogeneous wireless power transmission. *Power Delivery, IEEE Transactions on*. 2007;22(1):507–14.

[172] Guidelines for limiting exposure to time-varying electric, magnetic, and electromagnetic fields (up to 300 GHz). *Health Physics*. 1998;74(4): 494–522

[173] Guidelines for limiting exposure to time-varying electric, magnetic, and electromagnetic fields (1 Hz–100 kHz). *Health Physics*. 2010;99(6):818–36.

Chapter 7

Economic, Social and Environmental Dimensions of PHEV in the Smart Grid

Christopher J. Bennett, Markos Katsanevakis* and Rodney A. Stewart**

Abstract

Overcoming the technical hurdles to implementing plug-in hybrid electric vehicles (PHEV) technology into the smart grid is only one aspect of this disruptive transitional process. To ensure the rapid diffusion and efficient integration of PHEV in the smart grid, a range of governance, economic, social and environmental dimensions must also be considered and challenges addressed. Providing a robust governance framework is paramount, as it will drive both positive and perverse industry behaviours. Such frameworks must provide a set of rules and incentives to promote a stable market environment for PHEV roll-out over the long-term. Importantly, a well-designed governance framework will underpin the necessary economic thrust for a PHEV market to get established and grow. Such business drivers are sometimes not immediately obvious and are hard to quantify under current market conditions, such as quantifying the monetary benefits of distributed PHEV for the purpose of grid peak demand management and control. The economic drivers of PHEV are largely related to the capacity and related cost of energy storage and the provision of distributed power systems for resupplying them as required. Social dimensions are often multi-faceted and complex, but without convincing consumers that PHEV is a necessary transformative technology that is also economically and environmentally superior to traditional transportation methods, PHEV will never gain sufficient traction. Moreover, many people are still not convinced that the battery systems used in PHEV, which are mostly composed of Lithium, are sustainable. Proven cradle-to-grave environmentally friendly sourcing and life cycle management strategies for PHEV batteries is essential to ensure that this technology is acknowledged as a better solution than traditional liquid and gas transport fuels. To seize the full suite of opportunities and benefits available from PHEV technology, all of these intertwined challenges must be addressed in an integrated manner. Untangling these issues and many others and then formulating

*School of Engineering, Griffith University, Queensland, Australia

multi-pronged strategies to overcome them in a concurrent fashion is a challenge, but one which must be undertaken in order to progress PHEV diffusion in society globally. This chapter seeks to unpack these four non-technical dimensions to PHEV diffusion. It will look at the opportunities and benefits related to each dimension along with the associated challenges and strategies to address them.

7.1 Introduction

While PHEV represent a significant opportunity for society, there are a range of economic, social and environmental barriers that need to be addressed before it becomes the predominant transport technology. The economic, social and environmental dimensions are complex, multi-faceted and interrelated.

In the economic dimension, decision makers need information on how PHEV will affect people's behaviours, what would influence the uptake of electric vehicles and what would be the impacts on infrastructure such as the electricity grid. With this information, decision makers can then craft implementation frameworks that better consider the regulatory requirements as well as price incentives to achieve the best outcome for society and owners of electric vehicles. To do so, questions needs to be asked, such as what effects will PHEV charging from the electricity network have? What incentives can be put in place to minimise adverse impacts? And, what benefits can be yielded through EV proliferation and how can this proliferation be achieved? Using economic theory, these questions are answered under two key concepts: *Smart Electric Vehicle Charging* and *Electric Vehicle as Distributed Energy Resources*.

The social dimension concerns itself with the attitudes and beliefs of individuals in society and PHEV. When individuals, potential owners and consumers of PHEV, concern themselves with a mode of transportation they consider their needs and wants against the knowledge that they have at hand. Questions of convenience, trust of the technology, safety and risk enter their minds. In turn, industry decision makers should take into account whether or not PHEV will fulfil the potential owner's want for autonomy (travel where and when they want to travel), which they have come to expect since the introduction of the car. Understandably, owners will be concerned about PHEV purchase costs, servicing and maintenance requirements and costs, resale value over time, and life cycle ownership costs.

The environmental dimension addresses the environmental impacts that PHEV have from cradle to grave in comparison to vehicles currently available on the market. It is essential that PHEV be weighed up against conventional and other new transport technologies, and its environmental credentials assessed in terms of air quality, greenhouse gas emissions, pollutant runoff, battery waste management, to name a few. If the rollout of PHEV will provide better long-term conditions for human health and the environment, then it may be considered over conventional transport technologies even if it has a higher life cycle cost. The first section of this chapter addresses the critical economic dimensions of widespread PHEV implementation.

7.2 Economic Dimension

At present, the greatest challenge to the societal uptake of PHEV is economic in nature. On an aggregated level, the direct cost of the vehicle to consumers, the costs endured by electricity network operators and the costs of managing waste are all factors. Analysis of the charging operations of PHEVs, the ancillary functions that PHEVs can provide and the associated control systems are required in order to reduce costs and, in turn, yield greater benefit. PHEV charging may have a range of impacts on the electricity network due to the anticipated variability in PHEV charging, how rapidly PHEVs charge and when they are charging. Potential ancillary functions are 'supporting the grid', which acts to distribute energy resources, potentially mitigating negative economic effects or providing net benefits to network operators. In the event that PHEVs became cost-neutral compared to conventional combustion engine vehicles, additional waste management infrastructure would be required to deal with the increased quantity of battery waste.

The residential LV distribution network's load has two main characteristic profiles: summer and winter (illustrated in Figure 7.1 [Bennett *et al.*, 2014a]). The load profile for summer is such that the lowest loads are experienced during late night and early morning. From early morning onwards, the network load gradually increases until the peak demand period is approached. There is a large increase in load during peak demand periods, typically occurring between 5 p.m. and 9 p.m., dependent upon the specific network. The winter load profile also experiences the least demand during late night and early morning. Conversely, two peak demand periods occur between 8 a.m. and 10 a.m., and 5 p.m. and 9 p.m. Peak demand periods have the greatest effect on the design and network augmentation schedules of the network, such that the network must accommodate these periods so no failures occur. In turn, significant investment is made to the network to ensure continuous operation for proportionally brief operational periods (Nelson *et al.*, 2011 & Simshauser *et al.*, 2011). The widespread uptake of PHEVs in residential areas would have a profound influence on network infrastructure. If PHEVs were to charge during peak demand periods, networks would run closer to full capacity, possibly warranting network augmentation (Lemoine *et al.*, 2008; Putrus *et al.*, 2009 & Qian *et al.*, 2011). As a significant portion of the cost of electricity derives from its transmission and distribution, a consequence would be an increased electricity cost to consumers. Thus, the utility of PHEVs to owners would decrease, and those not owning them would be subsidising network augmentation. To avoid this, the design and operation of PHEVs should be such that they are not economically negative. This can be achieved by PHEVs supporting the grid. A number of possibilities for PHEV support of the grid are pending quantitative research prior to implementation:

- smart charging of PHEVs
- PHEVs as distributed energy resources
- the use of PHEVs to utilise pre-existing solar PV generation

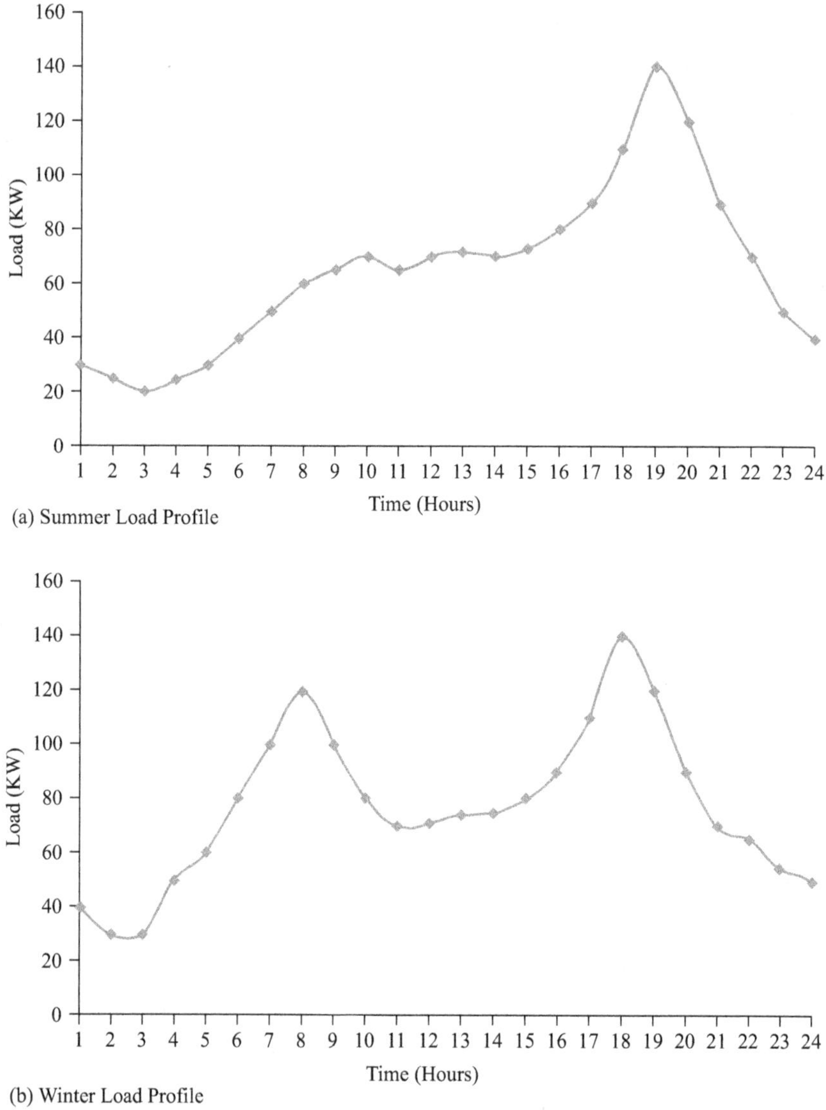

Figure 7.1 Summer and winter load profiles

7.2.1 Smart Electric Vehicle Charging

Smart charging of PHEVs entails charging in a way that does not detrimentally affect the grid. An example of smart charging would be restrictions on when PHEVs can be charged. A second example is charging PHEVs from energy sources not directly connected to the electricity grid. Curtailing PHEV charging periods could be enacted through tariffs, regulatory dicta or network operator communications systems.

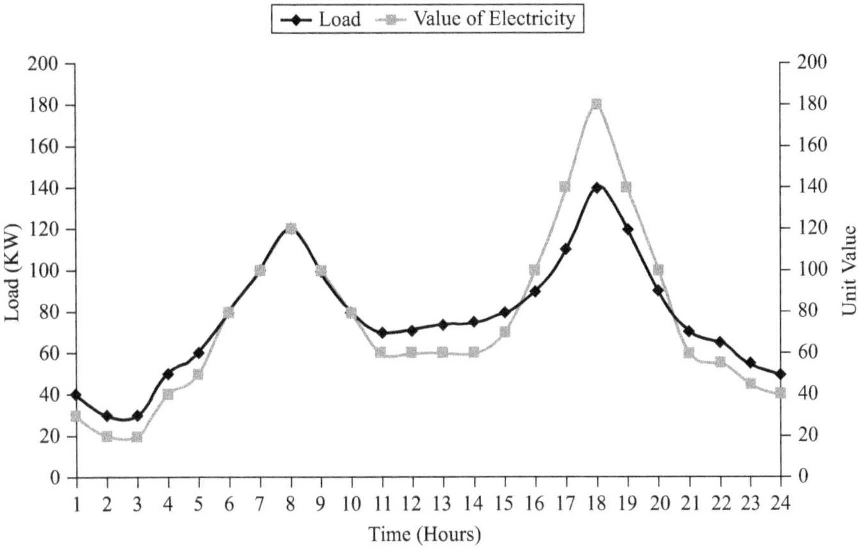

Figure 7.2 Value of electricity versus time

Time of use tariffs and a demand management technique require that the price of electricity is proportional to its value at different points throughout the day (Strbac, 2008; Deilami *et al.*, 2011 & Qian *et al.*, 2011). Figure 7.2 displays the value of electricity over time. As discussed, network design and augmentation are dependent on the load experienced during peak demand periods. As the load in peak demand periods approaches network capacity, network augmentation may be required. In turn, as supplying the capacity to meet peak demand dictates the price of electricity, time of use tariffs would set the highest price of electricity during this period. Late night and early morning would have the lowest price as these times experience the lowest demand. Price throughout the rest of the day would change to reflect the intersection of demand and supply curves. As consumers respond to price signals beyond the elastic region, they will alter their behaviour to consume electricity and charge PHEVs in accordance with their willingness to pay (Lemoine *et al.*, 2008). Price sensitive consumers will shift their non-essential electricity consumption to periods in which the price is low. Non-price sensitive consumers will pay for electricity according to its value, so will be paying for the capacity required of the network to supply electricity during peak demand periods.

Figure 7.3 presents the results of implementing a time of use tariff. Included are electricity supply and morning and evening electricity demand curves. The original price of electricity curve is set at 58 price units. The intersection of the demand curves and the original price curve establishes the original consumption quantities in the morning (Q_{m1}) and evening (Q_{e1}). In the scenario presented, the time of use tariffs are set at the value of electricity for the morning and evening periods – that is, the intersections of the demand curves with the supply curve. This sets the new consumption quantities for the morning (Q_{m2}) and evening (Q_{e2}).

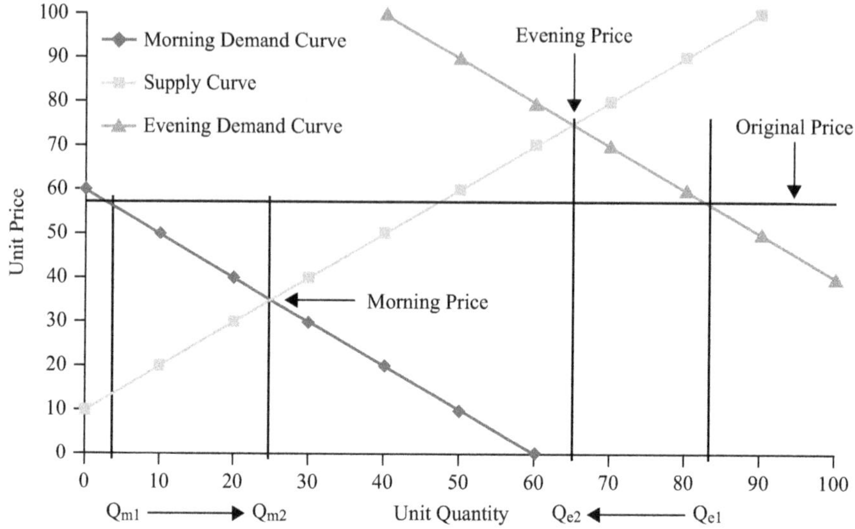

Figure 7.3 Time of use tariff

Allowing the price of electricity to reflect its value means that in the morning the quantity of electricity consumers will increase from Q_{m1} to Q_{m2}. In the evening, electricity consumed will decrease from Q_{e1} to Q_{e2}. Time of use tariffs are set to reflect the value of electricity at different times, flattening the load profile.

Regulating PHEV charging from the electricity network by restricting the periods in which they are allowed to charge could support the grid. If restrictions were implemented, they would be during peak demand periods such as 8 a.m. to 10 a.m. in the winter load profile and 5 p.m. to 9 p.m. during summer and winter load profiles. Consumers would be obliged to shift vehicle charging to periods of low demand. Benefits of supporting the grid would be the greater utilisation of electricity network infrastructure during non-peak demand periods, or customers finding alternative ways of charging PHEVs. These would mean that network operators and generators would be able to recover costs of network investments over greater periods, reducing the marginal cost of supplying electricity. Figure 7.4 presents the change in marginal cost (MC_1 to MC_2) for a given investment, in which total cost does not change and units produced increases. A neutral effect of the regulatory approach is that PHEVs will not contribute to load during peak demand periods. A drawback of this approach compared to time of use tariffs is that customers lose autonomy. Some consumers may prefer to pay a higher price for electricity during peak demand periods and charge their vehicles. Some instances may necessitate PHEV charging during these periods to fulfil important tasks or attend emergencies. The loss of autonomy may be mitigated by commercial charging stations.

Through the installation of a network communication system, the network operator would have to include EVs into its portfolio of load management practices. In some networks, this would include ripple control of electric hot water systems and

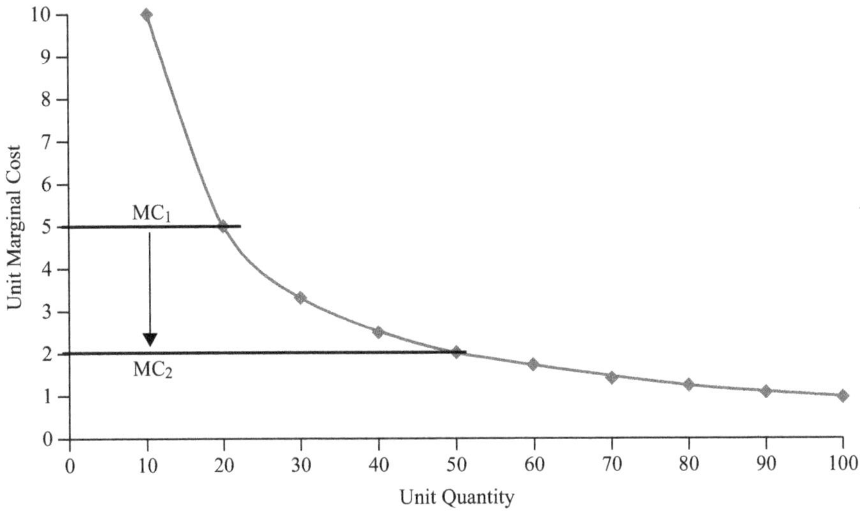

Figure 7.4 Marginal cost of supplying electricity

pool pumps (Deilami *et al.*, 2011). As the electricity network demand approaches capacity, the network operator may opt to engage in remedial measures such as load shedding. In turn, PHEVs would be prevented from charging and contributing to demand as the network approaches capacity. Additional uses of a network communication system may include remote scheduling of PHEV charging and the use of PHEVs to store excess generation. Similar to possible regulations, the network operator may dictate that PHEVs are charged during non-peak demand periods, reducing the marginal cost of generating and distributing electricity. Unlike the regulatory approach, the network operator would be free to experiment and discern the optimal scheduling arrangement, which produces the best outcomes for the customer and the electricity network. Instances occasionally occur in which there is excess generation over demand; in such circumstances, PHEVs can absorb excess generation. The inclusion of PHEVs in the portfolio of load management practices and their operation would be at the discretion of the network operator, and its obligation to ensure the stability of the electricity network and prevent brown or black outs.

7.2.2 PHEVs as Distributed Energy Resources

PHEVs have the potential to act as distributed energy resources (DERs) due to their energy storage capabilities. The use of PHEVs as DERs would yield functionality similar to more conventional grid-based energy storage technologies, such as fixed batteries (Brooks, 2002; Andersson *et al.*, 2010 & Peterson, 2010). Differences between the two technologies reside in the operational control and how economic value is created.

The conventional operation of energy storage is such that during low demand periods, energy storage systems absorb energy generated and discharge it during

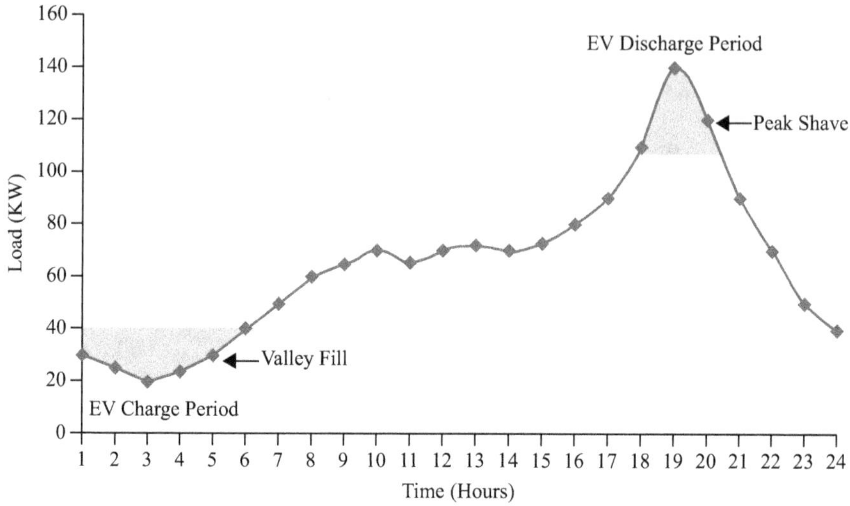

Figure 7.5 Peak shave and valley fill

peak demand periods (Mohd *et al.*, 2008). This conventional operation is known as 'peak shaving and valley filling', illustrated in Figure 7.5. This can be achieved by time-of-day heuristics or more advanced methods involving the use of load forecasts, and the optimisation of charge and discharge schedules with the use of this information. The use of PHEVs to perform this function adds a greater degree of complexity to the control operation, due to PHEVs' discharge of energy into the grid. The use of PHEVs as energy storage means that there will be no fixed energy storage capacity, and the network operator or control system will have no prior knowledge of when a customer will use their PHEV or when it would be available to discharge energy into the grid. To use PHEVs adequately for energy storage, a number of systems would need to be created, such as load forecast models, energy storage availability forecast models, discharge schedule optimisation and dynamic discharge optimisation.

The development of the necessary forecast models fall under two categories, depending on whether the system is deterministic or non-deterministic in nature. Load on the electricity network follows a deterministic pattern, with weather accounting, temporal and autoregressive variables accounting for much of the observed variance (Bennett *et al.*, 2014a & Bennett *et al.*, 2014b). In turn, load forecast models are usually deterministic, and can be developed through techniques such as regression, autoregressive models with exogenous variables and machine learning techniques (neural networks and support vector machines) (Bennett *et al.*, 2014a & Bennett *et al.*, 2014b). These techniques have benefits and drawbacks. Regression is the simplest technique to use, yet requires a priori data analysis to identify the type of relationship for which to programme a variable. Autoregressive models require information about future load on the network to be found prior to demand, through identifying periodic patterns. External variables can be included

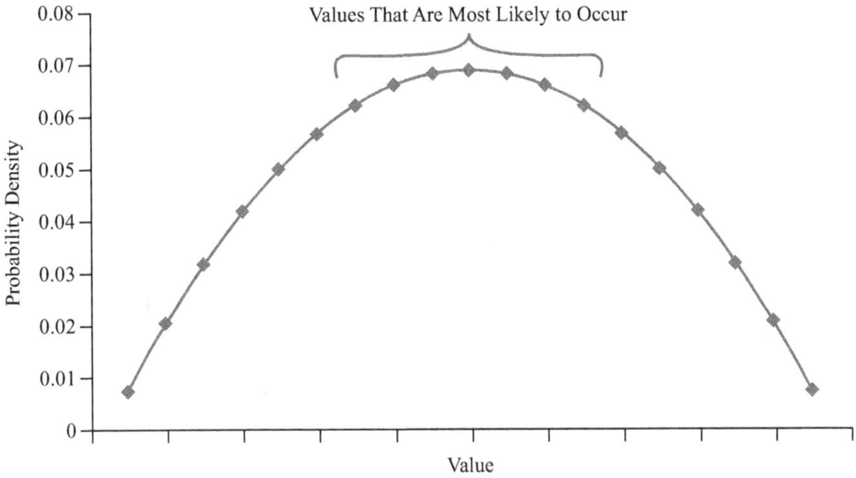

Figure 7.6 Stochastic forecast

with autoregressive models, to increase accuracy. Machine learning techniques are able to identify non-linear relationships between dependent and independent variables. A drawback of machine learning techniques is that they require extensive computational resources.

Energy storage capacity available as DERs would be categorised as a non-deterministic system (Fluhr *et al.*, 2010 & Qian *et al.*, 2011). There is no way to know the behaviour of customers in advance, or the availability of PHEVs to be used according to direct relationships with exogenous variables. Consequently, forecasting available energy storage capacity would rely on estimating the availability of a certain number of PHEVs and their capacity (Fluhr *et al.*, 2010 & Qian *et al.*, 2011). To do this, a stochastic modelling technique (or techniques that are similar) would be required, to develop energy storage capacity forecast models. As demonstrated in Figure 7.6, a stochastic model is used to produce a probability distribution of a random variable (availability of energy storage capacity) as a function of historical time series data, random variables and deterministic influences. The values in the probability distribution coinciding with the highest occurrence probabilities establish the range of likely values that a scheduling algorithm should consider and optimise therein. Variables in forecasting what constitutes available energy storage capacity include:

- the number of PHEVs available
- the charge state of the available PHEV
- the phase of the electricity line the PHEV will be attached to
- for how long the PHEV will be available

Optimising the discharge schedule and operating the system in real time adds complexity, requiring in-depth investigation before implementation.

The reduction in peak demand from the use of energy storage reduces the load on pre-existing infrastructure, in effect increasing network capacity. Economic value is created when installation of energy storage is more cost effective than network augmentation. The benefit of using PHEVs rather than installing fixed energy storage systems is that the costs of providing energy storage infrastructure is born by the PHEV owner rather than the network operator. The network operator yields value in the reduction of load via network augmentation deferral, without associated costs. The feasibility of PHEVs as DERs depends on the consistency of the PHEV's availability in supporting the grid. It is reasonable to expect that PHEV owners will oppose the use of their vehicle as a DER without compensation greater than the associated costs of allowing the use of their PHEV. To increase the reliability and availability of energy storage capacity, owners of PHEVs should be incentivised or compensated. Ways to do so include the creation of bidirectional time of use tariffs, or static feed in tariffs.

Bidirectional time of use tariffs require that the price offered throughout the day for both electricity consumption and power discharge (feed in tariff) reflects the value of electricity at the respected time (Brooks, 2002; Peterson, 2010 & Lopes *et al.*, 2011). There would be a trade-off for the PHEV owner between the use (or potential use) of the vehicle and earning revenue from the use of PHEVs as a DER (Lopes *et al.*, 2011). The use (or potential use) of the vehicle would be the owner's opportunity cost. If the discharging portion of the bidirectional tariff outweighs the opportunity cost and the cost of charging the vehicle, an increased energy storage capacity would be available for discharging. As the value of the opportunity cost to owners will fluctuate based upon external variables, the network operator may have to offer higher tariffs during different times of year, to achieve the desired energy storage capacity.

A static feed in tariff would operate similar to residential solar PV feed in tariffs. For a given amount of energy discharge into the grid, the owner would be receiving a fixed amount of money in return. The tariff should be offered for a specific duration, as an incentive to customers to invest in energy storage devices. A potential drawback of this tariff regime is that the tariff may not reflect the actual value of the system installed. In turn, it is more costly for this regime to exist than the conventional network augmentation scheme. Additional regulations would be required to ensure that energy storage systems discharge into the grid when load is high.

Figure 7.7 presents the result of a static feed in tariff, or the partial result of a bidirectional tariff implementation. The figure includes the electricity supply curve, the evening electricity demand curve and the original fixed price. The intersection of the original fixed price represents the quantity of electricity PHEV owners that are willing to supply, Q_{s1}. The intersection of the evening demand curve and the supply curve represents the willingness of PHEV owners to supply a quantity of electricity, Q_{s2}. The increase in quantity supplied from Q_{s1} to Q_{s2} suggests that the benefit yielded to owners was greater than the opportunity cost of not supplying electricity to the grid. Based on the principles of supply and demand, the implementation of a compensation arrangement will increase the reliability of energy storage capacity.

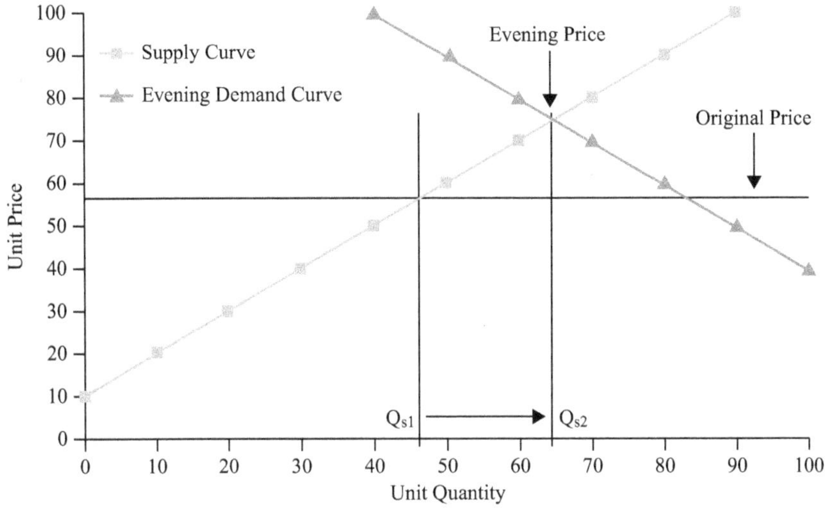

Figure 7.7 Compensation to EV owners

7.3 Social Dimension

A number of social challenges need to be addressed before the widespread diffusion of PHEV can be realised in advanced economies. Outlines of some of the key challenges follow.

7.3.1 Convenience

Since the advent of the car, urban dwellers have utilised the convenience of owning one or more vehicles per household, available for use at any time of day. The Internal Combustion Engine (ICE) and its reliance on liquid petrol or gas has meant that this transport option is always available, as long as fuel supplies (and the finances to purchase them) are available, almost always the case in advanced economies. The proliferation of petrol stations means that even in a geographically large country such as Australia, with one of the lowest global population densities, rarely will 500 km be travelled without the presence of a petrol station. Further, fuel tanks deplete after one to two weeks (based on average consumption and filling times of five to ten minutes). For workers and families in advanced economies, the ICE vehicle is a convenient means of travelling within and between destinations at any time.

The PHEV offers the same opportunity for travel but possesses some technological limitations at present, which reduce the convenience of the PHEV if compared to ICE technology. First, the price, frequency and length of charging are much higher. Long charging times are particularly inconvenient when travelling long distances as they force motorists to stop frequently, after travelling around 300–500 km. Further, motorists must wait while their vehicle charges. Conversely, motorists of conventional vehicles occasionally drive for up to 14 hours per day in one sitting, with only short rest breaks. 'Battery swap' stations have been considered and could result in much quicker stops for long-distance travellers, but the costs and warranty aspects of such systems requires further assessment and testing (Mak *et al.*, 2013). Further, rapid battery charge technology is being researched and could reach develop such that charging can occur at high c-rates (i.e. ratio between capacity of the battery bank and rate of charge) (Mak *et al.*, 2013).

Installing PHEV charging infrastructure is a commonly discussed barrier to the widespread adoption of PHEV. This 'chicken and egg' scenario will continue to prevent a greater adoption of PHEVs in many cities. If there is no market for PHEV charging, there is no incentive to build charge stations; if there are not sufficient PHEV charge stations, consumers will be reluctant to transition to driving PHEVs. Only strong believers in the benefits of PHEVs will suffer the inconvenience, with the majority only moving to PHEVs if they offer similar convenience as ICE vehicles. Alternatively, the average person will transition if the price of petrol reaches a point where ICE vehicles are no longer financially viable. For the building of charging stations, initial government support is required, in the form of subsidies, taxation relief and technology standardisation enforcement (such as charging plugs and electrics) (Steinhilber *et al.*, 2013). A worst-case scenario would be if the power electronics of different vehicles and stations were incompatible (Khan & Khan, 2013).

7.3.2 Uncertainty, Trust and Equity

For all major technology transformations, the designer and merchant must be able to dispel consumer uncertainty and generate a level of trust (Ziefle *et al.*, 2014). This is particularly difficult when a new technology competes with a proven technology (such as the ICE vehicle) largely viewed as safe, reliable and understandable. Before trust and loyalty can be established for PHEV technology, uncertainty must be reduced for the following issues:

- *After sales service and cradle-to-grave maintenance of PHEVs.* Vehicles require continual maintenance and repair. PHEV manufacturers and dealers should provide buyers with long-term warranties and access to service centres. As the greatest area of uncertainty for PHEVs is the 'true' reliable life of batteries, long-term warranties of eight or more years are required, to instil trust in consumers as well as trained and competent technicians to service PHEVs reliably (Heymans *et al.*, 2014).
- *PHEV battery technology lifespan, redundancy and refurbishment costs.* If battery technology has a reliable life of only five to eight years, and the replacement of battery packs after this period is substantial (30% of a new vehicle purchase price), the second-hand market for PHEVs will be very poor. The costs of installing new or refurbished battery packs should be low enough to ensure that PHEVs have a life in the second-hand car market, a key feature of the traditional ICE vehicle market, in which sellers would expect around half of the original car value back after approximately five years.
- *Proliferation of charge stations and points.* Purchasers of PHEVs will want to see that they can charge their vehicles reliably for a range of circumstances.

This includes short-, medium- and long-distance travel scenarios. Charge stations will not be constructed until there are a viable number of PHEVs to support them, and PHEVs will not be purchased until there are sufficient charge stations and points for customers. Consumers also require sufficient information on where charge points are located, and mobile applications – such as PHEV Charging Station Locator and *Plugshare* – are being offered in the United States.

Further, to avoid public backlash against PHEV technology, there must also be social equity considerations. A key aspect of PHEVs is that they require charge points/stations, at home or work. This means that a range of urban planning considerations for cities and regions must be considered requiring incorporation of either short or long-charge technology. With existing multi-residential buildings, how will limited charge points be assigned to residents? Within the workplace, who will have access to parks with charge points? How are these assigned – on a first-in basis, according to level in the organisation, the necessity to charge or other considerations? A range of equity issues regarding charging access will present should PHEVs become commonplace.

7.3.3 Perception of Poor Safety

Consumers have a few safety concerns about PHEVs. Such concerns can be addressed easily through appropriate testing and compliance. However, although isolated, safety incidents involving PHEVs are more newsworthy than ICE vehicle safety incidents, so are more likely to influence perceptions of PHEVs negatively. Below are consumers' main safety concerns:

- *Battery system catching fire.* A number of cases have been reported in which the Lithium Ion batteries commonly used in PHEVs catch on fire (Lu *et al.*, 2013). These have often been blamed on vehicle accidents that compress the space in which batteries are contained (Lu *et al.*, 2013). Incidents are not necessarily frequent but they are reported widely in the media, as this is a relatively new technology, meaning that public perception is affected. Extensive laboratory and field testing is being undertaken to ensure the long-term safety of PHEV battery systems.
- *Safety related to PHEV low noise.* Early PHEVs and hybrid vehicles were criticised for their lack of noise (their engines are silent), as pedestrians could not hear them coming, thereby leading to pedestrian incidents (Cocron & Krems, 2013). This was especially problematic for the visually impaired who rely on noise to determine whether a vehicle is coming (Cocron & Krems, 2013). A number of standards have been introduced to ensure that PHEVs and hybrid cars emit noise so that they can be heard by pedestrians. This issue is being worked on, and should not be an ongoing impediment to the adoption of PHEVs.

7.3.4 Public Perceptions and Overcoming Resistance to Change

If charge facilities and technologies (especially the charge rate) improve, a PHEV could provide a similar or better travel experience for most daily commuters. However, the perception still entrenched in many peoples' minds is linked to earlier PHEV technologies that were expensive to purchase, had poor resale value and limited range. A successful PHEV market will flourish when these perceptions diminish through successful city-scale PHEV roll-outs, rapid improvements in technology and through educating the risk-averse consumer. A key platform for improved perceptions of PHEV technology is information and social marketing of the benefits of PHEVs. Independently produced information on PHEV performance compared to ICE for a range of key performance indicators (charge time, cost of ownership, environmental aspects, travel distance etc.) relevant to consumers should be provided.

People are often resistant to change, so it is important that PHEV promoters transition people in a subtle manner. PHEVs are still considered a revolutionary technology by the mainstream population, not just a logical transition for vehicles. Manufacturers have used a number of strategies to transition people from one technological generation to the next.

7.4 Environmental Dimension

Motivation for replacing conventional cars with PHEVs can originate in ever-increasing fuel prices, as well as environmental concerns associated with sustainable living, such as the desire to improve air quality in urban areas by replacing conventional vehicles and their pollutants. Gas emissions from hybrid or purely electric vehicles can be significantly reduced or totally absent, a fair statement when only the exhaust pipe of the PHEV is considered, excluding harmful emissions resulting from the thermal electricity generation necessary for charging PHEV batteries. The aforementioned example is indicative of narrow examination of the matter; therefore, to study the environmental impact thoroughly, the effects of batteries, fuel, lightweight materials, electricity generation and so on should be included. Further, the examination should include the life phases of the vehicle and the outcome should be compared to the environmental impact caused by vehicles of different technologies. The above are taken into consideration during a life cycle assessment (LCA); therefore, this principle is discussed in this section.

7.4.1 LCA for PHEVs and Consideration of Further Automotive Technologies

For a comprehensive assessment of PHEVs as entities, beyond the utilisation period of the vehicle, phases such as extraction of raw materials and recycling, reuse or waste management should be included in the LCA. This is so a full account of the lifetime environmental effects resulting from components incorporated in vehicles can be taken. Such components (apart from batteries) can be electric motors and power electronics. Common to different automotive technologies are the chassis, wheels, car interior and so on. The phases of a complete LCA, from cradle-to-grave, are presented in Figure 7.8 (Renaldi *et al.*, 2013; Messagie *et al.*, 2010; Humphreys *et al.*, 1996).

The examination of the LCA phases and the comparison of the results between pure PHEVs, hybrid and conventional vehicles are of great importance in ascertaining the environmental impact for each technology. Another outcome from the

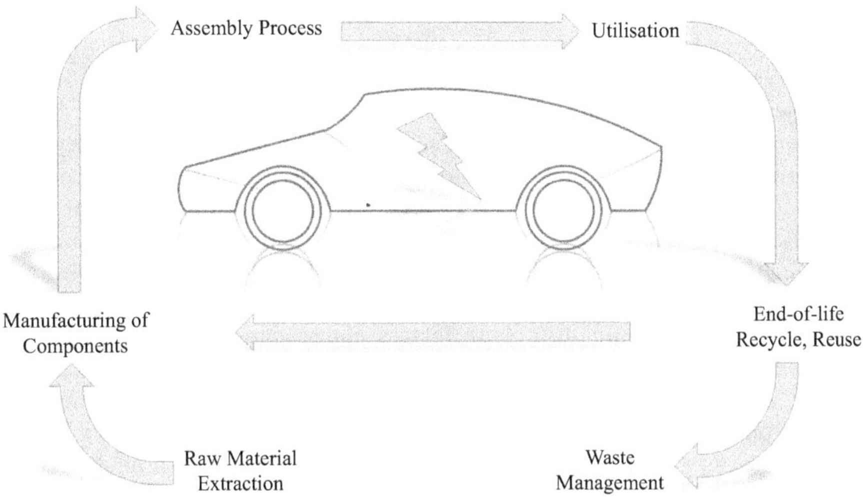

Figure 7.8 Life cycle assessment phases

LCA is to identify whether negative environmental effects have shifted from the utilisation phase to other life cycle periods. In terms of information to include in the assessment, a cradle-to-grave approach should be considered (rather than a cradle-to-gate, which does not account for the utilisation and end-of-life phases), containing material consumption, energy consumption and emissions during periods such as those presented in Figure 7.8. In terms of fuel, the well-to-wheel process should be followed to include impact related to fuel production, distribution and consumption in the vehicle. In addition, the environmental effect from the generated power intended for charging PHEV batteries should also be factored into any assessment. A more detailed description of the information can be found in Humphreys *et al.*'s study (1996). The stages concerning electrical energy production for PHEVs, namely, fuel production and transportation, electricity generation, transmission and distribution, battery charging and PHEV operation should be included. For a conventional vehicle, the aforementioned stages are adapted to crude oil extraction and transportation, oil refining, gasoline transportation, vehicle fuelling and operation (Renaldi *et al.*, 2013; Messagie *et al.*, 2010; Humphreys *et al.*, 1996).

7.4.2 Environmental Impact Comparative Assessment

The outcome of the LCA for different vehicle technologies can be expressed through the following (Renaldi *et al.*, 2013; Messagie *et al.*, 2010; Humphreys *et al.*, 1996):

- air acidification
- eutrophication
- human health
- greenhouse effect

For batteries – and particularly for the Lithium Ion (Li-ion) type found in cutting edge technology PHEVs – the extraction of Lithium can be harmful to the environment, and in the case of the Nickel Metal Hydride (NiMH) type, mainly available in obsolete PHEVs. Therefore, recycling must be performed to mitigate the extent to which nickel influences air acidification. As long as battery recycling is performed, PHEVs have a lower acidification impact than hybrid and conventional vehicles, but without recycling, the impact is significantly increased. By end-of-life, a hybrid vehicle without battery recycling would pollute more than a conventional vehicle. In general, Li-ion technology is more environmentally friendly and less threatening to the human health than NiMH technology. In the case of extensive aluminium use, there is a higher threat of ozone depletion, mainly due to the extraction and production phases. Conversely, when a battery reaches end-of-life it can be reused or recycled, so its overall environmental impact from cradle-to-grave can be reduced significantly. The assessment should also include total materials utilised, energy consumed and environmental residuals, for a comparison between PHEVs and conventional vehicles (Renaldi *et al.*, 2013; Messagie *et al.*, 2010; Humphreys *et al.*, 1996).

In terms of eutrophication, conventional types of vehicles (diesel and petrol) have the highest impact, while PHEVs have the lowest, originating mainly from the production of the Lithium battery. To compensate for the weight increase caused by the batteries, the lightweight materials used in PHEVs achieve their target only if the utilisation phase of the vehicle is concerned, as their usage balances energy consumption and related greenhouse gas emissions. However, to create a thorough image of the environmental impact of the lightweight materials, the production/manufacturing phase should be included. If magnesium and carbon fibre materials have been used, they have a more detrimental effect on human health than conventional steel. PHEVs' impact on human health is still lower than conventional vehicles, and it is a combination between raw material extraction, manufacturing process and vehicle assembly, along with tires' and brakes' wear and tear. On contrary, the ICE vehicle's negative effect on human health is related to fuel production, distribution and consumption, along with raw material extraction, vehicle assembly and so on (Renaldi *et al.*, 2013 & Messagie *et al.*, 2010).

Petrol and diesel vehicles have the highest impact on greenhouse gas emissions, mainly due to their fuel consumption. Petrol produces the highest amount of greenhouse gases, reflected in the well-to-wheel phases – fuel production, distribution and consumption. In hybrid vehicles, emissions are reduced due to improved fuel consumption, but the impact is still considerable. Messagie *et al.* (2010) discusses that PHEVs have the lowest negative impact.

Further components necessary for the operation of PHEVs – such as electric motors and power electronics – contribute to the total environmental impact of the vehicle in terms of greenhouse emissions. These are estimated to be approximately 2 and 3 per cent of total emissions (Renaldi *et al.*, 2013).

Another aspect associated with PHEVs is the generation of electricity to be used in the vehicles, and its environmental impact, depending on its origin. For instance, in the future where energy with be sourced from a range of non-renewable and renewable sources, PHEVs are more environmentally friendly than petrol vehicles. The more that renewable energy sources (RES) are integrated and utilised to power PHEVs, the more emission free the transportation will become. Conversely, a small conventional vehicle with improved fuel efficiency can have a lower environmental impact (such as on the greenhouse effect) than an oversized PHEV powered solely by coal- or oil-fired thermal units. Further, in the case of thermal generation, capacity should be added by the utility, in order to supply the excess demand of PHEVs. The additional emissions must be registered in the LCA of the PHEVs, and also included should be emissions accruing from the utilisation of different types of units and variation in dispatch to supply the demand of PHEVs (Messagie *et al.*, 2010; Humphreys *et al.*, 1996).

7.4.3 Global Metal Reserves, Production and Pricing Data Related to Vehicle Batteries and Electric Motors

Another dimension related to the environment which needs to be explored is the adequacy of metal reserves required in the manufacturing process of EVs. Such metals are mainly related to the batteries and electric motors of the vehicles and in order to estimate the longevity of the different battery technologies, the resources and in turn reserves have to be examined as well as the treatment of batteries by the end of their life cycle has to be discussed.

The reserves of the three main metals utilised in batteries are presented in Table 7.1 (Chatzivasileiadi *et al.*, 2013) while the remaining period for each of them based on the reserves over the current annual demand for each is also presented. It can be seen that the reserves of Nickel and Lead are running out but to the contrary, reserves of Lithium are expected to last for a considerable period of time. The values illustrated in Table 7.1 may change in the future particularly in case of Lithium and Nickel due to factors such as technological improvements, discovery of new reserves and recycling of the metals. These factors are expected to significantly affect the remaining period of the materials towards an increase as well as to limit their environmental impact.

The evidence provided in Table 7.1 regarding the extended period of Lithium reserves is confirmed by Kesler *et al.* (2012) where it is deemed that resources are adequate to cover the current demand and even beyond the twenty-first century

Table 7.1 Global metal reserves (Chatzivasileiadi et al., 2013)

Metal	Reserves (ktons)	Annual demand (ktons)	Remaining period (years)
Lithium	13,000	34	382.35
Nickel	80,000	1,800	44.44
Lead	85,000	4,500	18.88

provided that recycling of the batteries takes place. In the same work it is stated that the existing global Lithium resources are maintained in four main forms as follows:

- Pegmatite
- Brine
- Unusual rocks
- Unusual fluids

According to Kesler *et al.* (2012), the Lithium resource in pegmatites is identified at 3.9 Mt, the Lithium resource in brine is expected to be 21.6 Mt while the Lithium in unusual rocks and fluids adds up to 5.4 Mt. At this stage the attention needs to be brought on the fact that the expected resources have to be processed in order to be converted into reserves so Lithium can be produced at a reasonable cost. This observation has been made by Kesler *et al.* (2012), and it has been supplemented by the statement that examination of a number of other parameters has to be performed in order to find out about the feasibility of the conversion, for instance in case of brine deposits, brine constituents such as potassium and magnesium among others would prevent from processing.

Another challenge associated with the Lithium resources in brine which has to be addressed, is that 70 per cent of the resources are located in four countries namely Argentina, Bolivia, Chile and China translated to a strategic advantage on their behalf. On the other hand, the Lithium resources in pegmatites are more scattered among different countries Kesler *et al.* (2012), this could be translated into reduced market domination of a small number of countries over others which is however partially true when considering that the expected resources in brine account for the two-third of the total global resources. According to Goonan (2012) the issue of market domination of a few countries is even more intense as it is reported that 75 per cent of the global Lithium resources exist in just three countries i.e. Argentina, Bolivia and Chile while low-cost resource is also available in China and the United States.

In terms of Lithium based battery chemistries, the chemistries of Lithium (Li) Nickel (Ni) Manganese (Mn) Cobalt (Co) Oxide (O_2) i.e. $LiNiMnCoO_2$ and Lithium Cobalt Oxide i.e. $LiCoO_2$ present the highest energy density as it is presented in Goonan (2012), with 100–170 Wh/kg and 100–150 Wh/kg in accordance indicating that apart from Lithium, the expensive Cobalt and Nickel are present for the first battery type and Cobalt for the later battery type.

In terms of recycling, the main motivation in order to be performed is the expectation of retrieval of the high cost Cobalt and Nickel while secondarily comes the retrieval of Lithium which accounts for <3 per cent of the production cost for most of the Li-ion batteries. Due to the high price of Cobalt, Sony since 1992 has established a collaboration with another entity in order to retrieve it from Li-ion batteries which have reached their end-of-life (Goonan, 2012).

Battery recycling data from the United States indicated that the recycling level was proximate to 20 per cent while up to 2006 Li-ion batteries were not considered to be a hazardous waste. Batteries which have reached their end-of-life in the United States are likely to be still in possession of their owners, e.g. at homes, or disposed through the municipality's waste management system, e.g. landfills (Goonan, 2012). The expectation for higher recycling level of Li-ion batteries is considerable as it is related to the increase of utilisation of vehicles equipped with batteries as it should follow the effective recycling system applied to lead-acid batteries. If comprehensive recycling of EV batteries takes place, the recovered Lithium could be sufficient for 50 per cent of the Lithium demand in batteries by the year of 2040 (Goonan, 2012).

The alternative battery type which has been used in vehicles, i.e. the NiMH type in the Toyota's hybrid vehicle, the Prius, contains four different rare earth elements which are Lanthanum (La) 60 per cent, Praseodymium (Pr) 20 per cent, Neodymium (Nd) 10 per cent and Cerium (Ce) 10 per cent (Eriksson *et al.*, 2011). Furthermore it also contains Ni, Co, Mn and possibly Aluminium (Al).

So far the discussion focused on batteries; however, EVs are also equipped with permanent magnet electric motors. These motors also contain rare earth metals which namely are Nd in 75 per cent, Pr in 25 per cent as well as Dysprosium (Dy) and Terbium (Tb) in very small quantities. The very small quantities in Dy and Tb are used in order to improve characteristics of the magnet such as temperature stability and coercivity (Eriksson *et al.*, 2011). There can be found two common types of permanent magnets one is with Neodymium, Iron (Fe) and Boron (B) i.e. NdFeB and the other is with Samarium (Sm) and Cobalt i.e. SmCo, however the NdFeB type outperforms the later type in terms of magnetic energy density therefore it is preferable in vehicle applications (Eriksson *et al.*, 2011).

In order to find out about the global reserves and production in rare earths, the unit of rare earth oxide (REO) is utilised which is the total amount of rare earth elements existing in a mineral (Eriksson *et al.*, 2011). According to Figure 7.9, China is the world leader in REO reserves with 55 million metric tons while the second and third places in the world, for individual countries, are occupied by Brazil and the United States with 22 million metric tons and 13 million metric tons, respectively, in accordance

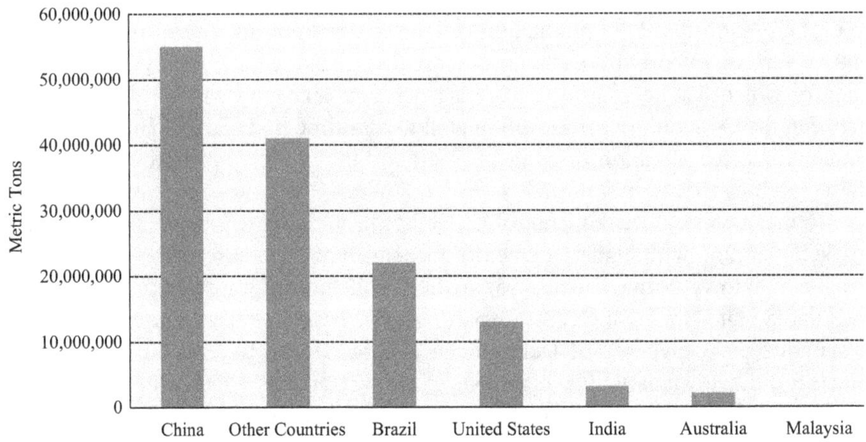

Figure 7.9 REO global reserves in 2013 (U.S. Geological Survey, 2014)

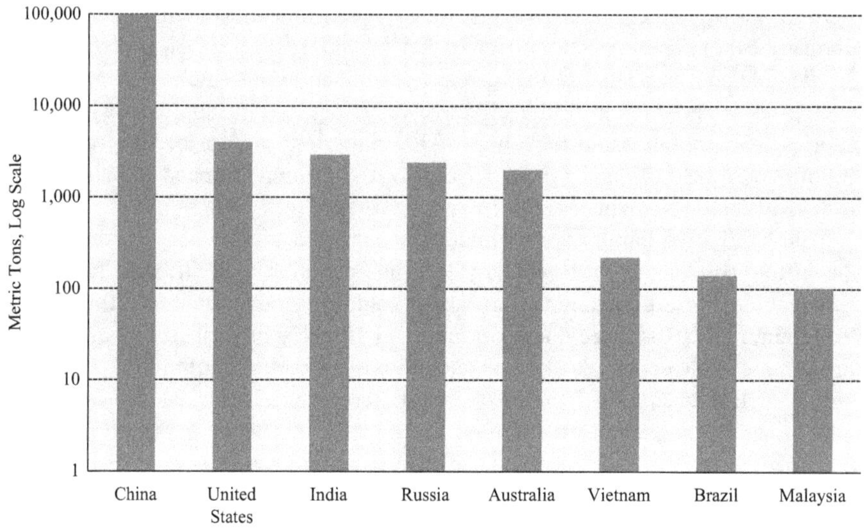

Figure 7.10 REO global production per country in 2013 (U.S. Geological Survey, 2014)

(U.S. Geological Survey, 2014). The size of the reserves as it has been presented for 2013 is determined by two main criteria, the first is related to the production cost and if the process is cost-effective, while the other criterion is the extraction/process potential provided by the today's technology (Eriksson *et al.*, 2011).

When production is concerned, once again China has dominated over the second country as it can be seen in Figure 7.10 with production to have reached 100,000 metric tons against 4,000 metric tons produced by the United States. In turn the countries India, Russia and Australia can be identified to be in a production range between 2,900 and 2,000 metric tons which is vastly lower than the production leader (U.S. Geological Survey, 2014). Considering the nearly exclusive role of China in the REO production, it is in the position to control market prices and to vary its exports by enacting quotas; this gives a strategic advantage to the country which has been used to exercise pressure on Japan when a disagreement between the two countries occurred related to maritime boundaries. This action has concentrated the attention of companies around the globe whose products rely on these supplies (Massari *et al.*, 2013).

In the context of market control which has been discussed above and towards a market operation with higher degree of independency, several research centres have taken action towards the discovery of artificial substitutes for rare earth elements. In the United States for example, the Department of Energy (DOE) through its funding, it supports the University of Delaware in a joint effort with Hitachi and General Electric (GE), towards the development of strong magnets which will be based on nano-composites instead of rare earths. In Europe, although mineral resources are very low, still it is attempted a reduced dependency on the market by intensely supporting the 'urban mining' concept which allows significant amounts of rare

Table 7.2 Prices for rare earth metals (Metal-pages, 2014)

Metal	Price ($/kg)
Lanthanum (La)	9.6
Cerium (Ce)	10
Praseodymium (Pr)	150
Neodymium (Nd)	83
Samarium (Sm)	25
Terbium (Tb)	825
Dysprosium (Dy)	475

earths to be retrieved by performing recycling of hi-tech wastes with the significant advantage that this source of rear earths is already there (Massari *et al.*, 2013).

Regarding the current market prices (December 2014) for the different rare earth metals, which concern this work, are provided in Table 7.2 (Metal-pages, 2014). Furthermore, in the U.S. DOE report (2010) and U.S. DOE report (2011) are presented two figures with the short-term and medium-term criticality related to the supply risk for each metal. Combining the two sources, the highly priced metals of Table 7.2 can be correlated with the metals subjected to high supply risk. Particularly Tb and Dy which are priced in 825$/kg and 475$/kg in accordance are also identified to be the metals with the highest supply risk in the U.S. DOE report (2010) and U.S. DOE report (2011).

7.5 General Method for Developing PHEV Scheduling Systems

7.5.1 Introduction

This following section outlines a general method for developing scheduling systems for PHEV in the electricity network. The aim of scheduling systems is to coordinate the charging and discharging of PHEV. To minimise the impacts that PHEV have on the electricity network, scheduling systems should aim to promote charging during low load and charge during high load (peak demand periods). In the residential low voltage distribution network low load periods occur late at night, early morning and the middle of the day. Daily peak demand occurs during the early evening. In commercial networks low load occurs during late night and early morning, and peak demand occurs during the middle of the day to early afternoon. The load in this circumstance is defined as what the load would be absent the effects that charging and discharging has on the network. For the scheduling systems to operate efficiently a number of subcomponents are required including load forecasting (LF), energy storage availability forecasting (ESA), initial scheduling (IS) and an online control system (OCS). The LF provides information about future load states such that low load and peak demand periods can be identified and quantified. The amount of PHEV connected to the network is up to the discretion of PHEV owner entailing that the precise amount of energy from the PHEV battery banks available for grid support

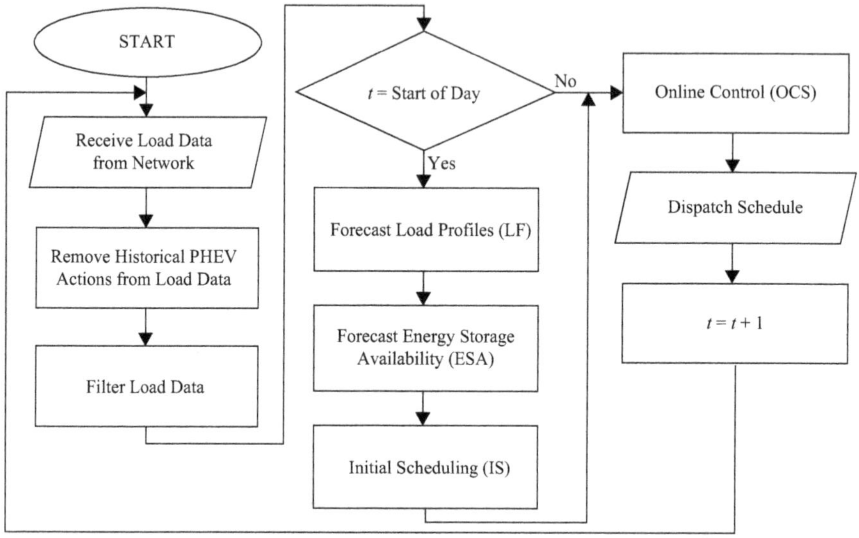

Figure 7.11 General PHEV scheduling system flow chart

is unknown. Since the actions of people in the aggregate are able to forecast with a reasonable degree of accuracy, the amount of energy storage available may also be able to be forecast. The ESA estimates how much energy storage will be available (number of electric vehicles connected to the electricity network) for supporting the grid. The IS uses the load profile forecasts and the energy storage availability forecasts to create an initial schedule for the day. The OCS receives the schedule at the start of the day and throughout the day analyses recent historical load and the activity of PHEV to mitigate scheduling error.

Figure 7.11 displays a general flow chart for forecast based PHEV scheduling systems. For each time interval (t) the scheduling system receives historical load data. Previous PHEV charge and discharge actions are removed from the load data. The load data is then filtered to remove noise from the underlying moving average to prevent scheduling errors. Filtering can be conducted by a variety of techniques such as Kalman flitering, Fourier Transform filtering and seasonal exponential smoothing. If the time interval is the start of the day, the system calls the LF, ESA and IS subcomponents. The forecasted load profiles and the forecasted energy storage availability are passed to IS in order for an initial schedule to be calculated. The calculated schedule is then passed to the OCS that adjusts the schedule in real time throughout the day in response to local conditions to mitigate scheduling error. As an example: if there is less energy storage available than what was initially anticipated, the OCS will reduce the amount of energy discharged into the electricity network. If there is more energy storage available, the OCS may increase the amount of energy discharged into the network. After schedule adjustments have been made, the schedule for the current time interval is dispatched to the PHEV connected to the electricity network.

7.5.2 Developing Load Forecast Models

The time series modelling technique is one most common technique used in the electric power industry to forecast load. Time series models can either be univariate or multivariate. The structure of univariate models is such that the variable being forecast is a function of temporal lags of the same variable. Multivariate models are structure such that the dependent variable being forecast is a function of both temporal lags of the same variable and independent variables. In the load forecast space independent variables may include temperature, humidity, day of the week and time of the year (Bennett *et al.*, 2014a; Bennett *et al.*, 2014b). Equation (7.1) displays a general univariate model and (7.2) displays a general multivariate model.

$$y_t = \sum_{i=1}^{m} \beta_i y_{t-i} + \beta_0 \tag{7.1}$$

$$y_t = \sum_{i=1}^{p} \beta_i y_{t-i} + \sum_{j=1}^{q} \alpha_j x_j + \beta_0 \tag{7.2}$$

where y is the variable being forecast, β is a coefficient, β_0 is a constant, t is the time interval, i is the temporal lag, m is the number of lagged y variables in the function, x represents independent variables, j is the particular independent variable and q is the total number of independent variables.

Univariate time series models fall under the general model called autoregressive integrated moving average ARIMA(p,d,q). The 'p' represents the number of discrete differences the time series has undergone. Discrete differencing is engaged in when the presence of a unit root is detected by a unit root test such as the Dicky-Fuller test. The presence of a unit root means that if an exogenous shock occurs its effect permeates through the rest of the time series. The 'd' is the number of lagged variables in the model. The 'q' represents the number of lagged error terms are in the model. Multivariate models fall under the ARIMA with exogenous variable definition (ARIMAX). From the establishment of the definitions, the first stage in the development of a univariate model is to determine the significant periodic patterns in the time series. This can be done using the autocorrelation function, partial autocorrelation function or a Fourier transform. The autocorrelation function is presented in (7.3):

$$r_l(Y_t, Y_{t-l}) = \frac{1}{m-1} \times \frac{\sum (y - u_t)(y_{t-l} - u_{t-l})}{\sigma(Y_t)\sigma(Y_{t-l})}, \quad \text{for } 1 \leq l \leq M \tag{7.3}$$

where r is the correlation between time series vectors Y_t and Y_{t-l}, Y_t is the original time series, Y_{t-l} is the time series with l lags, u_t is the mean of Y_t, u_{t-l} is the mean of Y_{t-l}, $\sigma(Y_t)$ is the standard deviation of Y_t, $\sigma(Y_{t-l})$ is the standard deviation of Y_{t-1}, m is the length of Y_t and M is the maximum number of lags being investigated. From here a threshold of significance is required to be calculated (7.4) and lags greater than the threshold are to be included in the model.

$$\frac{z_{1-\frac{\alpha}{2}}}{\sqrt{m}} \tag{7.4}$$

where z is a statistic derived from the Student's t-distribution, α is the level of significance and m is the length of Y_t.

With the significant lags in the time series identified and a function established the coefficients of the model are now required to be calculated. The most common approach is to use least means squared regression as seen in (7.5):

$$\hat{\beta} = \left(X^T X\right)^{-1} X^T Y \tag{7.5}$$

$$X = \begin{bmatrix} x_{1,1} & \cdots & x_{1,p} \\ \vdots & \ddots & \vdots \\ x_{m,1} & \cdots & x_{m,p} \end{bmatrix} \tag{7.6}$$

$$Y = \begin{bmatrix} y_1 & \cdots & y_m \end{bmatrix}^T \tag{7.7}$$

where $\hat{\beta}$ is a column vector of coefficients of length p, X is a matrix of observations where each is row is a new observation and each column pertains to a different variable, Y is a vector of dependent variable observation, m is the number of observation and p is the number of variables in the function. The output of the model is as follows:

$$\hat{Y} = X\hat{\beta} + \epsilon \tag{7.8}$$

$$\hat{Y} = \begin{bmatrix} \hat{y}_1 & \cdots & \hat{y}_m \end{bmatrix} \tag{7.9}$$

where $\hat{Y}$ is the vector of predicted values and ϵ is the error term. The calculation of $\hat{Y}$ allows for the determination if there is autocorrelation in the error term (Durbin-Watson test) and the calculation of accuracy statistics. The Durbin-Watson test is calculated using (7.10):

$$d = \frac{\sum_{t=1}^{m} \left(e_t - e_{t-1}\right)^2}{\sum_{t=1}^{m} {e_t}^2} \tag{7.10}$$

where d is the Durbin-Watson statistic and e is the error for forecast at time t where the vector of error terms from the model is calculated by $\hat{Y}$ minus Y. d_l and d_u are critical Durbin-Watson statistics. If d is less than d_l then the error is positively autocorrelated and if (4-d) is less than d_l the error is negatively autocorrelated. If d is greater than d_l and less than d_u there is likely to be no autocorrelation. If autocorrelation is identified the model calculated is required to be altered. Running the autocorrelation function on the error vector will indicate whether there are repeating patterns. If the presence of repeating patterns is identified the model can be adjusted to take the patterns into account.

Once the model has been adjusted to take into account autocorrelation in the error term and non-stationary processes accuracy statistics can be calculated. Common accuracy statistics include the coefficient of determination (R^2), standard deviation of model error or root mean square error ($RMSE$), mean absolute error

(*MAE*) and mean absolute percentage error (*MAPE*). These accuracy statistics are calculated by using the following set of equations:

$$R^2 = 1 - \frac{\sum_{t=1}^{m}(y_t - \hat{y}_t)^2}{\sum_{t=1}^{m}(y - u)^2} \tag{7.11}$$

where u is the mean of Y. The R^2 statistic ranges from 0 to 1. A value of 1 entails the model has a perfect fit with the observed time series. Conversely, a value of 0 entails the model does not represent to the observed time series. The closer the value is to 1 the better the fit.

$$\text{RMSE} = \sqrt{\frac{\sum_{t=1}^{m}(y_t - \hat{y}_t)^2}{m - p}} \tag{7.12}$$

The *RMSE* statistic can be used to add a confidence interval to forecasts made using the model. The greater the *RMSE* the less accurate the model is. The confidence interval can be calculated using (7.13):

$$\text{CI} = \pm z\left(1 - \frac{\alpha}{2}, m - 2\right) \times \text{RMSE} \tag{7.13}$$

where CI is the confidence interval. For a forecast $\hat{y}$ and for a level of significance α, the forecast is between the upper and lower limits of $\hat{y} - \text{CI}$ and $\hat{y} + \text{CI}$.

$$\text{MAE} = \frac{\sum_{t=1}^{m}\text{abs}(y_t - \hat{y}_t)}{m} \tag{7.14}$$

where *MAE* is the mean absolute error and *abs()* is the absolute function. The greater the *MAE* the less accurate the model is.

$$\text{MAPE} = \frac{\sum_{t=1}^{m}\text{abs}((y_t - \hat{y}_t)/y)}{m} \tag{7.15}$$

where *MAPE* is the mean absolute percentage error. The greater the *MAPE* the less accurate the model is.

Further testing can be conducted by splitting the time series into a model development set used to estimate the coefficients and a validation set. The validation set is not used in the model development stage. The validation set's input variable data are inputted into the developed models and resulting in a set of forecast for the validation set. The new forecasts can be compared against the validation set's dependent variable observations and accuracy statistics can be calculated. The new accuracy statistics give a better indication of the performance of the model.

To forecast a range of values in the future recursive or rolling forecasting is used. To demonstrate this, a hypothetical model is established in (7.16):

$$\hat{y}_t = \beta_1 y_{t-1} + \beta_2 y_{t-2} + \beta_3 y_{t-24} + \beta_0 \tag{7.16}$$

The model is univariate with three significant lags identified by the autocorrelation function. The autocorrelation function identified a significant period being a 24 hour cycle. In order to forecast the next time interval $t+1$ without knowledge of y_t the forecast $\hat{y}_t$ is rolled into the model:

$$\hat{y}_{t+1} = \beta_1\hat{y}_t + \beta_2 y_{t-1} + \beta_3 y_{t-23} + \beta_0 \tag{7.17}$$

The forecast $\hat{y}_t$ replaced the observation y_{t-1}. To forecast further into the future the pattern of rolling forecasts into the model is repeated. As an example, forecast up to 24 hours ahead using this model, the final forecast in the series is in the form of

$$\hat{y}_{t+23} = \beta_1\hat{y}_{t+22} + \beta_2\hat{y}_{t+21} + \beta_3 y_{t-1} + \beta_0 \tag{7.18}$$

For the purpose of forecasting load profiles and energy storage availability the recursive forecasting method can be used. A drawback of this method is that if significant forecast error is made, the error will permeate throughout the entire forecast period resulting a high likelihood that there will be a significant deviation from the actual load that is to occur in the future. To mitigate this, more forecast can be made throughout the time period allowing for the incorporation of new observations.

7.5.3 Initial Scheduling

For a IS to be developed the power electronics that the PHEV scheduling system applies to must be first emulated. Thus models of PHEV battery banks, inverters and the local electricity network are to be included within the IS. The models are not required to be exact replications; they only are required to approximate the system. The models constrain the system such to prevent issues including the exceedance of design ratings and operation of battery banks such that the charge drops below the depth of discharge (DoD) or breaches maximum charge limits (MC). From here, the goals of the scheduling system, and subsequently the IS, govern the rest of the structure surrounding the power electronics components. As an example, a goal may be that the scheduling system only dictates the discharge from battery banks and does not coordinate the charging. PHEV owners may charge their vehicles at will during the day. Using the assumption from the example an algorithm for initial scheduling is provided:

1. The LF and ESA provide load profile forecasts to the IS.
2. Peak demand periods are identified in the load profile forecasts.
3. For the peak for each peak demand period is reduce according to the amount of energy that is available for the specific period. This provides an initial discharge schedule.
4. The initial schedule is dispatched to the OCS.

Referring to Figure 7.5 the area between the peak and the level at which the peak is reduced to is equivalent to the amount of energy storage available constrained by the specifications of the power electronics of the system.

7.5.4 Online Control System

The goal of the OCS is to mitigate scheduling error that may occur due to forecast inaccuracy. There are two main types of scheduling error. The first is a mismatch between the actual peak load and the observed peak load. This would lead to an inefficient discharge schedule that may not sufficiently reduce peaks. The second is an occurrence of insufficient energy to discharge into the electricity network across the peak demand period. A method of mitigating scheduling error is

1. For each time interval the previous time interval's load and available energy storage data is collected and used to update load and energy storage availability forecasts.
2. A simulation is run with the new data to determine if the initial schedule is not sufficient.
3. If the schedule is not sufficient the amount of energy being discharged into the network is either increased or decreased.

7.6 Conclusion

The chapter provides a summary of the barriers and opportunities of widespread PHEV. Specifically, it examined the economic, social and environmental dimensions of PHEV. The economic dimension discussed how PHEV could have a potential negative impact on the electricity network if there was unregulated battery charging and discharge. The major problem being the requirement for additional distribution infrastructure to cope with the increased demand for energy created through PHEV operation as well as the associated power quality issues. Social issues could arise if it was perceived that PHEV owners were being subsidised by electricity consumers. One solution recommended herein was to restrict PHEV charging by technical or regulatory means. A second type of solution was proposed through the use of price incentives which would allow owners to use their PHEV at will but also pay according to their use. The social dimension covered individuals' likely take-up of PHEV according to their attitudes and beliefs. For the rollout of PHEV to be successful, car designers through to city planners need to make PHEV and the necessary infrastructure to support it as equally convenient to current transport options. Some strategies to achieve this include making PHEV servicing and maintenance convenient, providing infrastructure for PHEV owners (i.e. charging or battery swap stations), addressing safety concerns and providing the right information to society to change perceptions. The environmental dimension alluded that PHEV have the potential to yield environmental gains over conventional vehicles. PHEV tend to perform better in terms of reducing local air pollution. PHEV proliferation means that there will be substantial battery waste issues over time, which mandates the proper reuse and recycling of batteries and the invention of more advanced battery technology that will minimise environmental impact. PHEV charging from the electricity network means that there is still a reliance on fossil fuels in most cities; hence, there are minimal effects on reducing or abating greenhouse gas emissions unless city electricity generation has a reasonable proportion of renewables. Addressing the complete spectrum of economic, social and environmental dimensions of PHEV is paramount before this technology will push aside the age of the ICE.

References

Andersson, S. L., Elofsson, A. K., Galus, M. D., Göransson, L., Karlsson, S., Johnsson, F. *et al.* (2010). Plug-in hybrid electric vehicles as regulating power providers: Case studies of Sweden and Germany. *Energy Policy*, *38*(6), 2751–2762.

Bennett, C. J., Stewart, R. A., & Lu, J. W. (2014a). Forecasting low voltage distribution network demand profiles using a pattern recognition based expert system. *Energy*, *67*, 200–212.

Bennett, C., Stewart, R. A., & Lu, J. (2014b). Autoregressive with exogenous variables and neural network short-term load forecast models for residential low voltage distribution networks. *Energies*, *7*(5), 2938–2960.

Brooks, A. N. (2002). *Vehicle-to-grid demonstration project: Grid regulation ancillary service with a battery electric vehicle*. California Environmental Protection Agency, Air Resources Board, Research Division.

Chatzivasileiadi, A., Ampatzi, E., & Knight, I. (2013). Characteristics of electrical energy storage technologies and their applications in buildings. *Renewable and Sustainable Energy Reviews*, *25*, 814–830.

Cocron, P., & Krems, J. F. (2013). Driver perceptions of the safety implications of quiet electric vehicles. *Accident Analysis & Prevention*, *58*, 122–131.

Deilami, S., Masoum, A. S., Moses, P. S., & Masoum, M. A. (2011). Real-time coordination of plug-in electric vehicle charging in smart grids to minimize power losses and improve voltage profile. *Smart Grid, IEEE Transactions on*, *2*(3), 456–467.

Eriksson, T., & Olsson, D. (2011). The Product Chains of Rare Earth Elements Used in Permanent Magnets and NiMH Batteries for Electric Vehicles. Available at: http://studentarbeten.chalmers.se/publication/142592-the-product-chains-of-rare-earth-elements-used-in-permanent-magnets-and-nimh-batteries-for-electric [Accessed 12 July 2014].

Fluhr, J., Ahlert, K. H., & Weinhardt, C. (2010). A stochastic model for simulating the availability of electric vehicles for services to the power grid. In *System Sciences (HICSS), 2010 43rd Hawaii International Conference on* (pp. 1–10). IEEE.

Goonan, T. G. (2012). Lithium use in batteries: US Geological Survey Circular *1371*, 14 p. Available at: http://pubs.usgs.gov/circ/1371/ [Accessed 12 July 2014].

Heymans, C., Walker, S. B., Young, S. B., & Fowler, M. (2014). Economic analysis of second use electric vehicle batteries for residential energy storage and load-levelling. *Energy Policy*, *71*, 22–30.

Humphreys, K. K., Placet, M., & Singh, M. (1996). Life-cycle assessment of electric vehicles in the United States. In *31st Intersociety Energy Conversion Engineering Conference (IECEC)*. IEEE 11–16 August, pp. 2124–2127.

Kesler, S. E., Gruber, P. W., Medina, P. A., Keoleian, G. A., Everson, M.P., & Wallington, T. J. (2012). Global lithium resources: Relative importance of pegmatite, brine and other deposits. *Ore Geology Reviews*, *48*, 55–69.

Khan, R. H., & Khan, J. Y. (2013). A comprehensive review of the application characteristics and traffic requirements of a smart grid communications network. *Computer Networks*, *57*(3), 825–845.

Lemoine, D. M., Kammen, D. M., & Farrell, A. E. (2008). An innovation and policy agenda for commercially competitive plug-in hybrid electric vehicles. *Environmental Research Letters*, *3*(1), 014003.

Lopes, J. A. P., Soares, F. J., & Almeida, P. M. R. (2011). Integration of electric vehicles in the electric power system. *Proceedings of the IEEE*, *99*(1), 168–183.

Lu, L., Han, X., Li, J., Hua, J., & Ouyang, M. (2013). A review on the key issues for lithium-ion battery management in electric vehicles. *Journal of Power Sources*, *226*, 272–288.

Mak, H-Y. Rong, Y., & Shen Z-J. M. (2013). Infrastructure planning for electric vehicles with battery swapping. *Management Science*, *59*(7), 1557–1575.

Massari, S., & Ruberti, M. (2013). Rare earth elements as critical raw materials: Focus on international markets and future strategies. *Resources Policy*, *38*(1), 36–43.

Messagie, M., Boureima, F., Matheys, J., Sergeant, N., Turcksin, L., Macharis, C. *et al.* (2010). Life Cycle Assessment of conventional and alternative small passenger vehicles in Belgium. In *Vehicle Power and Propulsion Conference (VPPC)*. IEEE 1–3 September, pp. 1–5.

Metal-pages, (2014). Available at: http://www.metal-pages.com/metalprices/rareearths/ [Accessed 5 December 2014].

Mohd, A., Ortjohann, E., Schmelter, A., Hamsic, N., & Morton, D. (2008). Challenges in integrating distributed energy storage systems into future smart grid. In *Industrial Electronics, 2008. ISIE 2008. IEEE International Symposium on* (pp. 1627–1632). IEEE.

Nelson, T., Simshauser, P., & Kelley, S. (2011). Australian residential solar feed-in tariffs: Industry stimulus or regressive form of taxation? *Economic Analysis and Policy*, *41*(2), 113.

Peterson, S. B., Whitacre, J. F., & Apt, J. (2010). The economics of using plug-in hybrid electric vehicle battery packs for grid storage. *Journal of Power Sources*, *195*(8), 2377–2384.

Putrus, G. A., Suwanapingkarl, P., Johnston, D., Bentley, E. C., & Narayana, M. (2009). Impact of electric vehicles on power distribution networks. In *Vehicle Power and Propulsion Conference, 2009. VPPC'09*. IEEE (pp. 827–831). IEEE.

Qian, K., Zhou, C., Allan, M., & Yuan, Y. (2011). Modeling of load demand due to EV battery charging in distribution systems. *Power Systems, IEEE Transactions on*, *26*(2), 802–810.

Renaldi, Yuanditra, W., Raharno, S., & Martawirya Y. Y. (2013). Environmental consideration in materials selection for electric vehicles. In *Joint*

International Conference on Rural Information & Communication Technology and Electric-Vehicle Technology (rICT & ICeV-T). IEEE 26–28 November, pp. 1–4.

Simshauser, P., Nelson, T., & Doan, T. (2011). The boomerang paradox, part II: policy prescriptions for reducing fuel poverty in Australia. *The Electricity Journal*, *24*(2), 63–75.

Steinhilber, S., Wells, P., & Thankappan, S. (2013) Socio-technical inertia: Understanding the barriers to electric vehicles. *Energy Policy*, *60*, 531–539.

Strbac, G. (2008). Demand side management: Benefits and challenges. *Energy policy*, *36*(12), 4419–4426.

U.S. DOE report (2010). U.S. Department of Energy: Critical Materials Strategy.

U.S. DOE report (2011). U.S. Department of Energy: Critical Materials Strategy.

U.S. Geological Survey (2014). Annual publications: Mineral Commodity Summaries: Rare Earths: 2014. [online] Available at: http://minerals.usgs.gov/minerals/pubs/commodity/rare_earths/ [Accessed 4 December 2014].

Ziefle, M., Beul-Leusmann, S., Kasugai, K., & Schwalm, M. (2014). Design, user experience, and usability. User experience design for everyday life applications and services. Lecture notes in computer science, *8519*, 628–639.

Index

AC conductance method 126–7
air gap 188
alkaline fuel cells (AFCs) 51
all-electric motorbikes 15
Australia
  motor vehicle types in 23
  passenger vehicle ownership in 23–4
  private motor vehicle use 18
  public transportation services 21
automobile technologies, history of 160–2

backup generators 55
backup power supplies 54–6
battery capacity dependency 139
battery electric vehicles (BEVs) 54, 107, 158–9
  see also electric vehicles (EVs)
battery energy storage plants (BESP) 137
battery equivalent circuit with parasitic branch 138
Benz Patent-Motorwagen 52
biodiesel 56
biological fuel cells (BFCs) 51
BMW Active E 17

ChargeIQ smartphone application 31–2
Coulomb-counting technique 125–6
coupling factor (*k*) 188
current source inverter (CSI) 135

Davidson, Harley 15
DC/AC inverter 53
DC/DC converter 47, 53
DC load method 127–8
DFIG 132, 143
direct methanol fuel cell (DMFCs) 51
distributed energy resource (DER) controllers 34
distributed energy sources/distributed generations (DGs) 2, 43–4, 63–4
  connecting 43–4
  fuel cells 50–2
  solar 45–8
  types 44–5
  wind 48–50
distributed power conversion schemes 83
distributed-STATCOMs (D-STATCOMs) 167
distributed storage (DS) technologies 2
doubly fed induction generator (DFIG) 62
DRCT class with electric vehicle addition 34–5
Durbin-Watson test 246
dynamic wireless charging system 204–7

electric power systems (EPSs) 3
electric vehicle control (EVCT) 35
  class 35
  class for EV communication and control 37
  generic nature of 38
  logical node (LN) 36
  V2Gstatus and EconStatus 36

electric vehicles (EVs) 11, 52–4, 158
  batteries of 53
    circuit diagram of 84
  battery electric vehicles (BEVs) 54
  charging options 16–18
  charging power levels for 112–14
  cumulative impact of 12
  drivers and the smart grid, impact on 30–2
  dynamical modelling of 84–8
    connected to single-phase smart grid node 85–7
    connected to three-phase smart grid node 87–8
  electrical network, impact on 25–30
  electrical specs of 16
  interface between the smart grid applications and 31–2
  pure 54
  in the smart grid
    control, monitoring and management strategies 117–29
    distributed controller design for EVs 92–4
    dq-frame 86–8
    feedback linearisation and feedback linearisability 89–92
    impacts of EV penetration on grid power profile 108–11
    parasitic branch models 85
    performance of designed controller 94–100
    power conversion problem formulation 88–9
    single-phase smart grid node 85–7
    three-phase smart grid node 87–8
  smartphone application of EV-charging 32
  static model of 82
  types 12–17
  V2G and G2V modes 108
  see also battery electric vehicles (BEVs); plug-in electric vehicles (PEVs); plug-in hybrid electric vehicles (PHEVs)
Electric Vehicle-Smart Grid Interoperability Center 33
energy storage systems (ESS) 44
EV battery charging/discharging controller 144
EV energy storage system (EV-ESS) modelling 137–42
  battery capacity dependency 139
  main branch resistances 139–40
  parasitic branch 141–2
  thermal model 141
  third-order battery model constant parameters 139, 142

feedback linearisation technique 89
Ferdinand Porsche 158
four-quadrant topologies 6–7
fuel-cell electric vehicle (FCEV) 164
fuel cells (FCs) 50–2
  types 51
fuel cell vehicles (FCVs) 52
fuzzy-based frequency controller 82

high-frequency magnetic-link based medium-voltage inverter 48
HomePlug 129–31
hybrid electric vehicles (HEVs) 13, 52–3, 107
  parallel power trains 13–14
  powertrain architectures 111–16
    combination of series and parallel 115
    features 116
  series power trains 13
  see also plug-in hybrid electric vehicles (PHEVs)
hybrid fuel cell vehicle 54
hybrid ICG/EV vehicles 52
hybridization of cars 13
hydrogen–air fuel cell system 52

IEC 61850 Substation Communication Standard 33–4
  IEC 61850-7-420 extension 34–8
    universal electric vehicle modeling 38

inductive power transfer (IPT) 186
insulated-gate bipolar transistor (IGBT) 176
intentional islanding
  active power of wind generator 150
  DC bus voltage before and after EV ESS 148
  power consumed by EV ESS 149
  reactive power of wind generator 150
  supplied power from PV before and after EV ESS 149
interconnection switches 2
internal combustion engines (ICE) 157
  ICE-powered vehicles 158–9
internal dynamic of system 91

Kirchhoff's current law (KCL) 85
Kirchhoff's voltage law (KVL) 86
Kyoto Protocol 157

large magnitude pulse resistance (LMPR) 128–9
Lightning bikes 16
linear battery model 137
load forecast models for PHEV 245–8
  autoregressive integrated moving average (ARIMA) 245–6
  coefficient of determination (R^2) 247
  mean absolute error (MAE) 247
  mean absolute percentage error (MAPE) 247
  standard deviation of model error or root mean square error (RMSE) 247
  see also plug-in hybrid electric vehicles (PHEVs)
logical node (LN) 34–6

Mauer, Michael 158
maximum power point (MPP) 46
maximum power point tracking (MPPT) systems 46, 134
  direct
    forced oscillations 57
    incremental conductance method 57
    parasitic capacitance 57
    perturb-and-observe method 56–7
  indirect
    curve-fitting method 57
    look-up table method 57
    open-circuit voltage and short-circuit current method 57–8
medium-frequency transformer-link 47
metal-oxide semiconductor field-effect transistors (MOSFETs) 176
Microgrids 2, 165–7
  applications 2
  basic technologies for operation 2–3
  comprise low voltage (LV) distribution systems of 5
  considerations for planning and operating 3
    to achieve smart operation 63
  control strategies 63–7
    based on voltage droop method 66
    droop characteristics 65–6
  DC 83
  defined 2–3
  with doubly fed induction generator (DFIG) 62
  equivalent circuit of parallel-inverters based 65
  flow of active power 65
  grid-connected and islanded modes 60
  grid-connected operation 69–72
  for households 61
  IEEE 1547 series of standards 3
  islanded operation 72–3
    response of common AC bus voltage 73
  modes of operation 3
  MPPT strategies for PV based 56
  with multi-DGs and a control center 61–3
  planned 3
  power coupling 67

power flows within 68
power quality comparison 73
IEEE Standard 1547 73
simulation results and discussion 67–9
slopes selection 67
smart transformer for charging station in 180–1
static transfer switch (STS) 67, 69–70
synchronization of utility grid and 70
system parameters using voltage droop method 68
topologies 60–3
see also smart grids
micro-hybrids 13
mild hybrids 13
modular multilevel cascaded (MMC) inverter topology 47
molten carbonate fuel cells (MCFCs) 51
motor vehicle ownership and migration to electric vehicles 17–25
multilevel converters 47

net-zero energy transportation 4
nonlinear feedback linearising power conversion technique 91
nonlinear power conversion or switching control schemes 83

Ohmic capacity measurement 126–9

parallel-hybrid vehicles 159
partial feedback linearisation of smart grids with EVs 91–2
peaking power source (PPS) 53
permanent magnet synchronous generators (PMSGs) 49
perturb-and-observe method 56–7
phosphoric acid fuel cells (PAFCs) 51
photovoltaics (PVs) 108
PLL synchronization angle 136
plug-in electric vehicles (PEVs) 1
benefits of 7
charging diversity and utility interfaces 167–8
charging standards
limiting factor for power delivery 171
sockets 171–2
voltage ratings and frequencies 171
charging systems for 6–9
comparison of 8
CO_2 footprint of 5
communication between an EV and a utility network 168
communication requirements 163–5
contact-based charging methods 172–9
DC/DC converters 176
full-bridge converters with STATCOM capabilities 176–8
internal STATCOM 178
inverter topologies 173–5
on-board charging designs 178–9
rectifier topologies 173
transformer-based inverters 175
transformerless inverters 175–6
contact charging safety considerations
AC charging 181
DC charging 182–3
distributed generation 165–7
dynamic loads 158
formation of a microgrid or a smart grid 165–7
impact on the smart grid demand response 4–6
connection of batteries 4–5
uncontrolled connection in grid 5
load power characteristics alongside PV production and 165
on-board energy storage 166
overview 157–70
power discharging hardware and software 7

small-scale, network 165
storage configurations, central and local 168–70
storing of PV energy 6
V2G-enabled 163–5
wireless charging systems
applications 201–7
safety considerations 208–9
standards of wireless charging systems 197–201
wireless charging methods 198
wireless power characteristics 183–5
wireless power transfer 185–97
plug-in hybrid electric vehicles (PHEVs) 1, 14, 53–4, 107, 158
economic dimensions 225–33
as distributed energy resources (DERs) 229–33
smart charging of 226–9
environmental dimensions
environmental impact comparative assessment 237–9
life cycle assessment (LCA) 236–7
vehicle batteries 239–43
impact of 224
scheduling systems 243–9
initial 248
load forecast models 245–8
online control system 249
with series-parallel power train 15
social dimensions
convenience 233–4
public perception 235–6
safety aspects 235
uncertainty, trust and equity 234–5
point of common coupling (PCC) 144
Porsche Spyder 918 158
power conversion in smart grids with EVs 81–2
problem formulation 88–9
power factor correction (PFC) 167
unidirectional chargers 6
power-line communications 129–31
power systems, drawbacks of existing 58
proton exchange membrane fuel cells (PEMFCs) 51
protonic ceramic fuel cells (PCFCs) 51
public and private transport use in different countries 19
pure electric vehicle 15
PV power plants 45–7
PV system 134–6
basis of controlled parameter 135
current source inverter (CSI) 135
current–voltage relationship 134–5
PLL synchronization angle 136
voltage source inverter (VSI) 135

quality factor (*Q*) 188
quasi-Z source inverter 47

R-C blocks in battery model 138, 140
renewable energy resources (RES) 166–7, 239
powered smart grid 110
renewable sources of energy 4
residential LV distribution network 225
resonant inductive power transfer (RIPT) 186

series-hybrid vehicles 159
series-parallel hybrid vehicles 159
series-parallel power train 13–14
Singapore, public transport services 19
single-phase smart grid node 85–7
skin effect 196–7
smart grids 2, 165–7
feedback linearisation and feedback linearisability
overview of 90–1
partial 91–2
key technologies and challenges of information collection and energy management 59

power electronics control technology 59
security and privacy issues 59
semiconductor device 59
linearised model 83
using PEVs 2
impact on distributed energy resources 4–6
with V2G 6
vision 2
see also microgrids
the smart grid service providers (SGSPs) 82
the smart grid with EVs
control, monitoring and management strategies
battery capacity monitoring 124–9
battery management system (BMS) 119–24
Coulomb-counting technique 125–6
EV induction motor (IM) control 117–18
field-oriented control (FOC) of induction motor (IM) 118–19
industrially used 117
Ohmic capacity measurement 126–9
speed vector control of induction motor (IM) 118–19
V2G communication system 129–31
voltage and frequency control of induction motor (IM) 117–18
voltage profiling 124–5
distributed controller design for EVs 92–4
dq-frame 86–8
feedback linearisation and feedback linearisability 89–92
exactly linearised 90
generalised nonlinear systems 89
Lie derivative and relative degree 90
mathematical model 90
impacts of EV penetration on grid power profile 108–11
frequency and voltage regulation 109
supporting and balancing of intermittent RES powered smart grid 110
via charging and discharging 110–14
parasitic branch models 85
performance of designed controller 94–100
during the charging mode of EVs 97
during the discharging mode of EVs 97–100
line-to-ground fault 99–100
at a power factor other than unity 98–9
at unity power factor 98
power conversion problem formulation 88–9
power system models
EV energy storage system (EV-ESS) modelling 137–42
PV system 134–6
wind turbine 132–4
problem formulation and control strategy 143–5
simulation results
intentional islanding, case of 147–50
severe three-phase fault, case of 146–7
single-phase smart grid node 85–7
three-phase smart grid node 87–8
smart homes 2
smart meters 2, 31–2
solar energy sources 45–7
transformer data 46
solid oxide fuel cells (SOFCs) 51

stabiliser, static synchronous compensator (STATCOM) 167
stacked large pulse resistance (SLPR) method 128–9
starting, lighting, and ignition (SLI) in automotive area 44
static wireless charging system 201–4
supervisory control and data acquisition (SCADA) system 62
surface area and population densities of major cities 20
synthetic or artificial fuel 55

Tesla Roadster 158
third-order battery
- aggregate model 138
- constant parameters 142
- electrical network 139

three-phase fault
- active power of wind generator 147
- DC bus voltage before and after EV ESS 148
- power supplied by EV ESS 146–7
- reactive power of wind generator 148

three-phase six-pulse full bridge converter 87
three-phase smart grid node 87–8
Toyota Prius 158
trans-esterification process 56

uninterruptable power supplies (UPS) 137
United States
- motor-vehicle ownership per capita 21–2
- private motor vehicle use 18
- public transport use 20
- transportation modes for daily commute 21
- urban and rural travel in 24

urban and rural travel 24

vehicle-to-grid (V2G) technology 2, 4, 82, 163–5
- communications 7
- with four-quadrant topologies 6–7

V2G communication system
- HomePlug 129–31
- interference management and cognitive radio 131
- power-line communications 129–31
- wireless personal-area networking 131
- ZigBee 131

voltage profiling 124–5
voltage source inverter (VSI) 135

Western European Union (WEU) countries, motorized public mode and nonmotorized mode in 18–19
wind farms
- installed capacity 48
- multiple-cascaded modules for 50
- wind turbine generator system for direct medium-voltage grid integration 49–50

winding resistance 197
wind power generation 48–50
- modular high-power density wind generation system 51
- regional and local phenomenon 48–9

wind turbine generator system (WTGS) 132
wind turbine modelling 132–4
- algebraic–differential equations 133–4
- relationship between mechanical torque and torsional angle 133
- shafts 132–3
- squirrel cage induction generator (IG) 133
- turbine rotor 132
- two-mass drive train model 132

wireless personal-area networking 131
wireless power transfer (WPT) 7, 185–97
  capacitive 185
  coil structures for 191–2
  efficiency 188
  electromagnetic field–based 186
  ferro-magnetic core material and structures 192–201
  inductive 186–7
    resonant 187–91
    series-parallel resonant 187
  international standards 199
  magnetic gear and permanent magnet (PM)–based 185–6
  near-field-based 186
  SAE TIR J2954 international standards of 200
  safety certificates or compliance for industry-based 200

ZBAT 35
ZBCT 35
zero emission vehicles 44
Zero Motorcycles 15–16
ZigBee-based communication 31, 131
zinc-air fuel cells (ZAFCs) 51